Designing Green Networks and Network Operations

Reducing Enterprises' and
Carriers' Carbon Footprint and
Saving Run-the-Engine Costs with
Green Networks and Data Centers

Designing Green Networks and Network Operations

Reducing Enterprises' and
Carriers' Carbon Footprint and
Saving Run-the-Engine Costs with
Green Networks and Data Centers

DANIEL MINOLI

CRC Press
Taylor & Francis Group
Boca Raton London New York

CRC Press is an imprint of the
Taylor & Francis Group, an **informa** business

Cover photo: Eastern Shore, Virginia

CRC Press
Taylor & Francis Group
6000 Broken Sound Parkway NW, Suite 300
Boca Raton, FL 33487-2742

First issued in paperback 2017

© 2011 by Taylor & Francis Group, LLC
CRC Press is an imprint of Taylor & Francis Group, an Informa business

No claim to original U.S. Government works

ISBN-13: 978-1-4398-1638-7 (hbk)
ISBN-13: 978-1-138-11795-2 (pbk)

Library of Congress Cataloging-in-Publication Data

Minoli, Daniel, 1952-
    Designing green networks and network operations : saving run-the-engine costs/
author, Daniel Minoli.
        p. cm.
    Includes bibliographical references and index.
    ISBN 978-1-4398-1638-7 (alk. paper)
    1. Computer networks--Equipment and supplies. 2. Computer networks--Energy efficiency. 3. Electronic data processing departments--Energy efficiency. 4. Sustainable engineering. I. Title.

TK5105.53.M56 2010
004.6--dc22   2010001526

Visit the Taylor & Francis Web site at
http://www.taylorandfrancis.com

and the CRC Press Web site at
http://www.crcpress.com

Contents

Preface

This is the first book on the market to address Green networks and Green network operations. Electrical costs and cooling represent up to 35% to 45% of a data center's TCO (total cost of ownership). According to estimates, the 2-year cost of powering and cooling a server is equal to the cost of purchasing the server hardware, and the energy cost to operate a routers or switch over its lifetime can run into tens of thousands of dollars. In recent years the confluence of socio-political trends toward environmental responsibility and the pressing need to reduce Run-the-Engine (RTE) costs have given birth to a nascent discipline of Green IT. Some recent literature has emerged on Green Data Centers. This text extends and applies the concepts to Green Networks.

Firms are looking to minimize power, cooling, floor space, and online storage, while optimizing service performance, capacity, and availability. A variety of techniques are available to achieve these goals, including deploying more efficient hardware, and blade form-factor routers and switches, and pursuing consolidation, virtualization, network computing, use of Web Services, and outsourcing, among others.

Previous studies by this author have shown that, depending on the industry, the yearly networking costs typically are in the 1-2-4% of the top revenue line of a firm. It follows that a company with $1 billion in revenue could save between $1 million and $7 million annually and up to $21 million in 3 years. While the discrete value of individual activities may not be "spectacular," the aggregate value may be significant, particularly over a multi-year window. Also, while the potential savings will depend on the type of business the firm is in, network operators, carriers, teleport operators, web-hosting sites, financials, and insurance companies with large data centers will benefit the most.

This text extends the techniques that have been identified for data centers in the recent past to the networking environment and looks at opportunities for incorporating Green principles in the intranet and extranet—in fact, in the entire IT infrastructure. After an introduction and overview in Chapter 1, Chapter 2 provides a primer on networking. Chapter 3 looks at basic consumption concepts and analytical measures for Green operations. Chapter 4 provides a basic tutorial on power management while Chapter 5 explores basic HVAC concepts. Chapter 6 surveys regulatory guidelines and best practices. Chapter 7 covers approaches for green networks and data center environmentals (cooling, power). Chapter 8 discusses network computing and Web services as approaches to support green networks and data centers, and also discusses virtualization and cloud computing.

The results presented in this book are based on the dozens of networks and data nodes deployed by the author while at AT&T, Leading Edge Network Inc. (incubator for InfoPort Communications Group, Global Wireless Inc.), and SES Engineering, along with network and data center virtualization work at Capital One Financial.

This book can be used by practitioners, network engineers, IT personnel, vendors and manufacturers, wireline and wireless service providers, systems integrators, consultants, regulators, and students.

About the Author

Daniel Minoli has many years of technical hands-on and managerial experience in planning, designing, deploying, and operating IP/IPv6, telecom, wireless, and video networks, and data center systems and subsystems for global Best-in-Class carriers and financial companies. He has worked at financial firms such as AIG, Prudential Securities, Capital One Financial, and service provider firms such as Network Analysis Corporation, Bell Telephone Laboratories, ITT, Bell Communications Research (now Telcordia), AT&T, Leading Edge Network Inc., and SES Engineering, where he is Director of Terrestrial Systems Engineering (SES is the largest satellite services company in the world). At SES, Minoli has been responsible for the development and deployment of IPTV systems, terrestrial and mobile IP-based networking services, and IPv6 services over satellite links. He also played a founding role in the launching of two companies through the high-tech incubator Leading Edge Network Inc., which he ran in the early 2000s: Global Wireless Services, a provider of secure broadband hotspot mobile Internet and hotspot VoIP services, and InfoPort Communications Group, an optical and Gigabit Ethernet metropolitan carrier supporting data center/SAN/channel extension and cloud computing network access services. For several years he has been Session-, Tutorial-, and now overall Technical Program Chair for the IEEE ENTNET (Enterprise Networking) Conference; ENTNET focuses on enterprise networking requirements for large financial firms and other corporate institutions. He is also the founder and president emeritus of the IPv6 Institute, a certification organization for IPv6 networking technology, IPv6 global network deployment, and IPv6 security (www.ipv6institute.org).

Minoli has done extensive work in the network engineering, design, and implementation of carrier and corporate networks, as well as work in network/datacenter virtualization. The results presented in this book are based on work done while at AT&T, Leading Edge Network Inc., Capital One Financial, and SES Engineering. Some of the virtualization work has been documented in his book *A Networking Approach to Grid Computing* (Wiley, 2005) and in the book *Enterprise Architecture A to Z* (CRC Press, 2008).

Minoli has also written columns for *ComputerWorld, NetworkWorld*, and *Network Computing* (1985–2006). He has taught at New York University (Information Technology Institute), Rutgers University, and Stevens Institute of Technology (1984–2006). Also, he was a technology analyst at-large for Gartner/ DataPro (1985–2001); based on extensive hands-on work at financial firms and carriers, he tracked technologies and wrote CTO/CIO-level technical scans in the area of telephony and data systems, including topics on security, disaster recovery, network management, LANs, WANs (ATM and MPLS), wireless (LAN and public hotspot), VoIP, network design/economics, carrier networks (such as metro Ethernet and CWDM/DWDM), and e-commerce. Over the years he has advised venture capitalists for investments of $150 million in a dozen high-tech companies. He has acted as an expert witness in a (won) $11 billion lawsuit regarding a VoIP-based wireless air-to-ground communication system, and has been involved as a technical expert in a number of patent infringement proceedings.

Chapter 1

Introduction and Overview

1.1 Introduction

Networks, especially packet-based IP networks, including the Internet, have already made major contributions to the "greening" of the environment for three decades or more. *Greening* refers to minimizing energy consumption, maximizing energy use efficiency, and using, whenever possible, renewable energy sources; the utilization of eco-friendly components and consumables also plays a role. Some refer to greening as reducing the carbon footprint.* Just consider the carbon footprint saved by audio-teleconferencing, video teleconferencing/telepresence, Internet-based white-boarding/conferencing, telecommuting [MIN199502]; e-mail communication [MIN199901]; distance learning [MIN199601]; Web commerce, including on-line travel bookings, purchases, and other disintermediation [MIN199801]; Internet-based home businesses, multimedia applications such as music downloading (as contrasted to CD-based distribution) [MIN199401]; commercial-grade video enter-tainment such as video-on-demand (as contrasted to a trip to a theater) [MIN199501], and IPTV-based digital video recording (as contrasted with DVD-based distribu-tion), and other de-materialization efforts [MIN200801]; grid/cloud computing/virtualization (reducing space, power, and heating, ventilation, and air condition-ing (HVAC) costs) [MIN200501], [MIN200802]; remote sensing [MIN200701];

---

* "Carbon footprint" is a favored term de jour to describe a calculation of total carbon emissions of some system, namely, to describe the energy consumption in terms of the amount of green-house gases (GHG) produced to support a given activity (or piece of equipment); it is typically expressed in equivalent tons of carbon dioxide (CO_2), that is, CO_2e.

1

solar-powered satellite communications [MIN200901]; paperless news delivery, electronic libraries, remote operations (e.g., managing a function, a network, a data center, a site from a remote localized position); home banking and check-less banking with either credit card operations or scanned checks [MIN199402]; storage de-duplication; wireless/paperless remote order entry [MIN200201]; remote home-appliance monitoring; digital camera/picture transmission; Web-based services, and anticipated future nanotechnology-based breakthroughs [MIN200502]. Even more greening opportunities arise when one considers the entire Information and Communications Technology (ICT) space* [MIN200902].

Telecommunications is, in general terms, an environmentally friendly industry; however, carriers and service providers do use relatively large quantities of energy and exert a nontrivial impact on the environment; therefore, the time has come to introduce Green technologies and processing into the networking field itself. According to some sources,[†] the total contribution of ICT to global green house gases (GHG) emissions, including carbon dioxide (CO_2), is in the range of 2% to 2.5% of total emissions, with about 0.2% attributed to mobile telecom and about 0.3% to fixed telecom (the balance being with PCs, data centers, etc.). (See Figure 1.1, which provides two reasonably consistent views from two industry sources.) These percentages as well as the absolute values may grow as ICTs become even more widely deployed; in a business-as-usual (BAU) scenario, the ICT sector's own emissions are expected to increase from 0.53 billion tons (Gt) carbon dioxide equivalent (CO_2e) in 2002 to 1.43 $GtCO_2e$ in 2020. At the same time it should be noted that ICT contributes as much as 7% of global gross domestic product (GDP). Propitiously, there is a parallel recognition that ICT can enable significant reductions in emissions in other sectors of the economy, such as "smart" logistics, "smart" buildings,[‡] "smart" motor systems, and "smart" grids, to list just a few. By enabling other sectors to reduce their emissions, the ICT industry is positioned to be able to facilitate the reduction of global emissions by as much as 15% by 2020 [GES200801] (specific ICT opportunities—such as the replacement of goods and services with virtual equivalents [this being known as de-materialization] and the introduction of telecommand/telecommunication technology to enable energy efficiency—can lead to emission reductions five times the size of the sector's own footprint, namely, up to 7.8 $GtCO_2e$).

A discussion of ICT greening includes aspects, issues, and opportunities related to service providers (e.g., how a wireless carrier can reduce its energy consumption),

---

* Information Technology (IT) is a subset of the ICT field; in this text we use IT when focusing strictly on data processing and ICT when considering the entire field.

† European Telecommunications Standards Institute (ETSI) and Gartner (an IT research firm.)

‡ The concept of smart buildings already arose in the early 1980s, although the initial concept related to sharing telecom services. If one views this sharing as an example of "utility computing" (now called cloud computing—as discussed in Chapter 8), which does indeed save energy, then one can see that, at least in a limited way, the basic concept has some vintage.

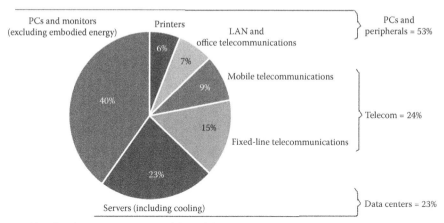

The ICT industry consumes 6–10% of the world's energy
The ICT emits 2–2.5% of CO_2 emissions

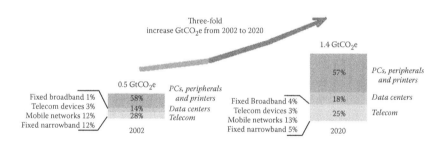

Note: Slight variation in percentages due to different time frames–results in general agreement

Figure 1.1   Approximate distribution of global CO_2 emissions from ICTs. (Top: Data from R. Kumar and L. Mieritz, Conceptualizing "GreenIT" and Data Centre Power and Cooling Issues, Gartner Research Paper No. G00150322, 2007. With permission. Bottom: Data from Alcatel-Lucent, Eco-Efficiency in Action—Alcatel-Lucent Sustainability Solutions for Access Networks, June 2009. With permission.)

as well as aspects, issues, and opportunities related to the enterprise (business and institutional) community: for example, how a large firm or institution can reduce its data center and networking use of energy or how a home consumer can do the same in reference to using Personal Computers (PCs), IT consumables, etc. This text addresses both sets of issues, and the concepts and principles discussed apply to

■ Data centers
■ Networking sections of data centers, supporting the intranet and/or extranet of the organization

- Telco rooms (aka telecom room, datacom rooms, networking rooms, equipment rooms)
- Carrier central offices
- Service provider points-of-presence
- Carrier hotels, meet-me-rooms, collocates (colos)
- Outdoor telecom shelters and remote wireless installations
- Cloud computing operations

To be fair, equipment vendors and service providers have been interested in reducing energy consumption, heat generation (which requires cooling), and the physical equipment footprint for a number of years for intrinsically beneficial reasons, specifically to increase operational efficiencies and, thus, reduce cost. The current renewed emphasis can be seen as an effort to accelerate these initiatives even further. The imperative with leading-edge enterprises is to increase the capabilities of their networks while (1) reducing data center and networking node energy consumption (both to run and to cool the Network Elements [NEs]), (2) reducing collateral environmentally unfriendly by-products, and (3) fostering green building and sustainable design. Firms are looking to minimize power, cooling, floor space, and on-line storage, while optimizing service performance, capacity, and availability. Service providers are looking to do the same for their edge (access) and core networks and for their switching/routing/muxing nodes. When people talk about green, they often focus on renewable energy sources, such as wind, solar, biomass, and fuel cells; some of these technologies can be used for wireless networks in rural environments and/or developing countries. However, the short-term "low-hanging fruit" for green networks is the design of NEs that make efficient use of traditional energy, and the optimized design of telecom/IT rooms, particularly from a cooling perspective.

There has been some practitioner-level attention to data center issues related to green IT in the recent past (as noted in Appendix 1A*), but so far there has not been much formal, explicit practitioner-level attention to the networking/telecom/carrier/service-provider issue of green technology, although research, advocacy, and academic literature is emerging. Energy consumption is recognized as being the activity that has the largest environmental impact for a(ll) telecommunications operator(s); nearly all the energy use is from electricity consumption; in turn, electricity is used to power and cool data, Internet, video, voice, and wireless communication networks, and data centers. At the same time, as noted, ICTs can help the greening cause by promoting the development of more energy-efficient devices, applications, and networks; by encouraging environmentally friendly design; and by reducing the carbon footprint of its own industry. To make rational statements about how much is gained, the industry needs technical standards and performance benchmarks to consistently "quantify" conformity to, or levels of, being

---

* Also see Appendix 1B for other resources.

green; fortunately these are now emerging. In this text the focus of a greening initiative relating to IT data centers and networks (whether intranets or carrier infrastructure) covers two main areas:

1. *System Load:* This relates to the consumption efficiency of the equipment in the data center or telecom node (e.g., IT equipment such as servers, storage, and networking equipment such as enterprise switches and routers—or, for example, carrier NEs such as repeaters, switches, digital cross-connects, core routers, and optical terminals). It represents the IT/networking work capacity available for a given IT/networking power consumption. Note that System Load needs to be specified as a function of the utilization of said capacity.
2. *Facilities Load:* This relates to the mechanical and electrical systems that support the IT/networking electrical load, such as cooling systems (e.g., chiller plant, fans, pumps), air-conditioning units, uninterruptible power supplies, power distribution units, and so on.

Until recently the environmental impact of ITC in general, and of the data center, the network infrastructure, and the support systems in particular, has been largely ignored. Data centers are energy-intensive operations: Server racks are now designed for more than 25 kW for high-end applications*; however, the typical server in a data center is estimated to waste about 30% of its consumed power. A large data center may have 200, 500, or 1,000 racks. There is also increasing energy demand for information storage systems. It is estimated that a high-availability data center that has a 1-MW demand will consume $20M of electricity over the data center's lifetime. Specifically, research shows that electrical costs for operating and cooling represent up to 35% to 45% of a data center's Total Cost of Ownership (TCO), and that energy costs have replaced real estate as the primary data center expense. Calculations show that the 2-year cost of powering and cooling a server is equal to the cost of purchasing the server hardware, and that the energy cost to operate a router and/or switch over its lifetime can run into tens of thousands of dollars. According to the Uptime Institute, a consortium of companies devoted to maximizing efficiency and uptime in the data center, more than 60% of the power used to cool equipment in the data center is actually wasted [OVE200701]. Furthermore, at this time a typical desktop PC wastes almost half the power it consumes as heat.

At the macro level, energy experts estimate that data centers use between 1.5% and 3% of all electricity generated in the United States. Studies published by the U.S. Environmental Protection Agency (EPA) indicate that in 2006 network equipment consumed 5% of the estimated 64 billion kilowatt-hours (kWh) of energy used in the United States by data centers (namely, 3.2 billion kWh).

---

* Average data center applications require 5–10 kW and mid-range data center applications require 15 kW. When averaged over the entire data center floor, the watts per square foot (WPSF) equate to 100–200 for average applications and up to 600 for high-end applications.

Network equipment ranked fourth on a list of six in terms of actual use, just behind storage. At $0.15/kWh that usage translates into $0.5 billion annually for the networking gear used in enterprises and $9.6 billion for the total. The issue is that in recent years, data center energy consumption has been growing at a Compound Annual Growth Rate (CAGR) of 14% [EPA200701] (basically doubling every 6 years, at least up to now*). That would make the 2010 figures at $780 million and $16 billion, respectively, as seen in Table 1.1. Worldwide, calculations developed by market research firm IDC indicate that companies spent over $26 billion to power and cool servers in 2005.

The confluence of socio-political trends toward environmental responsibility and the pressing need to reduce Run-the-Engine (RTE) costs has given birth to a nascent discipline of green IT and green networks. It should be noted, again, that the quest for efficiency is not entirely new. In fact, one can make the case that efficiency improvements have been sought throughout the Industrial Revolution. Up to the present, however, efficiency improvements have been driven almost invariably by the simple desire to optimize the cost-effectiveness of a process or manufacturing activity. (Energy) efficiency is part of a "greening" initiative, however, by itself it is not sufficient to optimize greenness of an operation; for example, the use of green (renewable) power sources, the reduction of waste byproducts, and the reduction of the carbon footprint per unit of manufactured goods are some of the other important factors.

Related to energy, some recent studies have shown that Best Practices can reduce data center energy consumption by upward of 45%. Optimizations are not limited to the data center itself, but the entire IT infrastructure: the data center, the intranet, the desktop, the extranet, and all the intervening networks (Wide Area Network, Metropolitan Area Network, Internet, wireless network, etc.). Consider this simple example: a 120-watt device (such as a PC or stationary laptop) not being used, but left on in sleep mode for (say) 16 hours a day for 365 days consumes about 700 kWh per year, which is about $100. A medium-sized company with 1,000 PCs obviously would spend $100,000 per year needlessly, and a company with 10,000 devices would waste $1 million per year. Hence, there are good reasons for pursuing green IT and green networking approaches.

This text extends and applies the concepts that have emerged for data center green operations to green networks; these include both corporate intranets as well as carrier/service-provider networks. According to some observers, "the topic of being and practicing green has captured the attention of every corporation, government entity, and standards developers in the information, entertainment and communications industry" [ATI200901]; hence the motivation in this work to address this topic in the context of networking technology and systems. Web hosting sites, networking-intensive "telco rooms," "collocates," networking sections within a data center, and switching centers can achieve measurable savings compared with a

---

* Doubling every 5 years, according to some [BRO200801].

Table 1.1   Power Usage (EPA Data) for Data Centers and Network Gear

	2000 Electricity use (billion kWh)	2006 Electricity use (billion kWh)	2006 % Total	2006 Rank	2000–2006 Electricity use CAGR	2010 Projection(*) Electricity use (billion kWh)	2010 Projection(*) Expenditures($B)
Site infrastructure	14.1	30.7	50%	1	14%	51.9	7.8
Volume servers	8.0	20.9	34%	2	17%	39.2	5.9
Storage	1.1	3.2	5%	3	20%	6.6	1.0
Network equipment	1.4	3.0	5%	4	14%	5.1	0.8
Mid-range servers	2.5	2.2	4%	5	–2%	2.0	0.3
High-end servers	1.1	1.5	2%	6	5%	1.8	0.3
Total	28.2	61.4			14%	106.5	16.0

(*) *This author*

BAU approach. A variety of techniques are available to achieve these goals, including deploying more efficient hardware, blade form-factor routers and switches, improved site cooling, consolidation, virtualization, Network Computing (also known as Cloud Computing), the use of Web Services (WS), and outsourcing, among others. Appendix 1C shows examples of how networking has already had a major greening impact.

The definition of *green** used in this book[†] is, as stated in the opening paragraph, the characteristic of (1) making (highly) optimized and/or economized use of energy, including the use of renewable energy; and (2) employing sustainable processes with a low carbon footprint and with an overall consideration for the "triple" bottom line (see below). Key terms used in this text are as follows (also refer to the Glossary) [GAL200801, SAE200801]:

Green: The tangible or physical attributes of a product or a property, in the context of energy efficiency/carbon footprint and/or eco-friendliness.

Greening: Minimizing energy consumption, maximizing energy use efficiency, and using, whenever possible, renewable energy sources. The use of eco-friendly components and consumables also plays a role. Succinctly: energy efficiency and renewable energy initiatives.

Green buildings: The implementation of design, construction, and operational strategies that reduce a building's environmental impact during both construction and operation, and that improve its occupants' health, comfort, and productivity throughout the building's life cycle. "Green buildings" is part of a larger trend toward sustainable design. (Some call these "smart buildings.")

Sustainable: Refers not only to green physical attributes, as in a building, but also to business processes, ethics, values, and social justice.

Sustainable design: System design where the objective is to create places, products, and services in a way that reduces the use of nonrenewable resources, minimizes environmental impact, and relates people with the natural environment.

Energy efficiency: Designing a system, a data center, a network node, a building, and so on, to use less energy for the same or higher performance as compared to conventional approaches. All data center subsystems, network subsystems, and buildings subsystems (e.g., HVAC, lighting) can contribute to higher energy efficiency.

Renewable energy: Energy generated from natural resources that are inexhaustible. Renewable energy technologies include but are not limited to solar power, wind power, hydroelectricity and micro-hydro, biomass, and biofuels.

---

* Green is used as an adjective.

[†] This is the first book on the market to address green networks and green network operations. Pike Research published a 64-page market report entitled "Green Telecom Networks" in June 2009, available at the time for a fee of U.S. $3,500 (U.S. $5,250 for a site license).

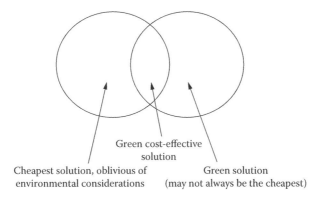

Green cost-effective
solution

Cheapest solution, oblivious of   Green solution
environmental considerations   (may not always be the cheapest)

Figure 1.2   Greening process versus economic optimization.

Carbon footprint: A calculation of the total carbon dioxide emission of a system
or activity.

Triple bottom line: A calculation of financial, environmental, and social performance. Often referred to as "profits, planet, and people." This calculation
method contrasts with the traditional business bottom line, which only considers profits.

While economic optimization does have an overlap with a greening process,
there are cases where the intersection of the two processes is empty (see Figure 1.2
for a graphical view). Greening in general and sustainability in particular span
three areas (see Figure 1.3):

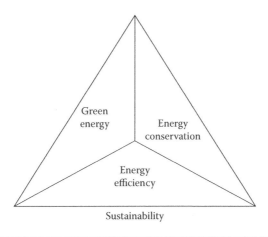

Green
energy

Energy
conservation

Energy
efficiency

Sustainability

Figure 1.3   Sustainability vectors.

- Energy efficiency
- Energy conservation
- Green energy

Table 1.2 enumerates some key industry segments related to green initiatives.

Data center/networking systems (equipment) of interest for greening include the following, among others:

- Routers, data switches, firewalls, fiber-optic systems and nodes, voice switches (traditional and those based on the Internet Protocol), video encoders, and radio transmitters including WiFi Access Points, WiMAX, cellular 3/4G, and satellite antennas
- Servers
- Storage systems
- Carrier network elements (e.g., fiber-optic terminals, digital cross-connect systems, multiplexers, core routers)
- Uninterruptible Power Supplies (UPSs)
- Power Distribution Units (PDUs)
- Power switch/transfer gear
- Emergency generators
- Batteries
- Chillers
- Computer Room Air Conditioners (CRACs)
- Direct eXpansion (DX) Air-Conditioning (AC) units
- AC pumps
- AC cooling tower
- Lighting systems

Data center power and cooling consumes 50% to 70% of the electrical power in a typical data center: 45% in HVAC, 25% in UPSs and PDUs and ancillary systems, and 30% in the actual IT and networking equipment (see Figure 1.4) [RAS200801, WOG200801]. The typical design load for the existing generation of 1 unit servers is around 13.25 kW per cabinet, but the load can be higher for high-density blade servers (blade servers can generate a heat load of 15 to 25 kW per cabinet*). Given these consumption levels, it makes sense to investigate how costs and usage can be better managed, especially for peak consumption, which is becoming a pressing issue in the power industry as a consequence of the fact that power companies find it increasingly difficult to build and deploy new generation plants.

Specific to the green IT/green networking issue, in the United States a recently enacted law requires the EPA to assess the power consumption in data centers,

---

* A data center with above-average kilowatt loading, typically greater than 10 kW/rack, is also known as a high-performance data center (HPDC).

Table 1.2 Key Green Segments

Segment	Stakeholders
Green power	Solar thermal, solar photovoltaics, wind, thermal energy, hydro, renewable heating, and clean coal, among others. There are three types of green power products, as defined by the EPA: renewable electricity, renewable energy certificates, and on-site renewable generation. Renewable electricity is generated using renewable energy resources and delivered through the utility grid; Renewable Energy Certificates (RECs) represent the environmental, social, and other positive attributes of power generated by renewable resources; and on-site renewable generation is electricity generated using renewable energy resources at the end-user's facility. In this context it should be noted that the U.S. American Recovery and Reinvestment Act of 2009 provides U.S.$59 billion (€ 45 billion) for green technologies, including U.S.$11 billion (€ 8 billion) for a smart electricity grid.
Green IT/green networks	Enterprise hardware and software technology vendors; data center vendors; semiconductor companies; virtualization, consolidation, and storage vendors; green IT compliance vendors and consultants; equipment take-back and disposal programs.
Green buildings	Integrated green communities, water and waste treatment, sustainable construction transport, energy-efficient insulation, rain water harvesting, fluid-applied membrane systems (for rooftop restoration), eco paints, materials recycling, energy audits and surveys, sound protection panels, low-voltage electric radiant floors, open- and closed-loop geothermal heating/cooling systems, green illumination, and smart buildings and controls.
Green fuels and transport	Alternative fuels, biodiesel, ethanol, hybrids, green batteries, flex fuel vehicles and conversions, synfuels, solar-assisted fuel synthesis, lightweight materials, mass transit, among others.
Government and regulatory bodies	Non-Governmental Organizations (NGOs), environmental organizations, renewable energy associations.

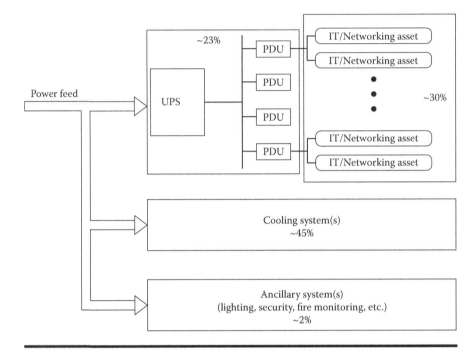

Figure 1.4 Approximate distribution of power consumption in a data center.

establish what manufacturers are doing to improve energy efficiency, and identify what incentives could be put in place to motivate organizations to adopt energy-efficient technology.* In general, organizations (carriers and enterprises) wishing to pursue a greening initiative should consider establishing energy targets, carbon targets, water targets, hazardous materials targets, and waste targets. There are international government-level efforts afoot to reduce GHG emissions†; in particular, the European Union (EU) is studying the issue of sustainability and has

---

* The EPA developed this report in response to the request from Congress stated in Public Law 109-431.
† Thirty-four countries have signed the Kyoto Protocol, the agreement negotiated under the United Nations Framework Convention on Climate Change (UNFCCC), which sets a target for average global carbon emissions reductions of 5.4% relative to 1990 levels by 2012. Individual regions and countries have also developed their own targets. In 2007, the European Union (EU) announced a 20% emissions reduction target compared to 1990 levels by 2020 and will increase this to 30% if there is an international agreement post-2012. The United Kingdom is aiming for a reduction of 60% below 1990 levels by 2050, with an interim target of about half that. Germany is aiming for a 40% cut below 1990 levels by 2020, while Norway will become carbon neutral by 2050. California's climate change legislation, known as AB 32, commits the state to 80% reductions below 1990 levels by 2050. China's latest 5-year plan (2006–2010) contains 20% energy efficiency improvement targets to try to reduce the impact of recent fuel shortages on its economic growth [GES200801].

published requirements for (among other things) 20% renewable energy in overall EU consumption by 2020. It follows that the move to green networks (and data centers) is driven by both financial and regulatory considerations. In fact, the interest in green data centers and networks is part of a recent, broader-based interest in environmentally friendly and socially conscious business operations. For example, the Leadership in Energy and Environmental Design (LEED®) accreditations of the Green Building Certification Institute (GBCI)/U.S. Green Building Council (USGBC®) reinforce the fact that during the past few years there has been increasing interest by businesses in green buildings and sustainable design. Commercially available proven technologies can cut buildings' energy use by 30% without a significant increase in investment cost, according to the U.N. Environment Program (UNEP) [SCH200901]. Having the best-in-class network components with respect to energy efficiency is not sufficient to build sustainable networks, but it is a necessary first step. It should be followed with responsible network design (which includes non-telecom infrastructure) and network operation [ECR200801, ECR200801, ECR200901].

Previous studies by this author have shown that, depending on the industry, the yearly networking costs typically are in the 1-2-4% of the top revenue line of a firm. As a pragmatic estimate, if a third of these expenditures are subject to firm-directed green networks optimizations, this would imply that the equivalent of 0.33% to 1.33% of the top-line dollars can be targeted. With 30% to 50% savings achievable by green technologies, a bottom-line improvement of 0.1% to 1% is achievable. That means that a company with $1 billion in revenue could save between $1 million and $7 million annually and up to $21 million in 3 years. Figure 1.5 depicts the theoretical savings achievable by firms by using green enterprise network principles. While the discrete value of individual activities may not be "spectacular," the aggregate value may be significant, particularly over a multi-year window. While the potential savings will depend on the type of business the firm is in, network operators, carriers, teleport operators, Web hosting sites, and financial and insurance companies with large data centers will benefit the most. It was forecasted recently that by 2011, about 25% of newly constructed data centers will be designed for maximum energy efficiency and minimum negative environmental impact. "Early-adopter companies" will find themselves in this group; other companies will follow. Datamonitor, a market research firm, recently found that 85% of IT professionals believe environmental factors are important in planning IT operations; however, just a quarter have written green criteria into their company's purchasing processes [DAT200901]. This text will hopefully sensitize corporate planners to the need for doing so.

1.2 Overview

Section 1.1 provided some motivations as to why some carriers and enterprises should consider greening their IT and network operations. This section provides

% in Networking Revenue		1		1		2		2		4		4
$	1,000,000	$	990	$	1,650	$	1,980	$	3,300	$	3,960	$	6,600
$	10,000,000	$	9,900	$	16,500	$	19,800	$	33,000	$	39,600	$	66,000
$	100,000,000	$	99,000	$	165,000	$	198,000	$	330,000	$	396,600	$	660,000
$	1,000,000,000	$	990,000	$	1,650,000	$	1,980,000	$	3,300,000	$	3,960,600	$	6,600,000
$	10,000,000,000	$	9,900,000	$	16,500,000	$	19,800,000	$	33,000,000	$	39,600,000	$	66,000,000
		Conservative		Aggressive		Conservative		Aggressive		Conservative		Aggressive

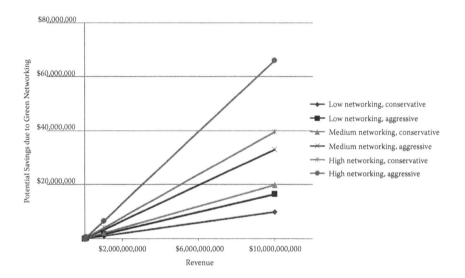

Figure 1.5 Theoretical savings achievable by firms using green network principles.

a summary and overview of possible greening strategies and solutions available to data center and networking practitioners, and can serve as a quick primer.

Modeling after EPA Executive Order 13101 for EPA's own greening, many firms and carriers are seeking to reduce their environmental footprint by increasing and promoting recycling, reducing materials entering the waste stream, promoting and achieving increased and preferential use of materials with recycled content, and emphasizing and increasing the purchase and use of environmentally preferable products, including green power, and greening their IT and networking operations. Typical broad opportunities include

- Green buildings
- Green janitorial/maintenance services
- Green copy paper/publication
- Green meetings
- Green office supplies
- Green electronics

- Green IT/networks
- Green fleets
- Green landscaping
- Green power
- Recycling/waste prevention

At a general level, some firms take these as their broad greening goals [ATI200901]:

- Promote energy efficiencies
- Reduce greenhouse gas emissions
- Promote "Reduce, Reuse, Recycle"
- Promote eco-aware business sustainability
- Support the potential for societal benefits

Some green initiatives seek savings in IT-related energy consumption and also achieve energy conservation through the use of IT. On the first topic, a DOE–Green Grid partnership has recently established a goal of 10% energy savings overall in U.S. data centers by 2011. This equates to a savings of 10.7 billion kWh (equivalent to the electricity consumed by 1 million typical U.S. households in a year), which would reduce greenhouse gas emissions by 6.5 million metrics tons of CO_2 per year. A variety of techniques will be needed by firms and organizations to achieve comparable goals on their own; some of these techniques are discussed in this work, especially from a networking and telecom perspective. This chapter section highlights some of the issues that will be discussed in the text.

The advancement in semiconductor technology over the past 50 years has enabled computers to become continuously more energy efficient. As computers themselves become more energy efficient, opportunities have arisen to utilize computers in ways that achieve improvements for the environment, including the displacement of carbon-intensive activities with lower carbon-emitting activities. The following trends have been noted, which we will also discuss [SKI200901]:

Trend #1: Continuous advancement of computational energy efficiency, in terms of computations per watt of energy used.

Trend #2: The consolidation of computers into powerful, large-scale computing utility centers that can be accessed remotely, including Grid/Cloud Computing installations. In a BAU mode, the worldwide deployment of servers is expected to reach 120 million by 2020, up from 20 million at press time. Enterprise virtualization and/or Cloud Computing are important to greening ICT.

Trend #3: The effective harnessing of computer technologies to achieve improved net environmental outcomes, including carbon reduction and de-materialization.* ICT's potential to reduce emissions extends across multiple sectors. "Eco-services" such as teleworking, telemedicine, home medical monitoring (e.g., with ZigBee devices connected to medical centers), home assistance for those with special needs and facilitating "smart," connected homes (in which heating, appliances, modems, set-top boxes, etc., are always available rather than always on) provide opportunities to reduce the consumption of energy and other resources [ALC200902].

We have already noted that the confluence of financial, environmental, and legislative concerns is now motivating, if not forcing, IT organizations to develop greener data centers. There is as yet no generally accepted, standardized reference framework for green data centers in terms of target architecture goals, benchmarks, and scope. Nonetheless, a set of pragmatic principles and approaches has emerged. These include the following [EPA200701]:

■ "Improved operation": Making use of energy-efficiency improvements beyond current trends that are operational in nature, and, as such, requiring little or no capital investment ("low-hanging fruit" scenario that calls for operating the IT assets more efficiently).
■ "Best practice operation": Seeking efficiency gains that can be obtained through increased adoption of the practices and technologies used in the most energy-efficient facilities in operation today.
■ "State-of-the-art operation": Seeking the maximum energy-efficiency savings that could be achieved using available technologies. This scenario assumes that servers and data centers will be operated at maximum possible energy efficiency using only the most efficient technologies and best management practices available today.

Figure 1.6, modeled after the EPA study, shows that new approaches may be advantageous [EPA200701]. For example, newer edge routers ("state-of-the-art operation") can incorporate a variety of functions, such as deep packet inspection (DPI), session border controller (SBC), and firewall, to run from a central processor (e.g., the Cisco QuantumFlow chip) rather than support such features on separate line cards of the router. Many of the features that traditionally were on a separate blade or a chassis can now be programmed on the central processor; this helps make these edge routers relatively small, resulting in space and power savings for a service provider.

The following observations are relevant to the discussion at hand [ECR200801]:

---

* For example, online bookstores utilize 1/14th of the energy than a traditional store to sell a book, and moving 50% of the DVD/video rentals to broadband distribution could save 4 million barrels of fuel and avoid 1.3 million tons of CO_2 emission annually [ACI200701, TIA200701].

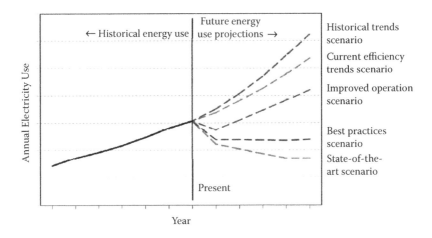

Figure 1.6 Unmanaged power consumption growth versus usage with best practices.

Rising energy costs and increasingly rigid environmental standards work in tandem to draw attention to the "power efficiency" aspect of data networking. Governments and corporations around the world are tightening energy and emission budgets, thus creating demand for new, energy-aware generations of telecom equipment. In response, telecom vendors are starting to label their offerings as "green" and "environment friendly." But one important detail about this theme is often missing: verifiable data to support these "green" marketing claims. As the first step, one must define an efficiency metric that supports informed decisions related to network equipment power consumption.

The evolving greening initiatives of telecom/networking equipment vendors and service providers fall into three general categories [ATI200902]:

■ Reducing energy consumption and heat generation (and/or carbon footprint), and reducing equipment footprint, thereby lowering energy costs in networks and user devices (using an accepted industry measurement metric)
■ Using renewable materials to manufacture products and recycled materials for packaging
■ Conforming to government regulations and recommendations for using renewable energy

Entities such as the Alliance for Telecommunications Industry Solutions (ATIS), the American National Standards Institute (ANSI), ETSI, the International Telecommunication Union (ITU), the Institute of Electrical and Electronics

Engineers (IEEE), the U.S. Department of Energy, the EPA, the European Telecommunications Network Operators' Association (ETNO), the Optical Internetworking Forum (OIF), among others, and a sizable number of equipment suppliers are now assessing approaches and technologies toward energy efficient networking. The motivation is that efficient power designs enable operational expenditure (OPEX) reductions and an increase in performance and reliability.

To obtain normalized benchmarks on efficiency that can be used across institutions, one needs to make use of analytical metrics. There is a need for standardized methodologies to measure and report energy consumption, and to uniformly quantify a network component's ratio of "work performed" to energy consumed. Power consumption in networking equipment and transmission channels (copper, fiber-optic, radio) is related to losses during the transfer of electric charges, which in turn are caused by imperfect conductors and electrical isolators. The exact rate of this consumption depends on the technology (operating voltage and fabrication process), as well as on the frequency of transitions and the number of gates involved. The latter is driven by the architecture and purpose of a network element [ECR200801].

In the context of the data center, one needs a metric that can be employed to compare the amount of electricity the data center consumes for power and cooling with the amount of power used by the data center's IT equipment. Two metrics are emerging as industry standards for measuring data center power consumption: Power Usage Effectiveness (PUE) and Data Center Infrastructure Efficiency (DCiE). These metrics are being advocated by The Green Grid, an industry consortium formed in 2007 to develop standards for measuring data center efficiency and productivity. The Green Grid defines these two metrics as follows [GRE200701]:

$$PUE = \text{Total Facility Power} \, / \, \text{IT Equipment Power}$$

This ratio should be less than 2; the closer to 1, the better.

$$DCiE = \text{IT Equipment Power} \times 100 \, / \, \text{Total Facility Power}$$

DCiE is a percentage; the larger the number, the better.

Clearly, DCiE = 1/PUE. The PUE for traditional data centers is generally between 1.8 and 3.0; the goal is clearly to design data and telecom centers that have much lower PUEs, say around 1.1. For example, containerized data centers manufactured by some vendors, including but not limited to Sun Microsystems, Rackable Systems, and Hewlett-Packard, are able to achieve a PUE of around 1.1 by using containers to house IT equipment compactly and water jackets to quickly remove heat from the equipment; similar concepts can be extended to high-density Internet routing and Web hosting nodes. In these equations, the Total Facility Power is defined as the power measured at the utility meter—the power dedicated solely to the data center; this is important in mixed-use buildings that house data centers as one of a number of consumers of power. The IT Equipment Power is defined as the equipment that is used to manage, process, store, or route data within the data center. Nonetheless, additional work remains to be done in the industry, to develop methodologies to collect power-consumption data and to apply these metrics in such a manner that they can be used to compare the efficiency of data centers in different organizations [MAR200801].

In the context of networks, note that different vendors typically make use of different technologies and architectures for various NEs. As a result, there could be unequal power consumption between equipment belonging to the same NE class, even for the same traffic-handling capacity. The goal of a network architect (now) must be to deploy the most efficient equipment. In principle, to compare devices, one would place two or more NEs under the same load and measure their respective power consumptions; however, this is not practical in the real world. Therefore, developers have sought to use calculated metrics to obtain a sense of what the efficiency is. In order to define a useful efficiency metric for NEs, one needs to normalize energy consumption E by effective full-duplex throughput T, namely,

$$\text{Efficiency metric} = E/T$$

where E is the energy consumption (in watts) and T is the effective system throughput (in bits per second [bps]). The values of E and T may come from either internal testing or the vendor's data. The efficiency metric is typically quoted in watts/Gbps, identifying the amount of energy (in Joules) required to move a block of data (in bits) across the device.

The Energy Consumption Rating (ECR) Initiative is a framework for measuring the energy efficiency of network and telecom devices. In 2008 in response to growing interest from national and international standard bodies and businesses to lower the environmental and operational footprint of networking, the ECR Initiative developed a methodology for reporting, measuring, and regulating the energy efficiency of network and telecom components. The ECR metric provides a common energy denominator between different network and telecom systems operating within a single class. The ECR methodology defines the procedures and conditions for measurements and calculations, and can be readily implemented with industry-standard test equipment. The ECR draft specification defines rules for classifying network and telecom equipment into classes and a methodology for measuring energy efficiency within each class. The final "performance-per-energy unit" rating can be reported as a peak (scalar) or synthetic (weighted) metric that takes dynamic power management capabilities into account. This rating can be further utilized to optimize energy consumption for telecom and network equipment [ECR200901]. The primary metric is a peak ECR metric, which is calculated according to the following formula:

$$\text{ECR} = E_f / T_f \text{ (expressed in watts per 10 Gbps)}$$

where T_f is the maximum throughput (Gbps) achieved in the measurement, E_f is the energy consumption (watts) measured during running test T_f, and ECR is normalized to watts/10 Gbps and has a physical meaning of energy consumption to move 10 Gbits worth of user data per second. This reflects the best possible platform performance for a fully equipped system within a chosen application and relates to the commonly used interface speed. The second metric is a weighted (synthetic) metric called ECRW; this metric takes idle mode into account. It is used in addition to the primary metric to estimate power management capabilities of the device. ECRW reflects the dynamic power management capabilities of the device, which match energy consumption to the actual work accomplished:

$$ECRW = ((\alpha \times E_f) + (\beta \times E_h) + (\gamma \times E_i)) / T_f \text{ (dimensionless)}$$

where T_f is the maximum throughput (Gbps) achieved in the measurement, E_f is the energy consumption (watts) measured during running test T_f, E_h is the energy consumption (watts) measured during half-load test, E_i is the energy consumption (watts) measured during idle test, and α, β, γ are the weight coefficients to reflect the mixed mode of operation (ECR specifies $\alpha = 0.35$, $\beta = 0.4$, and $\gamma = 0.25$).

Along these lines and further giving emphasis to the timeliness of this topic, in early 2009 the ATIS published three standards used to determine telecommunication equipment's energy efficiency. These standards introduce the Telecommunications Energy Efficiency Ratio (TEER) as a measure of network-element efficiency. The standards provide a comprehensive methodology for measuring and reporting energy consumption, and uniformly quantifying a network component's ratio of "work performed" to energy consumed. As highlighted by these observations, it is important to include power management requirements into network specifications as part of a network/IT equipment procurement process. Other metrics have also been defined.

In summation, organizations need a comprehensive and standardized methodology for measuring and reporting energy consumption. The methodolgies defined above are a first step in that direction.

Organizations have shown interest of late in undertaking greening initiatives such as, but not limited to, the following:

■ Deployment of socially responsible corporate policies and procedures
■ Reduction/conservation of energy use
■ Increased energy efficiency solutions
  – Replacement of "power-hungry" ICT elements
  – Carbon footprint reduction/conservation
■ Reduction, reuse, and recycling of waste
■ Reduction of water usage
■ Reduction of hazardous materials

- Disposal of hazardous material
- Increased energy-efficiency solutions, both in the infrastructure and in the systems (e.g., servers, NEs)
- Adherence to hot-aisle/cold-aisle concepts
- Support of end-user conservation
  - Home energy management (equipment manufacturers, power companies, carriers)
  - Smart appliances (equipment manufacturers)
  - Simple auto-configuration, self-repair customer solutions (equipment manufacturers, carriers)
  - Time-of-day electric usage metering (power companies)
- Improved operations:
  - Advanced customer service provisioning, configuration, and fault management (carriers)
  - Zero "truck–rolls" (carriers)
- Make use of "Smart Logistics," "Smart Transportation," and "Smart Grids"
- Support consumer dematerialization (move bits not matter, e.g., consumer paper reduction, consumer DVD reduction)
- Use alternative/renewable electrical energy storage systems
- Improve electrical power distribution (within ICT facility, such as data center or central office)
- Reduce peak use; shift loads when possible
- Deployment of data center and Central Office environmental management systems and controls
- Deployment of newer central cooling systems (including possibly technologies such as compressive refrigerant, absorption/geothermal cooling, air-to-air heat exchangers absorption cooling, steam-jet refrigeration, natural chilled water)
- Satellite operators could replace older vacuum-tube-based high-power amplifiers (HPAs) with newer, more efficient solid-state power amplifiers (SSPAs)
- Green processes for corporate operations, such as fleet management, facilities management, telecommuting, virtual meetings/telepresence (reduce business travel using videoconferencing/telepresence)

Specifically for IT and networking, some simple guidelines to improve efficiency are as follows [JUN200801]:

- Make a commitment to purchase energy-efficient systems and use power management; consider local energy costs and trends.
- Include power management requirements into network specifications as part of a network/IT equipment procurement process.
- Select equipment with the best efficiency ratings (when using an industry-accepted efficiency metric).

- Plan for good system and path utilization; avoid excessive capacity reserves.
- Include rack space, cooling, and power conversion in the assessment, as these items can make a significant difference in monetary and environmental impact.
- Do not multiply entities beyond necessity. Optimized core, edge, and access infrastructures provide the best in energy efficiency.

Design considerations that impact cooling efficiencies include the following [BOT200701]:

- The building layout
- How racks are configured
- How racks are spaced
- How racks are diversified with different equipment to avoid hotspots (provisioning certain zones in the data center with different types of cooling technologies where very hot, high-density servers can be located)
- The type/architecture of the equipment used

Of late, changes in the quantity, configuration, and density of servers are reshaping the data center environment. Higher heat emissions can create hotspots that ultimately threaten server reliability; clearly, heat is created by increased power consumption. Recent advancements in platform (processor) and communications hardware are driving changes in how data centers and high-density telecom sites are cooled [LIE200701]. Servers in high-density arrangements, such as blade servers and carrier-grade communication equipment, typically exacerbate the power/cooling problem because stacking servers into a small footprint requires more power per rack and, consequently, more cooling. There are a number of strategies that network facilities planners can pursue to better manage power and cooling requirements. For example, warehouse-type buildings on one floor with high ceilings are better able to use their geometry for cooling the space than traditional "in-high-rise" data spaces because high-bay, warehouse-like buildings result in more efficient rack layout and airflow.

Some of the possible pragmatic strategies include, but are not limited to, the following:

- Mix high-density with low-density equipment.
- Spread out racks.
- Use cold-aisle/hot-aisle rack configurations.
- Use cold-aisle containment, e.g., closing off the ends of aisles with doors, using ducting to target cold air, and installing barriers atop rows to prevent hot air from circulating over the tops of the racks [MIT2010001].
- Look to avoid air leaks in raised floor.

■ Use blanking panels.
■ Maintain vapor barriers in the data/telecom center perimeter.
■ Virtualize operations to reduce the number of physical servers and other products. In a communications environment, virtualization could imply the use of packet (IP) technologies in place of traditional dedicated time division multiplexing technologies; this will reduce the raw aggregate data rate needed to support networking requirements while ensuring reduction of supporting equipment.
■ Perform periodic (e.g., biannual) carbon footprint analyses.

Table 1.3 depicts some other approaches that have been explored of late, particularly for the telecom/service-provider space, as described in [ETNO200801]. In addition, efforts are underway to define green networking protocols; nearly all existing networking protocols are energy intensive, although many sensornetwork protocols, including ZigBee, do incorporate power control mechanisms [MIN200701]. Suggested selection criteria for a green (network management) protocol include reducing the carbon footprint by at least 10% and being backwards compatible with existing protocols.*

Energy efficiency and renewable energy initiatives ("greening") span a wide area that encompasses all industries, organizations, and consumers.† Appendix 1D provides a taxonomy of general greening initiatives; while the broader topic of Appendix 1D is well beyond the scope of this text, a perusal of the issues identified in that appendix may stimulate additional investigation on the part of the reader. Although the global recession of the end of the recent decade impacted IT budgets, hindered capital expenditure, and slowed IT development, green IT strategies and solutions have gained footing during the economic downturn. Datamonitor, a market research firm, found in 2009 that green IT investments were still high on the corporate agenda and more than ever they were driven by compliance with environmental legislation and cost-saving objectives. The global economic recession has spurred a paradigm shift in the way organizations evaluate, budget for, and deploy Green IT [DAT200901].

---

* In the specific area of network management [MIN198901], two main contenders for "green protocols" were available at press time: SIPP (Simple Integrated Protocol of Protocols) and SLURP (Simple Logical Ubiquitous Reconfigurable Protocol); according to some observers, this could be the biggest—and most audacious—undertaking to impact Network Management since "CMIP vs. SNMP wars of the late 1980s/early 1990s" [ZUK200901].

† Halving the worldwide CO_2 emissions by 2050 means that the worldwide energy consumption in 2050 would have to be reduced to 1960s levels. Because the current population is double that of the 1960s, it means dropping living standards by half. By 2050, the world's population is expected to increase by 3 billion beyond the current level. It will be impossible to realize the goal of halving CO_2 emissions without new ideas. The importance of living close to work, teleworking, and distance education will increase as people come to place more value on staying put and traveling less often [WOG200801].

Table 1.3 Greening Solutions for the Telecom/Service-Provider Space

☐ Energy solutions for cooling in FTT (fiber to the) cabinet or curb architecture
☐ Heat pipe application for outdoor cabinets
☐ Technology for fuel cells; solutions for mobile networks
☐ Technology for fuel cells; solutions for telecom networks
☐ Energy monitoring/management system
☐ Building efficiency rating systems
☐ Properly define power efficiency ratios
☐ Cost-effectiveness and energy efficiency of selected cooling options for data enters
☐ Borehole cooling
☐ Geothermal heat exchanger
☐ Operational actions & energy reduction
☐ Increase the temperature in equipment rooms (e.g., to 78°F)
☐ Use of frequency converters
☐ Use of bio-fuel
☐ Purchase of nature-made star eco-energy
☐ Mistral fresh air cooling (Swisscom)
☐ Fresh air cooling in telephone exchanges (BT)
☐ Simple fresh air cooling (France Telecom Orange)
☐ Purchase of renewable energy (BT, Swisscom, and KPN)
☐ Operation of equipment in an open-air environment
☐ Liquid pressure amplification
☐ Installation of combined solar panel and wind generator
☐ Energy optimization using switchable connector strips
☐ Energy optimization in digital exchanges
☐ DC power systems—energy saving routine
☐ Air conditioning system fan replacement project
☐ Air conditioning system electrically commutated fans
☐ Implementation of building optimization program
☐ Power saver plugs
☐ Use of rest heat
☐ High voltage DC architecture
☐ Use of cold-aisle/hot-aisle and/or cold-aisle containment

Stakeholders of "green concepts" include the following groups [GRE200901]:

- Industry/vendors/material and equipment manufacturers, including integrators and investors from solar, photovoltaics, wind, thermal energy, hydro, biofuels, renewable heating, and clean coal
- Green builders and architects, including the construction industry, integrated green communities, green building materials, materials recycling, and sustainable construction transport
- Green material and equipment manufacturers, including eco paints, sound protection panels, smart buildings and controls, wood-burning stoves, rainwater collectors, energy-efficient insulation, solar heating, fluid-applied membrane systems, and green illumination
- Law firms: energy audits and surveys
- Enterprise hardware and software technology vendors, including data center vendors; network equipment vendors; wireline and wireless service providers; semiconductor companies; virtualization, consolidation, and storage vendors; green IT compliance vendors and consultants; and equipment take-back and disposal programs
- Vendors from green fuel, batteries, and transportation sector, including biodiesel and ethanol manufacturers, automobile vendors, mass transit equipment manufacturers, lightweight material manufacturers, and hybrids
- Financial institutions, including banks, private investors, entrepreneurs, and venture capitalists
- Government and regulatory bodies, including non-governmental organizations, environmental organizations, and renewable energy associations
- Green lifestyle, including accessories, food manufacturers, and suppliers
- IT and network personnel in enterprises, along with consultants, system integrators, academics, researchers, and students

Legislation of the EU and others now requires vendors to ensure that the design of electrical and electronic equipment minimizes the impact on the environment during its entire life cycle. It should be noted, however, that increased power efficiency by itself is only the beginning of the process required to achieve truly green data centers and networks; for example, Gartner makes the point that [BRO200801]:

> 'Green' requires an end-to-end, integrated view of the data center, including the building, energy efficiency, waste management, asset management, capacity management, technology architecture, support services, energy sources and operations.

To that end, some are advocating the introduction of a Chief Green Officer (CGO) in the corporate structure. Table 1.4 identifies some areas where green concepts apply. Organizations (carriers and enterprises) may opt to undertake greening

Table 1.4: Areas Where "Green" Concepts Can Apply

Resource	Definition	Network	Data Center	Office
Energy Use	For all corporate sites, electricity purchased from energy suppliers; also the amount of each type of fuel consumed at the sites —e.g., natural gas, fuel oil, kerosene per year	x	x	x
Ozone-Depleting Substances	For all corporate sites, the number of refrigerant systems that use ozone-depleting substances (for example, air conditioning units), the type used (for example R22, R407C), and the amount of refrigerant leakage that occurred over the year	x	x	x
Paper and Toner	The total amount of paper consumed at all corporate sites and the amount of toner (all colors) consumed at all corporate sites per year		x	x
Media	The total amount of magnetic tapes and optical disks (eventually) discarded at all corporate sites, per year		x	x
Solid Waste	For all corporate sites, the total amount of waste produced at the site (excluding electronic equipment and batteries), and the amount sent to landfill, to incineration, or for recycling by material type (plastic, cardboard, aluminum cans, etc.) per year		x	x

Table 1.4: Areas Where "Green" Concepts Can Apply (Continued)

Resource	Definition	Network	Data Center	Office
Water and Sewer	For all corporate sites, the total amount of water consumed at the office and the amount of liquid sewer per year		x	x
Business Travel (and "truck rolls")	For all corporate sites, the amount of business travel made by employees based at the site, including fleet/hire car use, train, bus, and air travel. This includes "truck rolls" to handle remote installations. Data can be provided in terms of distance, fuel consumed, and/or cost per year	x	x	x
Electronic Equipment and Batteries Discarded	For all corporate sites, the total amount of electronic equipment discarded (without being donated to a cause) and batteries used for business purposes discarded per year	x	x	x

initiatives with the goal of eventually obtaining an industry certification, both to have a benchmark of reference and to get public recognition for their efforts. Environmental programs and certifications in the United States include but are not limited to the following*:

■ EPA's ENERGY STAR® Program
■ EPA's Resource Conservation Challenge (RCC)
■ EPA's Reduce, Reuse, Recycle Program
■ GREEN SEAL™ Program
■ USGBC's Leadership in Energy and Environmental Design (LEED®) Green Building Rating System™

---

* Other countries have appropriate/regional certification entities.

Solutions to energy challenges also require academic research. Some areas of nascent research in the green networking space include the following [IEE200901]:

- Energy-efficient protocols and protocol extensions
- Energy-efficient transmission technologies
- Cross-layer optimizations
- Energy-efficient technology for network equipment
- Energy-efficient switch and base station architectures
- Low-power sleep mode
- Exploitation of passive network elements
- Energy-efficient communications management
- Advanced architectures and frameworks
- Hierarchical and distributed techniques
- Remote power management for terminals
- Context-based power management
- Measurement and profiling of energy consumption
- Instrumentation for energy consumption measurement
- Energy-efficiency in specific networks
- Mobile and wireless access networks
- Broadband access networks
- Management of home and office networks
- Organic light-emitting diode (OLED) technology
- Green storage
- e-Recycling

For example, some proponents see OLED technology as affording a potentially important energy-saving opportunity due to its low energy consumption and emissions. OLED is a relatively new technology that can be used for lighting applications, allowing innovative design capabilities and energy savings for evolving lighting products; OLED is also usable for high-resolution displays; OLED-based systems are ultrathin and lightweight. As another example, green storage technologies can help businesses reduce their IT/data center footprints. Some techniques that can be employed include snapshots, de-duplication, writeable clones, compression, and thin-provisioning. Storage-specific aspects of "green" include the following topics [RIE200901, YOD200901]:

- Metrics and modeling—what matters for storage power and how it can be measured
- How idle is storage?—balancing energy efficiency with the various background activities that are part of the regular care and feeding of large-scale data storage
- Individual storage devices and associated media, including hard disk drives, tape drives, optical devices, and solid-state storage

- Aggregate devices, including disk storage arrays with varying redundancy (JBOD—just a bunch of disks—and RAID—redundant array of inexpensive drives)
- Comparison of tape libraries and virtual tape (VTL) systems
- Use of spin-down and similar techniques to save power for idle data
- Compression, de-duplication, and similar techniques to reduce the number of unintended or excess copies of data
- Storage virtualization

For yet another example in the area of ICT equipment manufacturing and e-recycling (e.g., computers and storage components), there are environmental regulations and initiatives that affect manufacturers and end users, including but not limited to, for example, ENERGY STAR, RoHS (Restriction of the Use of Certain Hazardous Substances), and WEEE (Waste Electrical and Electronic Equipment Directive). Specifically, the objective of WEEE/RoHS is to increase the recycling and/or reuse of such products; WEEE/RoHS also requires that heavy metals such as lead, mercury, cadmium, and chromium and flame retardants such as polybrominated biphenyls (PBBs) and polybrominated diphenyl ethers (PBDEs) be replaced by safer alternatives.

Vendors of data center equipment, as well as telecom and networking equipment, are now responding to the need of the community for green initiatives, of late also favored by new political/governmental interests. These initiatives are covered under the banner of eco-sustainable energy and/or eco-sustainable operations. There is work now underway in Europe and Asia to create an international standard on energy efficiency for telecommunications carriers, and some work is already being done in the United States. Industry sources make the claim that the U.S. telecommunications industry has made significant progress on "going green" in the recent past [WIL200901]. Telecom/networking companies that had announced green equipment development initiatives by press time include, but are not limited to,

- Alcatel-Lucent
- AT&T
- BetterWorld Telecom
- BT
- CableLabs
- China Mobile
- Cisco Systems
- Cricket Communications (Leap Wireless)
- France Telecom/Orange
- Hewlett-Packard (HP)
- Huawei
- IBM
- Innovazul

- Juniper Networks
- Motorola
- Nokia Siemens Networks
- PowerOasis
- Sprint Nextel
- Telstra
- Verizon

For example, Alcatel-Lucent recently stated the following principles related to their eco sustainability initiatives:

> As a responsible corporate citizen and leader in the telecoms industry, Alcatel-Lucent works to enrich people's lives—by transforming the way the world communicates and connects, and by acting in full recognition of its social and environmental responsibilities as a company. Alcatel-Lucent also recognizes the important role the telecommunications sector must play in the global effort to address environmental issues such as climate change. It is devoted to be an active community player and to involve suppliers and business partners in its sustainability approach.
>
> Alcatel-Lucent is deeply committed to ensuring that our customers are able to continuously reduce their carbon footprint while offering consistently high-quality services to their subscribers. Launching innovative and eco-sustainable technologies that really make a difference, is top of mind for us [ALC200901].

Eco-sustainable energy options for carriers, especially to support cellular/3G/4G/LTE (Long Term Evolution) systems, include*

- Solar power
- Wind power
- Hybrid power
- Fuel cell
- Biodiesel
- Compressed air
- Flywheel (more specifically, Flywheel Energy Storage [FES])

Some illustrative examples of how a number of these concepts are beginning to be implemented follow; while these examples illustrate only press-time activities, they provide a perspective of the growing interest on the part of suppliers and network operators to adopt green principles.

---

* These technologies are discussed in chapters that follow.

In the area of green data networks, for example, Verizon announced it would require vendors to cut energy usage by 20% for any products that Verizon purchased for the network after 1 January 2009. Verizon has also been looking at energy efficiency with the use of fuel cells. BT and Verizon were planning to incorporate energy efficiency into broadband architectures [WIL200901].

As an example of a vendor supporting green data networks, an upgrade to the Cisco Systems' IOS operating system announced in early 2009 allowed some of Cisco's enterprise routers to handle power management. The network is a pervasive platform and can provide the intelligence for a holistic approach to managing power across a complete organization. In the most ambitious cases, the router will interplay with facilities-maintenance software to turn down lights during the weekends or move office laptops into battery mode. Initially, Cisco is deploying the new software, called EnergyWise, onto fixed-function Catalyst switches, where it will control the power to Power-over-Ethernet (PoE) devices. These are easy targets because the router already interacts with those devices. EnergyWise can turn off one of the radios in each WiFi access point (AP) during the weekends, or turn IP phones into low-power mode [MAT200901]. The initial release placed EnergyWise on the Catalyst 2000 and 3000 lines and on Integrated Service Routers (ISRs). The departmental-size switches are typically in the wiring closet, touching the endpoints directly (such as the APs and the IP phones) where one can make localized decisions. Later in 2009, Cisco was planning to expand EnergyWise's reach to the Catalyst 4500 family and extend support to some non-PoE devices such as PCs. The flagship Catalyst 6500 core switch/router was expected to get EnergyWise in 2010. Cisco stated that its longer-term goal is to pull together the other major network components.

Cisco is positioning its power monitoring function as a way for corporations to save money. By 2010 Cisco was planning to have the features working with building-management software from partners, enabling Cisco's routers to interact with customers' air conditioning and turning off lights during weekends [MAT200901]. Cisco also reportedly plans to connect EnergyWise to network management systems. Press-time partners include vendors that sell administration software for network operating centers, and vendors that produce software for remotely monitoring and controlling power usage on a PC. Naturally, these extensions have significant security implications, because hackers who are able to penetrate the network nodes could be in a position to control the building's HVAC (and even the building's physical security) system.

As another example of green networks, Alcatel-Lucent was pursuing the issue at press time by developing solar-powered radio sites. Power consumption can contribute up to 35% of a radio site's total costs when the power supply equipment and the power bill are jointly calculated. Continually rising energy prices mean that the energy bill is an increasingly large cost factor in telecom network expansion and operation. Alcatel-Lucent reported that it had sold/installed more than 300 solar-powered radio sites at press time. Alcatel-Lucent offers approaches to reducing network costs related to power consumption of the base station, the cell site, and the total network. The Alcatel-Lucent GSM (Global System for Mobile communications) base transceiver station (BTS) family of products goes beyond reducing power consumption: It promises power efficiency and ensures reduced energy consumption with increased coverage and traffic flow capabilities. Power performance also includes new outdoor no-air-conditioning solutions, eco-sustainable power options, and fewer network cell sites:

■ *Power efficiency:* Lowering network energy consumption costs starts with the BTS itself. Key metrics to determining best-in-class power efficiency include (1) low power consumption per coverage and (2) traffic carried.
■ *No-air-conditioning outdoor solutions:* Removing the shelter and the need for air conditioning significantly reduces BTS cell site power consumption.
■ *High-coverage cell sites:* Reducing the number of network cell sites requires high coverage at each cell site; power efficiency allows for a network with the least number of sites to reduce the power consumption of the overall network.

Other illustrative examples from Alcatel-Lucent follow. Two products, the distributed digital subscriber line access multiplexer (DSLAM) and the passively cooled fiber-to-the-building (FTTB) DSLAMs, have been complemented by eco-sustainable indoor optical network terminals (ONTs) that are located at the customer's premise and convert optical signals back into electrical voice, video, or data stream. The new indoor ONTs reportedly consume up to 30% less power. The distributed DSLAMs, supported by the Intelligent Services Access Manager (ISAM) family of products, enable service providers to manage up to 24 times fewer nodes in FTTN or fiber-to-the-node architectures, resulting in more cost-effective operations. Additionally, through the use of simplified backplanes, power consumption is reportedly reduced by up to 20%. By introducing FTTB DSLAMs with "passive cooling" mechanisms, up to a 10% power savings can be realized at the home, when compared to traditional FTTB DSLAMs that are actively cooled. An additional benefit lies in the elimination of noise that traditionally accompanies an active cooling system.

As yet another example, Sony Ericsson has been unveiling handset models that it claims are more eco-friendly (devices use more recycled materials and consume less energy), and has stated that the company will push greener features across its product line in the next 2 years. Sony Ericsson has made public pronouncements that it targets a cut of 20% of its carbon dioxide emissions by 2015 [VIR200901].

An example (of many possible) of the deployment of green technology for a network operator is Ascent Media Group. The company provides media services to the media and entertainment industries that include motion picture studios, broadcast networks, and cable channels, including satellite uplinks. The company recently installed three magnetic-centrifugal chillers in collaboration with the Connecticut Energy Efficiency Fund (CEEF) and the Connecticut Light & Power Company (CL&P), as part of an incentive program. Frictionless centrifugal chillers provide energy efficiency and eliminate the high friction loss of conventional centrifugal compressors. The new systems increase the efficiency of the water-cooled air-conditioning system; with this technology, it is estimated that nearly 50% of the energy consumption associated with water cooling will be reduced. The new magnetic-centrifugal chillers use electromagnetic compressor technology; this avoids tasks that are carried out by moving parts, thereby eliminating the need for oil and bearings, and reducing friction and counterproductive heat generation during the cooling process. The chiller that recirculates cool water in turn cools the air that is distributed throughout the plant [ADK200801].

In the area of green data centers, for example, IBM recently announced a plan for increasing the level of energy efficiency in IT through its Green Data Center initiative. IBM has outlined a five-step approach designed to improve energy efficiency:

DIAGNOSE: Evaluate existing facilities—energy assessment, virtual 3-D power management, and thermal analytics

BUILD: Plan, build, or update to an energy-efficient data center

VIRTUALIZE: Virtualize IT infrastructures and special-purpose processors. Virtualization of a data center provides an abstraction layer between (1) physical servers and physical storage elements and (2) virtual machines and virtual storage. This virtualization increases efficiency via resource sharing by permitting a single server (or a small set of servers) to support many applications (appearing as individual servers to the various distinct applications).

MANAGE: Seize control with power management software.

COOL: Exploit liquid cooling solutions—inside and outside the data center.

The plan includes new products and services to reduce data center energy consumption. These principles can be extended and particularized to the network infrastructure.

This short illustrative survey shows that there will be development in this space in a going-forward manner.

1.3 Text Scope

The issues, approaches, and analytical metrics discussed in the previous two sections are covered in this text in the chapters that follow. The focus of the discussion is on networks and network equipment, but data centers are woven into the discussion. Chapter 2 provides a primer on networking. Chapter 3 looks at basic consumption concepts and analytical measures for green operations. Chapter 4 provides a basic tutorial on power management while Chapter 5 explores basic HVAC concepts. Chapter 6 surveys regulatory guidelines and best practices. Chapter 7 covers approaches for green networks and data center environmentals (cooling, power). Chapter 8 discusses network computing and Web services as approaches to support green networks and data centers; virtualization and cloud computing are also discussed.

This text should prove timely because some network equipment manufactures are already taking steps to develop green networking gear, as noted earlier. It follows that IT and networking managers also need to become familiar with these concepts. Other communities of interest include communications service providers, including fixed line network operators, mobile operators, and cable/multiple services operator (MSO) providers; systems integrators/VARs; green energy technology companies; green component technology companies; consultants; and regulatory staff and government agencies/state and local officials responsible for green initiatives. Observations such as these from Pike Research, an industry research firm, reinforce the timeliness of the content of this text [PIK200901]:

> Energy consumption is one of the leading drivers of operating expenses for both fixed and mobile network operators. Reliable access to electricity is limited in many developing countries that are currently the high-growth markets for telecommunications. At the same time, many operators have adopted corporate social responsibility initiatives with a goal of reducing their networks' carbon footprints, and network infrastructure vendors are striving to gain competitive advantage by reducing the power requirements of their equipment. All these factors will continue to converge over the next several years, creating significant market potential for greener telecom networks.
>
> The large equipment vendors are creating highly efficient network elements that consume far less power than in previous hardware generations. Operators and vendors alike are exploring innovative network architectures and topologies that will support more capacity with less infrastructure. And, the entire industry is working to incorporate renewable energy sources such as solar and wind power, particularly for off-grid mobile base stations in developing countries where the vast majority of subscriber additions will occur over the next 5 years.

Table 1.5 includes some other press-time observations that reinforce the timeliness of the content of this text.

Table 1.5   Some Press-Time Industry Observations

Observation	Reference
Many data centers operators [and service providers] are simply not aware of the financial, environmental, and infrastructure benefits to be gained from improving the energy efficiency of their facilities. Even awareness does not necessarily lead to good decision making, simply because there is no framework in place for the operators to aspire to. Making data centers more energy efficient is a multidimensional challenge that requires a concerted effort to optimize power distribution, cooling infrastructure, IT equipment and IT output [EUC200801].	EU Code of Conduct on Data Centers Energy Efficiency — Version 1.0, 30 October 2008

Table 1.5 Some Press Time Industry Observations (Continued)

Observation	Reference
While to date much of the discussion surrounding the "greening" of IT has been focused on data centers and equipment metering, advanced sustainability means taking a holistic view of the full business ecosystem. Green computing is a very hot topic these days, not only because of rising energy costs and potential savings, but also due to the impact on the environment. Energy to manufacture, store, operate, and cool computing systems has grown significantly in the recent years, primarily due to the volume of systems and computing that companies now heavily rely upon. Despite the huge surge in computing power demands, there are many existing technologies and methods by which significant savings can be made. [There are many] ways a typical organization can reduce their energy footprint while maintaining required levels of computing performance [GCI200901].	Green Computing Impact Organizations (GCIO)
The biggest threat to the telecom industry is energy use and the industry must recognize and address this as they develop and introduce new products and service solutions… Within the communications industry, there is increased competition so margins are under pressure; added pressure comes from the increased need to innovate and replace old technology with next generation networks capable of delivering new services and solutions. As a result, operators are starting to search for simple and cost effective methods to reduce maintenance costs, without impacting quality and network availability. One of the most significant costs for a majority of carriers is the energy bill; energy efficiency can deliver reductions in cost and CO_2 emissions [ETNO200801].	European Telecommunications Network Operators' Association
For a typical business desktop user, implementing advanced power-management policies alone—without compromising productivity—could save 60% of the electricity consumed. The wasted power is expended as heat, which means that on top of the cost of running the computer, you may also be spending more on air conditioning to cool the home or office. In offices, homes, and data centers, the added heat from inefficient	Climate Savers Computing Initiative (CSCI)

Continued

Table 1.5 Some Press Time Industry Observations (Continued)

Observation	Reference
computers can increase the demand on air conditioners and cooling systems, making the computing equipment even more expensive to run. Servers are typically more efficient than desktops, but still waste 30–40% of the input power. With proven technology that actually saves money in the long run, the vast majority of these energy losses can be eliminated. In addition, there is a significant opportunity to reduce overall energy consumption by putting systems into a lower power-consuming state when they are inactive for long periods of time. Even though most of today's desktop PCs are capable of automatically transitioning to a sleep or hibernate state when inactive, about 90% of systems have this functionality disabled.	
The data center has changed considerably through the decades as the evolution of information technology has enabled it to become the critical nerve center of today's enterprise. The number of data center facilities has increased over time as business demands increase, and each facility houses a rising amount of more powerful IT equipment. Data center managers around the world are running into limits related to power, cooling, and space— and the rise in demand for the important work of data centers has created a noticeable impact on the world's power grids. The efficiency of data centers has become an important topic of global discussion among end-users, policy-makers, technology providers, facility architects, and utility companies. When a standard set of measurements [is] adopted by the industry, it will be easier for end-users to manage their facilities and equipment to achieve optimal energy efficiency.	The Green Grid
The steadily rising energy cost and the need to reduce the global CO_2 emission to protect our environment are today's economical and ecological drivers for the emerging consideration of energy consumption in all fields of communications. Triggered by network operators experiencing energy cost as a significant new factor of their calculation, researchers have started to investigate approaches for reducing power consumption and manufacturers have started offering energy-efficient	IEEE

Table 1.5 Some Press Time Industry Observations (Continued)

Observation	Ref.
network components. Also standards bodies, such as the IEEE, are already developing standards for energy-efficient protocols. However, research and development in these areas is still at an early stage and the gamut of potential solutions is far from being explored. Particularly, there is interest to identify and address issues with a very high potential for significant energy saving, including access network infrastructures, home networks, and terminal equipment [IEE200901].	
The Obama Administration and the current Congress have shown commitment to new energy initiatives by upholding the Energy Policy Act of 2005 and Energy Independence and Security Act of 2007, and then crafting the AmericanRecovery and Reinvestment Act of 2009 (ARRA). In the energy arena, ARRA has battery grants, defense energy and efficiency programs, transmission congestion studies, investment tax credits (accessible to all renewable energy types, not just solar power), advanced energy manufacturing credits, credits for plug-in electric drive vehicles, and measures for alternative fuels and electrification of the transportation system [MEY200901].	Press
Governments and business have a wide range of large-scale initiatives on ICTs and their impacts on climate change and the environment. Initiatives concentrate on greening ICTs rather than tackling global warming and environmental degradation through the use of ICT applications Reducing the direct environmental impacts of ICTs is the most frequent objective of governments and businesses. Of 92 initiatives surveyed by the OECD, over two thirds focus on greening ICTs. Standards and labels such as U.S. Environmental Protection Agency's ENERGY STAR or the Electronic Product Environment Assessment Tool are examples. Encouraging R&D on resource efficient ICTs also ranks high (e.g., Japan's Green IT Project and the Climate Savers Computing Initiatives). In many cases, governments are taking the lead by greening their ICTs and implementing green procurement strategies (e.g., Denmark's Action Plan for Green IT and the United Kingdom's Green ICT Strategy).	Organization for Economic Co-operation and Development (OECD)

Continued

Table 1.5   Some Press Time Industry Observations (Continued)

Observation	Ref.
Less common is tackling global warming and environmental degradation by using ICT applications as enablers of change. Only one tenth of initiatives focus solely on using ICTs as an enabler, although almost one third of initiatives look at both greening ICT infrastructures as well as deploying ICT applications as enablers. Environmental information systems, smart transport and smart buildings are among the most frequently supported ICT-related applications. Very few business associations have strategies to apply ICTs outside of the ICT sector, although there are notable examples such as the Global e-Sustainability Initiative (GeSI) or The Digital Energy Solutions Campaign.	
Most initiatives aim at reducing energy consumption and increasing energy efficiency of ICTs or by using ICT applications. The aim of reducing energy consumption has been driven by high energy prices as well as environmental considerations. However, energy prices have fallen sharply (in 2009, oil prices were less than 40% of peak 2008 prices), and capital and credit have tightened to the choking point. Furthermore, venture capital investments in clean technologies in the United States were down by 87% in the first quarter of 2009 compared with the same quarter in 2008 [OEC200901].	
The ICT sector has transformed the way we live, work, learn and play. From mobile phones and microcomputer chips to the Internet, ICT has consistently delivered innovative products and services that are now an integral part of everyday life. ICT has systematically increased productivity and supported economic growth across both developed and developing countries. But what impact do pervasive information and communication technologies have on global warming? Is it a sector that will hinder or help our fight against dangerous climate change? To answer these questions, one needs to quantify the direct emissions from ICT products and services based on expected growth in the sector. Then one can look at where ICT could enable significant reductions of emissions in other sectors of the economy and has quantified these in terms of CO_2e emission savings and cost savings. Aside from emissions associated with deforestation, the largest contribution to	The Climate Group

Table 1.5   Some Press Time Industry Observations (Continued)

man-made GHG emissions comes from power generation and fuel used for transportation. It is therefore not surprising that the biggest role ICTs could play is in helping to improve energy efficiency in power transmission and distribution (T&D), in buildings and factories that demand power and in the use of transportation to deliver goods. In total, ICTs could deliver approximately 7.8 GtCO2e of emissions savings in 2020. This represents 15% of emissions in 2020 based on a BAU estimation. It represents a significant proportion of the reductions below 1990 levels that scientists and economists recommend by 2020 to avoid dangerous climate change. In economic terms, the ICT-enabled energy efficiency translates into approximately €600 billion ($946.5 billion) of cost savings. It is an opportunity that cannot be overlooked [GES200801].	

Appendix 1A: Green IT Bibliography

The following is a press-time listing of books on Green IT and related topics.

Green to Gold: How Smart Companies Use Environmental Strategy to Innovate, Create Value, and Build Competitive Advantage by D. Esty and A. Winston (Paperback, 9 January 2009).

The Green and Virtual Data Center by G. Schulz (Hardcover, 26, January 2009).

The Greening of IT: How Companies Can Make a Difference for the Environment by J. Lamb, IBM Press; 1st edition (29 March 2009).

Green IT For Dummies by C. Baroudi, J. Hill, A. Reinhold, and J. Senxian (Paperback, 27 April 2009).

Green IT: Reduce Your Information System's Environmental Impact While Adding to the Bottom Line by T. Velte, A. Velte, and R. Elsenpeter (McGraw-Hill, 8 September 2008).

Green Computing and Green IT Best Practices on Regulations and Industry Initiatives, Virtualization, Power Management, Materials Recycling and Telecommuting by J. Harris, Emereo Pty Ltd (Paperback, 21 August 2008).

Harvard Business Review on Green Business Strategy (Harvard Business Review Paperback Series) by Harvard Business School Press (Paperback, 1 November 2007).

The Clean Tech Revolution: The Next Big Growth and Investment Opportunity by R. Pernick and Clint Wilder (Hardcover, 12 June 2007).

Computers and the Environment: Understanding and Managing Their Impacts (Eco-Efficiency in Industry and Science) by R. Kuehr and E. Williams, Kluwer Academic Publ. (Hardcover, October 2003).

Appendix 1B: Partial List of Resources

This appendix provides a partial list of organizations now involved in green ITC technologies. The reader should be aware that there is an ever-growing universe of players in this space, and to track and document this universe can become an all-consuming effort. At the end of the day it is somewhat irrelevant what xyz-advocacy-entity is doing; the question really is: What is this planner actually doing (or able to do) for the firm with whom he or she is affiliated? The body of this text aims at assisting the practitioner in answering that question.

Entity	Notes
80 PLUS	http://www.80plus.org/ A program that promotes the development and purchase of energy-efficient power supplies for use in desktop, laptop, thin client, and workstation computers as well as servers. 80 PLUS is an electric utility-funded incentive program to integrate more energy-efficient power supplies into desktop computers and servers.
AISO.net	http://www.aiso.net
BOMA International	http://boma.org/
Build It Green	http://builditgreen.org/
CALARCH Benchmark 2.1	http://poet.lbl.gov/cal-arch/
Carbon Disclosure Project (CDP)	The CDP acts as an intermediary between shareholders and corporations on all climate-related issues, providing primary climate data from the world's largest corporations to the global marketplace.
Climate Counts	http://www.climatecounts.org/
Climate Savers Computing Initiative (CSCI)	http://www.climatesaverscomputing.org/ The CSCI promotes efforts to increase the use and effectiveness of power-management features by educating computer users on the benefits of these tools and by working with software vendors and IT departments to implement best practices. Member companies commit to purchasing energy-efficient PCs and servers for new IT purchases, and to broadly deploying power management. By publicly declaring their support for this important effort, companies

Entity	Notes
	demonstrate their commitment to the "greening" of IT and join other industry-leading companies and organizations blazing new trails in corporate social responsibility and sustainable IT.
Consumer Electronics Association (CEA)	http://www.ce.org The CEA's mission is to grow the consumer electronics industry. More than 2,000 companies currently make up the CEA membership. The consumer electronics industry is dedicated to the design, production, and marketing of energy-efficient products that provide consumers with a range of product features at competitive prices. The CEA advocates for market-driven solutions as the primary foundation for a national policy regarding the issue of electronic waste. CEA research on energy efficiency has been published in the following reports: • The Energy and Greenhouse Gas Emissions Impact of Telecommuting and e-Commerce, July 2007. • Energy Consumption by Consumer Electronics in U.S. Residences, January 2007.
CoStar Green Report	http://www.costar.com/
Deutsche Gesellschaft für nachhaltiges Bauen / German Sustainable Building Council	http://dgnb.de/
DOE Energy Efficiency and Renewable Energy	http://www.eere.energy.gov
DOE: Building Energy Software Tools directory	http://www.energytoolsdirectory.gov
DOE: Energy Plus Simulation Software	http://www.energyplus.gov

Continued

Entity	Notes
DOE: High-Performance Buildings	http://www.eere.energy.gov/buildings/highperformance
DOE's Federal Energy Management Program (FEMP)	http://www1.eere.energy.gov/femp/ FEMP's mission is to facilitate the U.S. Government's implementation of sound, cost-effective energy management and investment practices to enhance the nation's energy security and environmental stewardship. • Renewable energy: www.eere.energy.gov/femp/technologies/renewable_energy.cfm • Renewable purchasing: www.eere.energy.gov/femp/technologies/renewable_purchasepower.cfm • Design assistance: www.eere.energy.gov/femp/services/projectassistance.cfm • Training: www.eere.energy.gov/femp/technologies/renewable_training.cfm • Financing: www.eere.energy.gov/femp/services/project_facilitation.cfm
Eco-Structure Magazine	http://eco-structure.com/
Electronic Product Environmental Assessment Tool (EPEAT)	http://www.epeat.net/ EPEAT is a system that helps purchasers evaluate, compare, and select electronic products based on their environmental attributes. The system currently covers desktop and laptop computers, thin clients, workstations, and computer monitors. Desktops, laptops, and monitors that meet 23 required environmental performance criteria may be registered in EPEAT by their manufacturers in 40 countries worldwide. Registered products are rated Gold, Silver, or Bronze, depending on the percentage of 28 optional criteria they meet above the baseline criteria. EPEAT operates an ongoing verification program to ensure the credibility of the registry.

Entity	*Notes*
ENERGY STAR	http://www.energystar.gov/ The ENERGY STAR 4.0 specification, which took effect in July 2007, requires power supplies to be at least 80% efficient for most of their load range. It also puts limits on the energy used by devices when inactive, and requires systems to be shipped with power-management features enabled.
Global e-Sustainability Initiative (GeSI)	http://www.gesi.org/ The GeSI has been around for almost a decade and has the goal to foster sustainable development in the ICT sector via open cooperation and voluntary actions. The GeSI is currently focusing on supply chain and e-waste. GeSI works in conjunction with the United Nations Environment Program (UNEP) and the ITU.
Global System for Mobile (GSM) Association (GSMA) Development Fund Green Power Program	http://www.gsmworld.com GSMA is a global trade association of mobile phone operators and manufacturers across 218 countries with the goal of ensuring that mobile phones and wireless services work globally and are easily accessible, enhancing their value to individual customers and national economies while creating new business opportunities for operators and their suppliers. The Green Power for Mobile (GPM) program has been established to promote the use of green power to achieve two commercial objectives: (1) the expansion of mobile networks into regions currently lacking coverage—to bring coverage to the unconnected, and (2) the systematic reduction of reliance on diesel consumption by operators. GPM brings together a community of operators, vendors, and financers to collaboratively catalyze the uptake of green power technology to accomplish these goals. The GPM webpage has been created to guide operators through the process: from understanding the business opportunities, to deciding if green power is technically feasible in a specific geography, to highlighting how green power solutions can form part of corporate social responsibility [ATI200902].

Continued

Entity	Notes
Green Computing Impact Organization (GCIO)	http://www.gcio.org/
Green Electronics Council	http://www.greenelectronicscouncil.org/ The Green Electronics Council partners with environmental organizations, government agencies, manufacturers, and other interested stakeholders to improve the environmental and social performance of electronic products. In 2006, it received a grant from the U.S. Environmental Protection Agency to promote and implement the Electronic Products Environmental Assessment Tool (EPEAT) green computer standard. The EPEAT standard is an easy-to-use tool to rank computer desktops, laptops, and monitors based on their environmental attributes (see related entry in this table) [ATI200902].
The Green Grid	http://www.thegreengrid.org/ The Green Grid is a global consortium of IT companies and professionals seeking to improve energy efficiency in data centers and business computing ecosystems around the globe. The organization seeks to unite global industry efforts to standardize on a common set of metrics, processes, methods, and new technologies to further its common goals.
Green Power Network	http://apps3.eere.energy.gov/greenpower
Green Power Partnership	http://www.epa.gov/grnpower/ The Green Power Partnership is a voluntary program that supports the organizational procurement of green power by offering expert advice, technical support, tools, and resources.
Green Storage/SNIA	http://www.snia.org/forums/green/knowledge/tutorials/ The Storage Networking Industry Association's (SNIA) Green Storage Initiative is dedicated to advancing energy efficiency and conservation in all networked storage technologies and minimizing the environmental impact of data storage operations. Topics of interest include

Entity	Notes
	• Green Storage I - Economics, Environment, Energy and Engineering • Green Storage: Metrics and Measurements • Technologies for Green Storage • Building the Green Data Center: Towards Best Practices and Technical Considerations
Green.TMCnet	http://green.tmcnet.com/
GreenBuild 365	http://www.greenbuild365.org/
Greener World Media	http://www.greenerworldmedia.com
Greenhouse Gas Protocol (GHG Protocol)	http://www.ghgprotocol.org/ The GHG Protocol is a widely used international accounting tool for government and business leaders to understand, quantify, and manage greenhouse gas emissions. The GHG Protocol Initiative, a decade-long partnership between the World Resources Institute and the World Business Council for Sustainable Development, is working with businesses, governments, and environmental groups around the world to build a new generation of credible and effective programs for tackling climate change.
LEED™ (Leadership in Energy and Environmental Design)	A certification program, the primary U.S. benchmark for the design, construction, and operation of high-performance green buildings (certification program operated by the U.S. Green Building Council [USGBC]).
Lighting Design Lab Commercial Lighting Guides	http://www.lightingdesignlab.com/articles/coml_ltg_guides/coml_ltg_guides.htm
Mission Critical magazine	http://www.missioncriticalmagazine.com
National Data Center Energy Efficiency Information Program	http://www.eere.energy.gov/datacenters DOE Save Energy (Now data center Web page) EPA ENERGY STAR data center Web page http://www.energystar.gov/datacenters

Continued

Entity	Notes
National Governors Association	http://www.nga.org/portal/site/nga
NBI Advanced Buildings	http://www.advancedbuildings.net
Post Carbon Cities	http://postcarboncities.net/
Responsible Property Investing Center	http://responsibleproperty.net/
Revolution Events	http://www.greenitexpo.com/
RoHS (Restriction of the Use of Certain Hazardous Substances)	http://www.rohs.gov.uk/ Restriction of the Use of Certain Hazardous Substances in Electrical and Electronic Equipment Regulations 2008 (the "RoHS Regulations"). These regulations implement EU Directive 2002/95, which bans the placing on the EU market any new electrical and electronic equipment containing more than agreed-to levels of lead, cadmium, mercury, hexavalent chromium, polybrominated biphenyl (PBB) and polybrominated diphenyl ether (PBDE) flame retardants.
Standard Performance Evaluation Corporation (SPEC)	http://www.spec.org SPEC is an organization of computer industry vendors dedicated to developing standardized benchmarks and publishing reviewed results.
Sustainable ICT Forum	http://www.greenelectronicscouncil.org
TechSoup.org	http://www.techsoup.org/greentech/
The American Institute of Architects	http://www.aia.org
The Climate Conservancy	http://www.climateconservancy.org/
The Climate Group	http://www.theclimategroup.org/ The Climate Group is an independent, not-for-profit organization that works internationally with government and business leaders to advance climate change solutions and accelerate a low carbon

Entity	Notes
	economy. The Climate Group was founded in 2004 and has offices in the United Kingdom, United States, China, India, and Australia.
The Federal Partnership for Green Data Centers	http://www1.eere.energy.gov/femp/program/fpgdc.html
The Illuminating Engineers Society of North America	http://www.ies.org
The Silicon Valley Leadership Group: Data Center Energy Efficiency (DCEE) Summit	http://dcee.svlg.org
U.S. Department of Energy	http://www.doe.gov
U.S. Department of Energy, Energy Efficiency and Renewable Energy	http://www.eere.energy.gov/
U.S. Environmental Protection Agency (EPA) ENERGY STAR Program	http://www.energystar.gov/
UNEP Finance Initiative	http://www.unepfi.org/
Uptime Institute	http://www.uptimeinstitute.org The Institute created the Tier Classification System, widely accepted as the de facto industry standard, to provide a consistent means to compare typically unique, customized facilities to each other from the perspective of expected site infrastructure availability, or uptime. The Tier Classification System is a benchmarking system to effectively evaluate data center infrastructure in terms of business requirements for uptime.

Continued

Entity	Notes
U.S. Green Building Council (USGBC)	http://www.usgbc.org/ An organization that promotes sustainability related to how buildings are designed, built, and operated. The USGBC developed the Leadership in Energy and Environmental Design (LEED™) rating system.
Virtual Energy Forum	http://www.virtualenergyforum.com/
Waste Electrical and Electronic Equipment Directive (WEEE)	http://ec.europa.eu/environment/waste/weee/index_en.htm The WEEE Directive is the European Community directive 2002/96/EC on waste electrical and electronic equipment which, along with the RoHS Directive 2002/95/EC, became European Law in 2003, establishing collection, recycling, and recovery targets for electrical goods.
Windows for High Performance Buildings	http://www.commercialwindows.umn.edu
World Resource Institute (WRI)	http://www.wri.org/
World Wildlife Fund (WWF) Climate Savers Initiative	http://www.worldwildlife.org WWF is a global nature conservation organization.

Appendix 1C: Examples of How Networking Has Already Had a Major Greening Impact

The 2009 "ATIS Report on Environmental Sustainability" notes the value of using ICT to advance the concept of greening as follows [ATI200902]:

- ICT-enabled solutions could cut annual CO_2 emissions in the United States by up to 22% by 2020. This translates to gross energy and fuel savings of as much as $240 billion.
- A 7% increase in broadband adoption could result in $6.4 billion per year in mileage savings from unnecessary driving and 3.2 billion fewer pounds of carbon emissions in the United States. This is equivalent to taking 273,858 cars off the road for a year.

- If companies were to substitute videoconferencing for 10% of business air travel, it would reduce carbon emissions in the United States by some 35 million tons annually. This is equivalent to saving the amount of energy it takes to provide electricity to more than 4.2 million homes each year.
- Widespread use of broadband has the potential to reduce carbon emissions by more than 1 billion metric tons. This is equivalent to the annual CO_2 emissions of 215 coal-fired power plants.

The 2008 Report by the Global e-Sustainability Initiative (GeSI) entitled "SMART 2020: Enabling the Low Carbon Economy in the Information Age" notes the value of using ICT to advance the concept of greening as follows [GES200901]:

- *Smart motor systems:* A review of manufacturing in China has identified that without optimization, 10% of China's emissions (2% of global emissions) in 2020 will come from China's motor systems alone, and to improve industrial efficiency even by 10% would deliver up to 200 million tons (Mt) CO_2e savings. Applied globally, optimized motors and industrial automation would reduce 0.97 $GtCO_2$e in 2020, worth €68 billion ($107.2 billion).
- *Smart logistics:* Through a host of efficiencies in transport and storage, smart logistics in Europe could deliver fuel, electricity, and heating savings of 225 $MtCO_2$e. The global emissions savings from smart logistics in 2020 would reach 1.52 $GtCO_2$e, with energy savings worth €280 billion ($441.7 billion).
- *Smart buildings:* An analysis of buildings in North America indicates that better building design, management, and automation could save 15% of North America's buildings emissions. Globally, smart buildings technologies would enable 1.68 $GtCO_2$e of emissions savings, worth €216 billion ($340.8 billion).
- *Smart grids:* Reducing T&D losses in India's power sector by 30% is possible through better monitoring and management of electricity grids, first with smart meters and then by integrating more advanced ICTs into the so-called energy internet. Smart grid technologies were the largest opportunity found in the study and could globally reduce 2.03 $GtCO_2$e, worth €79 billion ($124.6 billion).

Appendix 1D: Taxonomy of General Greening Initiatives

The list that follows represents a taxonomy of general greening initiatives (that go beyond Green Networks and/or Green IT, but are worth considering in the broader context), as defined by U.S. Department of Energy's (DOE) documents. The DOE has a long-standing commitment to make government work better and cost less, to use the federal government's extensive purchasing power to stimulate markets for American energy and environmental technologies, and to save taxpayers money by reducing materials costs, waste disposal costs, and utility bills. Executive Order 13123, Section 403(d), instructs federal agencies to develop sustainable design principles and use them in planning and building new facilities. This order also instructs agencies to optimize life-cycle costs and other environmental and energy costs associated with the construction, life-cycle operation, and decommissioning of a facility. The order's chief goals are to reduce the greenhouse gas emissions associated with federal facility energy use by 30% by 2010 in comparison to 1990 levels, to reduce energy consumption by 35% between 1985 and 2010, and to increase water conservation and the cost-effective use of renewable energy. The DOE's Federal Energy Management Program (FEMP) has supported several federal facilities working to meet these goals through a process called greening [DOE200101]. A taxonomy of general opportunities for greening initiatives follows.

- **Site and Landscape Issues:**
  - Land-use planning and transportation
  - Site selection and site planning
  - Building placement and orientation on a site
  - Landscaping principles
  - Stormwater management
  - Plantings in the sustainable landscape
  - Water use in the landscape
  - Chemical use in the landscape

- **Building design:**
  - Integrated building design:
    - Passive solar design
    - Daylighting design
    - Natural ventilation
  - Building envelope:
    - Windows and glazing systems
    - Insulation

- **Energy Systems:**
  - Energy and conservation issues
  - HVAC systems:
    - Boilers
    - Air distribution systems
    - Chillers
    - Absorption cooling
    - Desiccant dehumidification
    - Ground-source heat pumps
    - HVAC technologies to consider

- Water heating:
  - Heat-recovery water heating
  - Solar water heating
- Lighting:
  - Linear fluorescent lighting
  - Electronic ballasts
  - Compact fluorescent lighting
  - Lighting controls
  - Exterior lighting
- Office, food service, and laundry equipment:
  - Office equipment
  - Food service/laundry equipment
- Energy management:
  - Energy management and control systems
  - Managing utility costs
- Electric motors and drives:
  - High-efficiency motors
  - Variable-frequency drives
  - Power factor correction
  - Energy-efficient elevators
- Electrical power systems:
  - Power systems analysis
  - Transformers
  - Microturbines
  - Fuel cells
  - Photovoltaics
  - Wind energy
  - Biomass energy systems
  - Combined heat and power

■ **Water and Wastewater:**
- Water management
- Toilets and urinals
- Showers, faucets, and drinking fountains
- Electronic controls for plumbing fixtures
- Reclaimed water
- Graywater collection and use
- Rainwater harvesting
- On-site wastewater treatment systems

■ **Materials, Waste Management, and Recycling:**
- Material selection:
  - Writing green specifications

- • Structural building components
- • Wood products
- • Low-slope roofing
- • Floor coverings
- • Paints and wall coverings
- • Contract furnishings
- − Operational waste reduction and recycling
- − Construction waste management
- − Deconstruction

■ **Indoor Environmental Quality:**
  - − Indoor air quality
  - − Controlling soil gases
  - − Controlling biological contaminants
  - − Productivity in the workplace
  - − Noise control and privacy

■ **Managing Buildings:**
  - − The role of Operations & Maintenance (O&M)
  - − Building commissioning
  - − Maintaining healthy indoor environments
  - − Leased buildings
  - − Measuring and monitoring benefits
  - − Setting standards and training

■ **Employee incentive programs**

References

[ACI200701]   American Consumer Institute (ACI), Broadband Services: Economic and Environmental Benefits, 31 October 2007.

[ADK200801]   J. Adkoli, Ascent installs eco-friendly HVAC system, 23 May 2008, *TMCnet,* online magazine, http://green.tmcnet.com.

[ALC200901]   Alcatel-Lucent expands portfolio with eco-sustainable broadband access gear that further reduces environmental impact of fiber-based networks, *FTTH Council Europe Conference,* Copenhagen, 12 February 2009, PRNewswire-FirstCall via COMTEX.

[ALC200902]   Alcatel-Lucent, Eco-efficiency in Action — Alcatel-Lucent Sustainability Solutions for Access Networks, June 2009.

[ATI200901]   Alliance for Telecommunications Industry Solutions (ATIS) Promotional information. ATIS, 1200 G Street, NW, Suite 500, Washington, DC 20005, 2009.

[ATI200902]   ATIS Report on Environmental Sustainability, A Report by the ATIS Exploratory Group on Green, ATIS, March 2009, 1200 G Street, NW, Suite 500, Washington, DC 20005.

[BOT200701]   B. Botelho, Gartner Predicts Data Center Power and Cooling Crisis, SearchDataCenter.com, 14 June 2007.

[BRO200801]   J. Brodkin, Gartner in "Green" Data Centre Warning, *Network World*, 30 October 2008.

[DAT200901]   Datamonitor, Can Green IT Bloom in an Economic Downturn? (Market Focus), Revolution Events Ltd., 24 June 2009, Hawkwell Barn, Hawkwell Business Center, Maidstone Road (A228), Pembury Kent, TN2 4AG, UK.

[DOE200101]   U.S. Department of Energy, Energy Efficiency and Renewable Energy, Greening Federal Facilities: An Energy, Environmental, and Economic Resource Guide for Federal Facility Managers and Designers, second edition, Pub. DOE/GO-102001-1165, NREL/BK-710-29267, May 2001.

[ECR200801]   ECR, Energy Efficiency for Network Equipment: Two Steps Beyond Greenwashing, 10 August 2008, Revision 1.0.2, 10-08-2008.

[ECR200802]   Network and Telecom Equipment - Energy and Performance Assessment, Test Procedure and Measurement Methodology, Draft 1.0.2, 14 October 2008.

[ECR200901]   The Energy Consumption Rating (ECR) Initiative, IXIA, Juniper Networks, http://www.ecrinitiative.org.

[EPA200701]   U.S. Environmental Protection Agency ENERGY STAR Program, Report to Congress on Server and Data Center Energy Efficiency, Public Law 109-431, 2 August 2007.

[ETNO200801]   Energy Task Team, European Telecommunications Network Operators' Association, The Hague, London, April 2008.

[EUC200801]   EU Code of Conduct on Data Centers Energy Efficiency — Version 1.0, 30 October 2008.

[GAL200801]   Galley Eco Capital LLC, Glossary, 2008, San Francisco, CA.

[GCI0200901]   Green Computing Impact Organizations (GCIO), http://www.gcio.org/.

[GES200801]   SMART 2020: Enabling the Low Carbon Economy in the Information Age, Report by the Global e-Sustainability Initiative (GeSI), 2008.

[GRE200701]   The Green Grid, The Green Grid Data Center Power Efficiency Metrics: PUE and DCiE, White paper, 2007.

[GRE200901]   The Green Energy Summit, India, http://www.greenenergysummit.com/.

[IBM200701]   IBM unveils plan to combat data center energy crisis; allocates $1 billion to advance "green" technology and services. IBM Press Room, 18 May 2007. http://www-03.ibm.com/press/us/en/pressrelease/21524.wss

[IEE200901]   *2nd IEEE Workshop on Green Communications,* 4 December 2009, Honolulu, HI.

[JUN200801]   Juniper, Energy Efficiency White paper, Juniper Networks, Inc., 1194 North Mathilda Avenue, Sunnyvale, CA 94089-1206, 2008.

[LIE200701]   Liebert Corp., Power Management Strategies for High-Density IT Facilities and Systems, White paper, 2007, Emerson Network Power, 1050 Dearborn Drive, P.O. Box 29186, Columbus, OH 43229.

[MAR200801]   C. Marsan, Two ways to measure data-center power consumption, *Network World,* 18 February 2008.

[MAT200901]   C. Matsumoto, Cisco's IOS goes green, *LightReading Online Magazine,* 27 January 2009.

[MEY200901]   P.E. Meyer, IEEE goes green: Coverage of the First Annual IEEE Green Technology Conference, *IEEE Today's Engineer On Line,* June 2009.

[MIN198901]   D. Minoli et al., *Expert Systems Applications in Integrated Network Management* (Artech House, 1989).

[MIN199401]   D. Minoli and R. Keinath, *Distributed Multimedia through Broadband Communication Services* (Artech House, 1994).

[MIN199402]   D. Minoli, *Imaging in Corporate Environments — Technology and Communication* (McGraw-Hill, 1994).

[MIN199501]   D. Minoli, *Video Dialtone Technology: Digital Video over ADSL, HFC, FTTC, and ATM* (McGraw-Hill, 1995).

[MIN199502]   D. Minoli and O.E. Eldib, *Telecommuting* (Artech House, 1995).

[MIN199601]   D. Minoli, *Distance Learning: Technology and Applications* (Artech House, 1996).

[MIN199801]   D. Minoli and E. Minoli, *Web Commerce Handbook* (McGraw-Hill, 1998).

[MIN199901]   D. Minoli and A. Schmidt, *Internet Architectures* (John Wiley & Sons, 1999).

[MIN200201]   D. Minoli, *Hotspot Networks: Wi-Fi for Public Access Locations* (McGraw-Hill, 2002).

[MIN200501]   D. Minoli, *A Networking-Approach to Grid Computing* (Wiley, 2005).

[MIN200502]   D. Minoli, *Nanotechnology Applications to Telecommunications and Networking* (Wiley, 2005).

[MIN200701]   D. Minoli, K. Sohraby, and T. Znati, *Wireless Sensor Networks* (Wiley, 2007).

[MIN200801]   D. Minoli, *IP Multicast with Applications to IPTV and Mobile DVB-H* (Wiley, 2008).

[MIN200802]   D. Minoli, *Enterprise Architecture A thru Z: Frameworks, Business Process Modeling, SOA, and Infrastructure Technology* (Auerbach, 2008).

[MIN200901]   D. Minoli, *Satellite Systems Engineering in an IPv6 Environment* (Taylor and Francis, 2009).

[MIN200902]   D. Minoli, Is there such a thing as a green network?, *IEEE Globecom/ IEEE ENTNET,* Honolulu, HI, December 2009.

[MIT201001]   R. L. Mitchell, Data Center Density Hits the Wall, Computer World Power and Cooling for the Modern Era, White paper, 2/2/2010.

[OEC200901]   OECD, Towards Green ICT Strategies. Assessing Policies and Programmes on ICT and the Environment, OECD, DSTI/ICCP/IE(2008)3/FINAL, May 2009.

[OVE200701]   S. Overby, How green data centers save money, *CIO,* 26 March 2007.

[PIK200901]   Pike Research, Green Telecom Networks, Market Report, 1320 Pearl Street, Suite 300, Boulder, CO, 80302.

[RIE200901] E. Riedel. Green Storage II: Metrics and Measurements, SNIA White paper, 2009, Storage Networking Industry Association. http://www.snia.org.

[RUS200801]   N. Rasmussen, Electrical Efficiency Measurements for Data Centers, White paper #154, American Power Conversion/Schneider Electric.

[SAE200801]   B. Saeger and H.W. Leppo, LEED® Prep: What You Really Need to Know to Pass the LEED NC v2.2 and CI v2.0 Exams (LDPR), 2008.

[SCH200901]   D. Schuettler and J. Boyle, "Green" Tech a Money Saver in Global Downturn: U.N., Reuters, 22 January 2009.

[SKI200901]   J. Skinner, Moore and Less: Moore's Law, Less Carbon, 25 September 2009, Climate Savers Computing Initiative, http://www.climatesaverscomputing.org

[TIA200701]   TIAX LLC, The Energy and Greenhouse Gas Emissions Impact of Telecommuting and e-Commerce, Final Report by TIAX LLC to the Consumer Electronics Association, 2007.

[VIR200901]   T. Virki, Sony Ericsson unveils two "greener" phones, Reuters, 4 June 2009.

[WIL200901]   C. Wilson, Green initiatives taking hold, Verizon Says, *TelephonyOnLine Magazine,* 9 March 2009.

[WOG200801]   2050 Working Group, Proposing a New Societal System for Cutting CO2 Emissions by Half, 10 October 2008, Ministry of Economy, Trade and Industry, Commerce and Information Policy Bureau, Information Economy Division.

[YOD200901]   A. Yoder. Technologies for Green Storage, SNI White paper, 2009, Storage Networking Industry Association. http://snia.org.

[ZUK200901]   D.N. Zuckerman, Workshop on green communications, *IEEE ICC 2009*, Dresden, Germany, June 2009.

Chapter 2

A Networking Primer

Networks, whether they are corporate intranets, wireline carrier networks, or wireless networks, are basically comprised of nodes and transmission channels. Nodes, also known as Network Elements (NEs), support a variety of communication functions, including multiplexing, switching, and routing. Many of these NEs reside at a particular layer of the Open Systems Interconnection Reference Model (OSIRM), while others support several layers of the protocol model (generally, it is more cost-effective to have an NE cover multiple layers, as compared to discrete devices; this way, there is no need for multiple chassis, multiple power supplies, multiple racks with interconnecting cables, multiple network monitoring systems, etc.). Regardless of the functional bundling, NEs are critical to the proper functioning of any network.

Several dozen different NEs exist for the above-named networks. This chapter looks at some of the basic NEs that are commonly used to construct communication networks. Readers who are familiar with these concepts may opt to skip or skim this chapter. Networking, telecom, and wireless products and services are developing at a rapid pace, and increased data rates for all sorts of networks (optical WAN, LAN, wireless, video, etc.) lead to new demands on resources. For example, the introduction of services such as broadband is challenging the communications industry to keep its energy consumption to a minimum [ETNO200801]. It is important, therefore, to have a good understanding of what network elements are and what function they support. While reading this chapter, keep in mind that greening initiatives of equipment vendors and service providers typically span the following goals [ATI200902]:

- Network efficiencies:
  - Size, Wattage, And Performance [SWAP] improvements

- Heat generation (e.g., fresh air cooling versus air conditioning)
- Energy efficiencies and power consumption
■ Data center efficiencies:
  - Power consumption of data centers, and measures to better manage energy use to achieve measurable results following the implementation of sustainability initiatives
■ Applications and services:
  - The role of applications to enable energy-conscious behavior and practices
  - Applications for the virtualization of server resources to reduce physical resource needs and power needs without sacrificing service
  - Applications for energy management to provide real-time usage data for homes and businesses
■ Life-cycle assessment:
  - Biodegradable materials (e.g., packaging and design)
  - Recycling
  - Disposal of hazardous material (best practices)
■ Corporate operations:
  - Reducing carbon footprint
  - Employee incentives and programs (e.g., telecommuting, virtual meetings/conferences, telepresence; telepresence applications allow for reduced carbon emissions from travel to support face-to-face meetings or worksites)
■ Equipment supply-chain (network/consumer):
  - Compliance requirements and associated programs (e.g., EPA ENERGY STAR® and Research Conservation Challenge [RCC] programs)
  - Equipment power-down and sleep modes
  - Power consumption
■ Alternative energy (renewable and sustainable):
  - Alternative energy sources for power to reduce dependency on power grids and on carbon-generating energy sources such as generators
■ Renewable energy such as wind and solar, especially as solutions for remote sites, in particular for wireless applications

2.1 Networking Functions

Fundamentally, a network exists to move information (voice, video, data, multimedia, instant messages, etc.) from a source point A to a remote sink point B (this being the case for unicast communication environments; in multicast environments, the sinks will be B1, B2, …, Bn; in broadcast environments, there will be, in principle, infinitely many sinks). See Figure 2.1. The transmission channel can operate as a guided medium, where the signal is contained within the physical medium itself (e.g., cable, fiber-optic link), or as an unguided medium, where the signal is sent without a physical medium but only as electromagnetic energy in the free-space

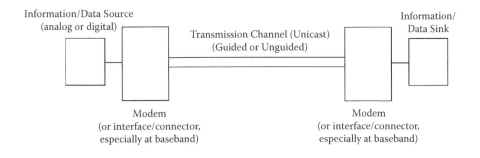

Figure 2.1   Basic communication framework.

field (e.g., radio/wireless communication, satellite communication). Furthermore, the signal can be carried in baseband mode, where the signal changes are coupled directly onto the channel (e.g., by direct connection to the medium itself), or in a broadband* mode where the information is modulated (added, superimposed) onto an adequate underlying carrying signal called the carrier. For example, Local Area Networks (LANs) operate at baseband; fiber-optic links also operate at baseband; cable modem communication, commercial radio, traditional over-the-air broadcast TV, among others, use modulation.

Almost invariably, additional requirements beyond this basic transmission function are imposed on a commercial network, such as, but not limited to, the information will be transmitted reliably, rapidly, with low latency/jitter/loss, cost-effectively, in an integrated fashion, and in a secure manner. Some of these requirements are met by the selection of the proper transmission channel; other requirements are met by employing appropriate NEs in the network, such as NEs supporting multiplexing, grooming, switching, and routing (many other functions are supported in a network, but our discussion here focuses on these key functions).

Networks can be classified in a number of ways. The classification can be in terms of (but not limited to)

- *Information-handling approaches:* circuit mode, packet mode, circuit-emulation over a packet mode; reservation mode (packet-mode reservation of a circuit-mode channel), and hybrid where the network combines packet switching and time division circuit switching in an integrated system[†]
- *Information media*[‡]: voice, data, video, multimedia/converged
- *Transmission class* (e.g., wireline, wireless, etc.)

---

* Note that in this traditional context, the term *broadband* simply implies the use of modulation, regardless of the actual bandwidth.

† This arrangement, however, is not common.

‡ Notice that the term *media* is used in two distinct ways: to describe the transmission medium (channel) and to describe the information stream—the intended meaning should be understood from the context.

- *Transmission mode:* physical or virtual (tunneled/encapsulated/overlay)
- *Transmission channel technology* (e.g., optical, cable, twisted pair, short-hop radio frequency, long-hop radio frequency, free-space optics)
- *Information-handling protocols*, particularly at a given OSIRM protocol layer
- *Administration:* private, public, and hybrid
- *Geographic technology:* Personal Area Network (PAN), Local Area Network (LAN), Metropolitan Area Network (MAN), Regional Area Network (RAN), and Wide Area Network (WAN)
- *Logical scope:* intranet, extranet, and Internet
- *Function/scope* (e.g., commercial, public safety, military)

Clearly, it follows that a comprehensive discussion of networking requires one to take a multidimensional view; in this chapter, however, we look at only a small subset of these dimensions.

One fundamental classification is related to information-handling approaches, as identified above. Communication started out to be circuit mode-based (also called connection-oriented communication), where a session (agreement to exchange information) is established before the information is transmitted between the source and the sink. To accomplish this, the network must maintain "state" information about the session. In connection-oriented systems, the content is almost always transmitted monolithically in its entirety between the source and the sink. This mode has been used in voice telephony since its inception in the nineteenth century and it is reasonably well suited for certain types of communication applications such as high-throughput, long-duration, fixed-bandwidth environments. However, this mode has limitations in terms of efficiency and cost-effectiveness for a number of media, particularly data.

More recently (since the late 1960s, but more so since the 1990s), designers have deployed packet mode communication (also called connectionless communication). Here, there is no requirement to establish a session (agreement to exchange information) before the information is transmitted between the source and the sink; instead, information is sent without formal pre-establishment of "terms" and the network does not maintain extensive "state" information.* The content is always segmented into (relatively) small packets (also called Protocol Data Units [PDUs] or datagrams), and the packets are transmitted between the source and the sink and re-assembled sequentially at the sink. This mode has been used for data communications and communication over the Internet; it is well suited for data communication applications, particularly for medium-throughput, short-duration, bursty-bandwidth environments. Recently, voice and video

---

* This is particularly true in the case of User Datagram Protocol (UDP) transmissions—while one could argue that Transmission Control Protocol (TCP) transmissions entail some sort of end-to-end handshake, it is still considered a connectionless service because the network does not maintain any state (although the endpoints maintain some state information).

applications have also seen a migration to the packet mode. In particular, packet mode is ideal for multimedia and/or streaming environments. Lately, carriers have seen a price erosion of their traditional services; and to compensate for this, they have been pursuing a strategy called Triple Play, where they aim to provide voice, data (Internet access), and entertainment-level video. To do this effectively, packet-mode communication is indispensable.

Circuit emulation over packet mode, reservation mode (packet-mode reservation of a circuit-mode channel), and hybrid has seen much less penetration than the circuit- and packet-mode systems just described. Hence, we focus here on the circuit-versus-packet classification. We also allude to the information media mode (voice, data, video, multimedia/converged). The other classifications are cited only parenthetically.

The subsections that follow examine five basic functions used to meet the above-named goals: regenerating, multiplexing, grooming, switching, and routing; information transfer (transmission) is only briefly discussed. All of these functions can occur at several layers of the OSIRM, particularly the lower layers. It should be noted, however, that some forms of multiplexing, grooming, or switching are more important, and/or well known, and/or supported by commercially available NEs than others; Figure 2.2 depicts some of the more well-known NEs. Figure 2.3 provides a generalized view of the functions of these key NEs. Also see Table 2.1 for a basic glossary of terms.

Next, we discuss in some detail the four key communication functions highlighted earlier.

2.1.1 Information Transfer and Regeneration Function

This function entails the bit transmission across a channel (medium). Because there are a variety of media, many of the transmission techniques are specific to the medium at hand. Functions such as, but not limited to, modulation, timing, noise/impairments management, and signal level management are typical in this context (these, however, are not covered here because it would require a lengthy discussion). Typical transmission problems (e.g., in optical communication) include the following:

- Signal attenuation
- Signal dispersion
- Signal nonlinearities
- Noise, especially due to simple repeaters and optical amplifiers used along the way
- Crosstalk and inter-symbol interference, especially occurring in optical switches and optical cross-connects

Some of these impairments can be dealt with using a regenerator. Figure 2.3 (top) depicts this signal regeneration function pictorially. Regeneration is an important

	Information Transfer	Multiplexing	Grooming	Switching
Transport Layer: end-to-end communication		L4 multiplexing	L4 grooming	L4 Switching
Network Layer: communication across a set of links	IP routing	L3 multiplexing	L3 grooming	IP switching MPLS VoIP (softswitch)
Datalink Layer: communication across a single link	Relaying	ATM Frame relay Ethernet	L2 grooming	ATM switching Frame relay switching Ethernet switching
Physical Layer: communication over a physical channel	Bit transmission Bit retransmission (repeater) (aka reamplification) 3R: Re-amplify, retime, reshape (optical systems)	T1 multiplexing SONET/SDH multiplexing SONET/SDH ADM Wavelength division multiplexing	Digital cross-connect systems SONET/SDH ADM All-optical cross-connect system	Matrix switching Voice/ISDN switching Optical switching Digital cross-connect systems ASON Video "router" (switch)

ADM = Add/Drop Multiplexer
ASON = Automatic Switched Optimal Network
ATM = Asynchronous Transfer Mode
ISDN = Integrated Services Digital Network
MPLS = MultiProtocol Label Switching

Figure 2.2 Well-known communication functions across OSIRM layers.

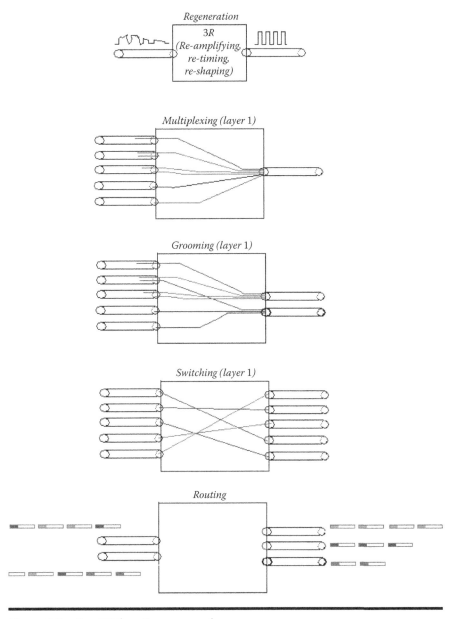

Figure 2.3   Key NE functions: examples.

Table 2.1   Basic Glossary of Terms

Grooming	Combining a relatively large set of low-utilization traffic streams to a higher-utilization stream or streams.
	Generally done at Layer 1 of the OSIRM, but can also be done at other layers.
	Differs from multiplexing in the sense that in the latter case, the incoming streams are usually considered fully loaded, and their entire content must be aggregated time slot by time slot onto the outgoing channel; in grooming, the incoming streams are usually lightly utilized and not every incoming time slot needs to be preserved.
Information transmission	Physical bit transmission (transfer) across a channel.
	Transmission techniques are specific to the medium (e.g., radio channel, fiber channel, twisted-pair copper channel, etc.).
	Often entails transducing, modulation, timing, noise/impairments management, signal-level management.
	Transmission is generally considered a Layer 1 function in the OSIRM.
Multiplexing	Basic communication mechanism to enable the sharing of the communication channel, based on cost-effectiveness considerations.
	Multiplexing allows multiple users to gain access to the channel, and do so in a reliable manner that does not impede or degrade communication.
	Multiplexing is achieved in a number of ways, depending on technology, channel characteristics, etc.
	Generally done at Layer 1 of the OSIRM, but can also be done at other layers.
Regeneration	Restoring the bit stream to its original shape and power level.
	Regeneration techniques are specific to the medium (e.g., radio channel, fiber channel, twisted-pair copper channel, etc.).
	Typical transmission problems include signal attenuation, signal dispersion, and crosstalk. Regeneration correctively addresses these issues via signal re-amplification, re-timing, and re-shaping.
	Regeneration is generally considered a Layer 1 function in the OSIRM.

Table 2.1 Basic Glossary of Terms (Continued)

Routing	The forwarding of packets based on the header information included in the Protocol Data Unit.
	Routing is a Layer 3 function in the OSIRM.
	The forwarding is based on topology information and other information such as priority, link status, etc.
	Topology information is collected (often in real-time) with the use of an ancillary routing protocol.
	Routing information is a "thin" type of "state" that is maintained in the network.
Switching	The mechanism that allows information arriving on any inlet (port) to be forwarded to/relayed to/interconnected with any outlet.
	The function is technology dependent.
	Furthermore, switching can be undertaken at Layer 1 in the OSIRM (e.g., voice switching), Layer 2 (e.g., cell switching), Layer 3 (e.g., MPLS), or even at higher layers.

function, especially in fiber-optic systems. Figure 2.4 depicts three versions of the repeating function, with increasing levels of sophistication. Figure 2.5 depicts the basic building blocks of various regenerators: a "high-end" regenerator includes the re-amplification, re-timing, and re-shaping (3R) functionality; and a "low-end" regenerator includes only the re-amplification (1R) function. 1R NEs are generically called repeaters. The functions of a 3R NE include

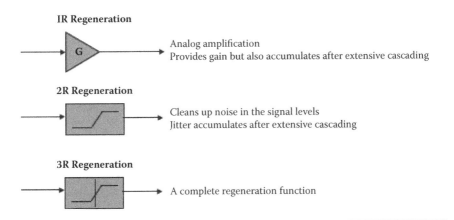

Figure 2.4 Examples of repeaters.

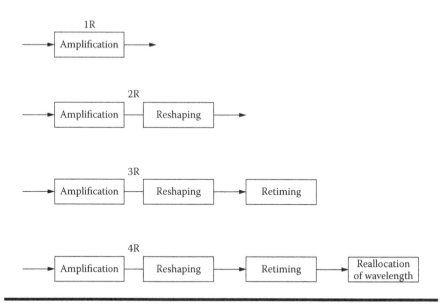

Figure 2.5   Basic functionality of various regenerators.

- Re-amplification: increases power levels above the system sensitivity
- Re-timing: suppresses timing jitter by optical clock recovery
- Re-shaping: suppresses noise and amplitude fluctuations by decision stage

Regenerators are invariably technology specific. Hence, one has, among others, LAN repeaters (even if rarely used), Wi-Fi repeaters, copper-line (T1 channel) repeaters, cable TV repeaters, and optical regenerators (of the 1R, 2R, or 3R form). Originally, optical repeaters required Optical-to-Electrical (O/E) followed by Electrical-to-Optical (E/O) conversion; this conversion added cost and performance limitations. An all-optical regenerator, now being deployed in many carrier networks, retimes, reshapes, and retransmits an optical signal (an all-optical scheme achieves the regeneration function without O/E conversion). Clearly, these NEs require power to operate and when one counts all the regenerators in a network, the aggregate consumption may be nontrivial.

2.1.2 Multiplexing Function

Figure 2.1 depicted a basic communications model where a source makes use of a channel to transmit information to a remote location. It turns out that, in most situations, the channel is relatively expensive. This is because the channel may require that a dedicated cable, wire pair, or fiber-optic link be strung between the two locations; this entails raw materials (e.g., copper, cable), rights-of-way, maintenance, protection, and so on. These deployments may entail a nontrivial carbon footprint over the life cycle of the link, starting with mining of the conductor, pole installation, ongoing powering, and ongoing repeating. Clearly, in the case of multicast or broadcast communication, this becomes even more expensive. (One can see why there is so much interest in wireless communication.)

It then follows that one of the most basic communication requirements is to use some mechanism to enable the sharing of the communication channel, based on cost-effectiveness considerations. This requirement is met using the mechanism of multiplexing. Figure 2.3 (second from top) depicts this function pictorially. Multiplexing, which can be achieved in a number of ways, depending on technology, goals, channel, and so on, allows multiple users to gain access to the channel, and do so in a reliable manner that does not impede or frustrate communication. See Figure 2.6. Generally, there will be a large set of incoming links operating at lower speed; the NE will collect traffic from these links and aggregate the traffic into a single, higher-speed bundle in such a manner that the traffic can be redistributed at the remote end in such a way that source-destination pairs are unmistakably identifiable.

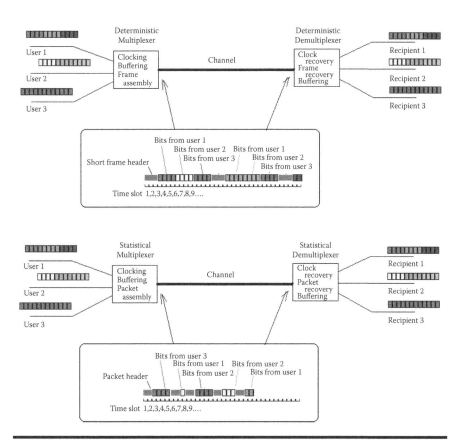

Figure 2.6 The concept of multiplexing.

Examples include a multiplexer that takes 24 DS0 channels and muxes them to a T1 line, or a multiplexer that takes 28 DS1/T1 channels and muxes them to a DS3/T3 line, or a multiplexer that takes 4 OC-3 channels and muxes them to an OC-12 line. What drives the need for muxing is the fact that there are consistent economies of scale in transmission systems. For example, traditionally, for the cost of six DS0 channels, one can use a T1 line that, in fact, can carry 24 DS0s. Hence, if one needed to remotely transmit eight DS0s, one would be much better off purchasing a T1 multiplexer and using a T1 line (the cost of the T1 mux is usually recoverable very quickly). These economies of scale are applicable at all speed levels; for example, for the cost of two OC-3 channels, one can use an OC-12 line, that in fact can carry four OC-3s. Synchronous Optical NETwork (SONET) Add/Drop Multiplexers (ADMs) can be seen as multiplexers or as grooming devices (see next section).

The multiplexing technique depends on a number of factors, including, but not limited to, underlying channel (e.g., radio, fiber-optic); technology (e.g., space-division multiplexing, frequency division multiplexing, wavelength division multiplexing, time division multiplexing, code division multiplexing, demand-assignment multiple access, or random access); discipline (e.g., deterministic multiplexing, statistical multiplexing); the protocol layer (e.g., physical layer, datalink layer, packet layer); and the purpose of the function.

Multiplexing is a fundamental network function that is indispensable to modern networking. Space-Division Multiplexing (SDM) is typical of satellite technology, cellular telephony, WiFi, WiMAX, over-the-air commercial radio, and TV broadcasting, to name a few. Frequency Division Multiplexing (FDM) is very typical of traditional radio and wireless transmission systems such as satellite technology, cellular telephony, WiFi, WiMAX, and over-the-air commercial radio and TV broadcasting (in all of these cases, it is used in addition to space-division multiplexing and optical systems). Wavelength Division Multiplexing (WDM) (a form of frequency division multiplexing) is now very common in optical transmission systems. During the past quarter century, the trend has been in favor of time division multiplexing as an initial first step and statistical multiplexing as a follow-on next step. Code Division Multiplexing (CDM) has been used in military applications and also in some cellular telephony applications. Demand-assignment multiple access has seen some limited applicability in satellite communication. Random access techniques were very common in traditional LANs, before the shift to switched LANs that has occurred since the mid-1990s.

In deterministic multiplexing, the sender and receiver of the information have some kind of well-established, time-invariant mechanism to identify the data sent; in statistical multiplexing, data ownership is generally achieved through the use of a label process (the label mechanism depends on the OSIRM layer); the mix of traffic is based on the arriving rate from different users, and not fixed a priori. Deterministic multiplexing is also called Time Division Multiplexing (TDM). See Figure 2.7.

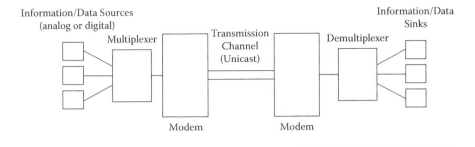

Figure 2.7 Deterministic multiplexing function versus statistical multiplexing function.

It is interesting to note that at the commercial level, physical layer multiplexing, such as that achieved with T1/T3 digital transmission lines, was important in the 1980s*; datalink layer multiplexing, such as that achieved in Frame Relay Service and in Cell Relay Service, was important in the 1990s; while in the 2000s, network-layer multiplexing (packet technology in IP and MultiProtocol Label Switching (MPLS)) has become ubiquitous. Not only is multiplexing done in the network, but it is also routinely done at the end-system level.

2.1.3 Grooming Function

"Network grooming" is a somewhat loosely defined industry term that describes a variety of traffic optimization functions. A basic description of grooming is the "repacking" of information from a (large) number of incoming links that are not fully utilized to a (smaller) set of outgoing links that are much better utilized. Grooming is the optimization of network traffic handling; Figure 2.3 (middle) depicts this function pictorially. One real-life example would be the "repacking" (say, grooming) of airline passengers arriving at an airport from various regional locations to be reloaded/repacked onto another airplane (of the same or larger size) to achieve higher utilization (seat occupancy). Clearly, the grooming of a network to optimize utilization of its traffic-carrying capacity has a significant impact on the network's cost effectiveness and availability. Optimization of the utilization can support a greening process. The availability figure-of-merit is improved by grooming the network in such a manner that traffic can easily be rerouted in case of a link (or other equipment) failure; this is often achieved by grooming capacity so that an available pool of alternate resources is guaranteed available in case of failure. Table 2.2 (based loosely on [BAR200501]) identifies the functionality of NEs at a more detailed level.

---

* It continues to be important at this time in the context of fiber-optic systems and traditional telephony.

Table 2.2   Grooming Functionality

Function	Definition	Example
Packing	Grouping lower-speed signal units into higher-speed transport units Operates on a digital signal hierarchy such as the traditional asynchronous Digital Signal (DS) hierarchy (embedded in each DS3 are 28 DS1 circuits; in turn, each DS1 carries 24 DS0 circuits), or the synchronous SONET/SDH hierarchy	Traditional telco cross-connect system
Assigning	Binding flows to transmission channels (e.g., time slots, frequencies, wavelengths) within a given transport layer	Assigning bit streams to SONET time slots Assigning Wavelength Division Multiplexing (WDM) lightpaths to specific wavelengths on each span of a given mesh or ring network
Converting	Altering signals between channels in the same transport layer	Reshuffle time slots of transiting traffic with Time Slot Interchange (TSI) within a SONET Add/Drop Multiplexor (ADM) Optical Cross-Connection (OCX) to convert lightpaths' wavelengths (also called Wavelength Cross-Connects [WXCs])
Extracting/inserting	Taking lower-speed bit streams to/from higher-speed units	Using an ADM to terminate a lower-rate SONET stream
Physical-level routing	Routing speed bit streams between their origins and destinations	Determining the path that each OC-3 needs to follow and creating a set of lightpaths in an optical network

Grooming makes use of the fact that there typically are multiple layers of transport within a carrier-class (or, at least, fairly complex) network. Generally, there will be a large set of incoming links operating at a given speed; the NE collects traffic from these links and aggregates it into a set (maybe as small as one, but could be more than one) of outgoing links (usually of the same speed as the incoming lines, but could also be a higher-speed line). The aggregation is done in such a manner that the traffic can be redistributed at the remote end such that source-destination pairs are unmistakably identifiable. The term *concentrator* is sometimes also used to describe this function (however, grooming is usually done in a "nonblocking" manner while pure concentration could be done in a blocking [read: overbooked] manner).

As implied above, grooming has come to encompass a variety of meanings within the telecommunications industry and literature. Grooming spans multiple distinct transmission channels or methods, and can occur within multiple layers of the same technology or between technologies: It can be performed when signals are bundled for extended-distance transmission and when cross-connection equipment converts signals between different wavelengths, channels, or time slots [BAR200501]. Some examples are shown in Table 2.3.

Grooming differs from multiplexing in the sense that in the latter case, the incoming streams are usually considered fully loaded and their entire content should be aggregated time slot by time slot onto the outgoing channel; in grooming, the incoming streams usually are lightly utilized and not every incoming time slot needs to be preserved. The example given previously about air travel is useful: in the airline grooming function, not every seat of the arriving planes needs to have a corresponding seat in the outgoing plane, because many of those seats (on the regional planes) were empty (in this example*). Note that sometimes (but not commonly) grooming entails the simple but complete reassignment of slots, one by one, but for an entire transmission facility.

The most well-known example of a grooming NE is a Digital Cross-connect System (DCS); see Figure 2.8 for an example. SONET ADMs can be seen as grooming devices or as TDM multiplexers. ADMs demultiplex a given signal (say, an OC-3) off the backbone for local distribution at a given point in the network (e.g., a metropolitan ring), and at the same time re-inject (re-multiplex) a different signal onto the ring that originates at that local point and needs to be transported over the backbone to some remote location. Network grooming methods are particular to the (multiplexing) technology in use, such as TDM links, SONET/Synchronous Digital Hierarchy (SDH) rings, WDM links, and WDM mesh networks.

---

* On the other hand, consider, for an illustrative example, a fully loaded plane at terminal that was ready to leave but was found to have a mechanical problem; if a new aircraft is brought to the terminal to replace the impaired aircraft there would have to be an exact seat-by-seat transfer from the old plane to the new plane.

Table 2.3   Various Flavors of Grooming

Core grooming	Traffic grooming achieved inside the network: combining low-utilization traffic streams to a higher-utilization stream over the carrier's high-speed core/backbone network. Typically done by the carrier.
End-to-end grooming	Traffic grooming achieved outside the network, i.e., beyond the edge of the network. Typically done directly by the end user. Also sometimes called bypass.
Grooming	Combining low-utilization traffic streams to a higher-utilization stream. A procedure of efficiently multiplexing/demultiplexing and switching low-speed traffic streams onto/from high-capacity bandwidth trunks in order to improve bandwidth utilization, optimize network throughput, and minimize network cost [ZHU200301].
	Term used to describe the optimization of capacity utilization in transport systems by means of cross-connections of conversions between different transport systems or layers within the same system [BAR200101].
	Grooming entails complex physical-level routing, and often implicitly assumes bundling or multiple capacities or multiple layers of transmission [BAR200501].
Grooming architectures	A strategy for the placement of intermediate grooming sites, routing of traffic, and rules for how often traffic is groomed as it traverses the network [WES200201].
Hierarchical grooming	The combination of end-to-end (sub-rate) and intermediate (core) grooming [CLI200301].
Next-generation optical grooming	Traffic grooming in the context of next-generation optical WDM networks to cost-effectively perform end-to-end automatic provisioning [ZHU200301].
Optical grooming	Grooming done on optical links, usually by utilizing Optical Switches (OSs) or Optical Cross-Connects (OCXs).
WDM grooming	Techniques used to combine low-speed traffic streams onto high-speed wavelengths in order to minimize the network-wide cost in terms of line terminating equipment and/or electronic switching [DUT200001].
	Bundling of low-speed traffic streams onto high-capacity optical channels [ZHU200302].

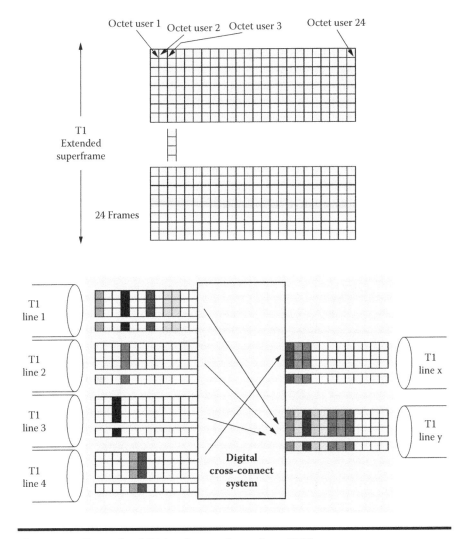

Figure 2.8   Example of T1-level grooming using a DCS.

2.1.4 Switching Function

As seen in Figure 2.2, switching is a commonly supported function: Thirteen types of switches are identified in the figure. The basic goal of switching is to allow information arriving on any inlet (port) to be forwarded/relayed/interconnected with any outlet. The function is technology dependent. Furthermore, switching can be undertaken at Layer 1 of the OSIRM (e.g., voice switching), Layer 2 (e.g., cell switching), Layer 3 (e.g., MPLS), or even at higher layers. Figure 2.3 (second from bottom) illustrates the basic concept.

Figure 2.9 shows the advantages of switching. Fundamentally, switching greatly reduces costs by allowing any-to-any connectivity without requiring that channels be deployed between any two pairs of entities that require to communicate: without switching or relaying, one would need $n \times (n-1)/2$ links to interconnect n users. Table 2.4 defines some of the key switching NEs.

Switches have line interfaces and trunk interfaces. A line interface is an unmultiplexed interface that supports information from/to a single transmitting user; a trunk interface is a multiplexed interface that supports information aggregated and proxied from/to the interface from a multitude of transmitting users. Switches typically incorporate call/session control mechanisms that allow rapid and dynamic call/session cross-connection to the outgoing line, trunk, or ultimate destination. Figure 2.10 depicts some interfaces of interest.

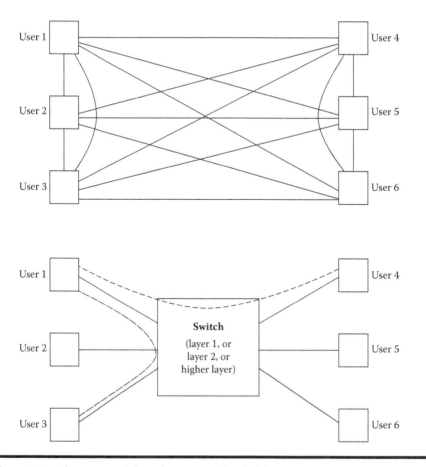

Figure 2.9 The connectivity advantages of switching.

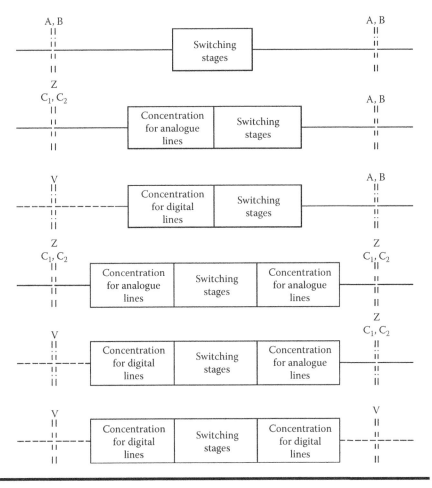

Figure 2.10 Interfaces as defined in ITU Q.551.

2.1.5 Routing Function

Routing is the forwarding of datagrams (packets) based on the header information included in the PDUs, also called packets or datagrams. Routing is a Layer 3 function in the OSIRM. Routed networks use statistically multiplexed packets to transfer data in a connectionless fashion.* A router is a device—or, in some cases, software in a computer—that determines the next network node to which a packet should be forwarded in order to advance the packet to its intended destination. Routing is usually done in a "best-effort" manner where there is

---

* Some older systems, such as ITU-T X.25, also support a connection-oriented, switched virtual connection mode,

Table 2.4   Key Switching Systems

Any-layer switch, at Layer x	An "any-layer" switch is a definition of a generic switch offered by the author of this book to define such a switch as an NE that (1) has protocol stacks on the line interface engine that support Layer x protocols with an equivalent Layer x protocol peer at a remote end, and (2) allows for the forwarding of any incoming PDU at Layer x from a line interface engine to an outgoing line/trunk interface engine by using/acting on information contained at the Layer x header. A line interface is an unmultiplexed interface that supports information from/to a single transmitting user; a trunk interface is a multiplexed interface that supports information aggregated and proxied from/to the interface from a multitude of transmitting users.
Any-layer switch, at Layer x, Protocol y	An "any-layer" switch where the protocol stack supports protocol y. An example is a VLAN switch: a Layer 2 switch that supports the IEEE 802.1p/1Q Protocol.
Matrix switch	Layer 1 switching system that transparently maps an incoming bit stream on a given port to a specified output port. The switching is typically done via a bus and/or internal memory infrastructure. It enables one to select an input source and connect it to one or more outputs. A matrix switch enables rapid reconfiguration under software control of many connections. A DCS is an example. Usually, such a switch does not have a sophisticated session (call) control mechanism. Circuit-switched technology.
Voice switch	Layer 1 traditional NE that allows incoming voice lines (or trunks) to connect with low- or no-blocking to outgoing lines (or trunks). Originally these switches handled analog voice streams, but since the 1970s, the technology has moved to digital Pulse Code Modulated (PCM) voice streams. Usually, such a switch has a sophisticated call control mechanism. Circuit-switched technology.
Integrated Services Digital Network (ISDN) switch	Layer 1 switch that allows incoming ISDN voice lines (or trunks) to be connected with low- or no-blocking to outgoing lines (or trunks). Switches support an out-of-band signaling protocol stack (specifically, Link Access Control – D) in its control plane, as well as an information-transfer stack in its user plane. Circuit-switched technology.

Table 2.4 Key Switching Systems (Continued)

Optical switch	Layer 1 switch that handles optical streams. Typically, it terminates optical-level interface, such as SONET, and either goes through an O/E/O conversion, or is completely O/O-based. More advanced systems switch optical-level wavelengths. Circuit-switched technology.
Digital cross-connect System (DCS)	A Layer 1 switching matrix optimized (standardized) for DS0, DS1, and/or DS3 facilities switching or grooming. Granularity of switchable entities varies but can be all of the above (DS0, DS1, or DS3). Circuit-switched technology. Typically does not have a sophisticated real-time control plane for session control, but the switching is accomplished by non- or near-real-time provisioning.
Automatically Switched Optical Network (ASON) switch	Layer 1 optical switch typically able to support incoming trunks supporting the ITU-T Rec. G8080/Y.1304 "Architecture for the Automatically Switched Optical Network (ASON)," November 2001. ASON has the objective of supporting a scalable, fast, efficient and simple transport network architecture. ASON introduces flexibility to the optical transport network by means of a control plane. It describes a reference architecture for this control plane and its interfaces specifying the basic relation among the three different defined planes: control plane, user (transport) plane, and management plane [ESC200501]. The control plane allows for real-time control of the connections through the switch. Circuit-switched technology. Switching typically occurs at the OC-12, OC-48, OC-192, or optical lambda level.
Video "router" switch	Layer 1 switching system that allows one to switch video signals, typically in Synchronous Digital Interface (SDI) format (or, in earlier days, analog video signals). For example, it allows one to connect more than one camera, Digital Video Recorder (DVR), video server and similar devices, to more than one monitor, video printer, and so on. An audio-video matrix switcher typically has several video and stereo/audio inputs, which can be directed by the user in any combination to various audio-video output devices connected to the switchers. Circuit-switched technology.

Continued

Table 2.4   Key Switching Systems (Continued)

Asynchronous Transfer Mode (ATM) switch	Layer 2 switch designed to support (medium-to-high-speed) switching of ATM cells. ATM cells are comprised of 48 octets of payload and 5 octets of header; the header contains a pointer to the destination (but not the destination itself). The stateful switch supports end-to-end virtual connections. ATM is defined by an extensive set of ITU-T standards. The fixed-length nature of the cells allows optimized, high-throughput inlet-to-outlet (input-port-to-output-port) forwarding (read: switching) of cells. Rigorous queue management allows the delivery of very well-defined Quality of Service (QoS) connectivity services. In theory, three planes are supported (the control plane, the user (transport) plane, and the management plane), enabling the support of *switched virtual connection;* in reality, only *permanent virtual connections* were actually supported by the majority of commercially available ATM switches. Typical line interface rates are at OC-3 and OC-12 rates, but in some cases, other rates are also supported. Packet (cell)-switched technology of the mid-1990s.
Frame Relay switch	Layer 2 switch designed to support megabit-per-second-level switching of Frame Relay frames. The stateful switch supports end-to-end virtual connections. Frame Relay is defined by an extensive set of ITU-T standards. Limited queue management supports a limited set of QoS-based services. In theory, three planes are supported (control plane, user [transport] plane, and management plane), enabling the support of *switched virtual connection;* in reality, only *permanent virtual connections* were actually supported by the majority of commercially available switches. Typical line interface rates are at DS0 and DS1 rates. Packet (frame)-switched technology of the early 1990s.
Ethernet (Local Area Network) switch	Layer 2 switch designed to support 10-, 100-, 1000-, and 10,000-Mbps switching of Ethernet connections (Ethernet frames). Ethernet is defined by an extensive set of IEEE standards. Packet (frame)-switched technology of the mid-1990s.
Virtual Local Area Network (VLAN) (Ethernet) switch	Layer 2 switch designed to support 10-, 100-, 1000-, and 10,000-Mbps switching of Ethernet connections (Ethernet frames), when the Ethernet frames support IEEE 802.1q/1p headers defining VLAN domains. Most Ethernet switches are, in fact, VLAN switches. Packet (frame)-switched technology of the mid-to-late 1990s.

Table 2.4   Key Switching Systems (Continued)

IP switch	Layer 3 switch that aims at treating (Layer 3) packets as if they were (Layer 2) frames. The goal is to expedite the packet forwarding time and simplify the forwarding process. Relies on switching concepts by using a label (at a lower layer) instead of a full Layer 3 routing function. Attempts to create a flat(er) network where destinations are logically one hop away rather than being logically multiple hops away. Supports the concept of "routing at the edges and switching at the core." Seeks to capture the best of each "ATM switching and IP routing." Started out as vendor-proprietary technology; now mostly based on MPLS. An "any-layer" switch operating at Layer 3. True packet-switched technology of the late 1990s/early 2000s.
MultiProtocol Label Switching (MPLS) switch	Layer 3 IP switching technology based on the IETF RFCs defining MPLS. True packet-switched technology of the late 1990s/early 2000s.
VoIP switch (softswitch)	Layer 3 IP switching technology optimized for handling Voice over IP (VoIP). In this environment, voice is digitized and compressed by an algorithm such as the one defined in ITU-T G.723.1; voice is then encapsulated in the Real Time Transport Protocol (RTP), User Datagram Protocol (UDP), and then IP. Replaces a traditional TDM (Layer 1) voice switch. Often based on software running on a general-purpose processor rather than being based on dedicated hardware. True packet-switched technology of the late 1990s/early 2000s.
Layer 4 switch	An "any-layer" switch operating at Layer 4. True packet-switched technology.

no absolute guarantee (at the network layer) that the packets will, in fact, arrive at the destination. The forwarding is based on topology information and other information (such as priority, link status, etc.). Topology information is collected (often in real-time) with the use of an ancillary protocol called a "routing protocol." The routing (forwarding) information is maintained in the routing table; this table contains a list of known routers, the addresses they can reach, and a cost metric associated with the path to each router so that the best available route is chosen. Routers are generally more complex (on a comparable interface basis) than other NEs.

Routing protocols can be of the Internal Gateway Protocol (IGP) type, or of the External Gateway Protocol (EGP) type. IGP is a protocol for exchanging routing information between gateways (hosts with routers) within an autonomous network—for example, a corporate network or an Internet Service Provider (ISP) network. Two commonly used IGPs are the Routing Information Protocol (RIP) and the Open Shortest Path First (OSPF) Protocol. OSPF is a link-state routing protocol defined in IETF RFC 1583 and RFC 1793; the multicast version, Multicast OSPF (MOSPF), is defined in RFC 1584 (some routing protocols are distance-vector type of protocols). Enhanced Interior Gateway Routing Protocol (EIGRP) is a well-known, vendor-specific network protocol (an IGP) that lets routers exchange information more efficiently than with earlier network protocols. See Table 2.5 for some basic routing concepts.

An EGP is a protocol for the distribution of routing information to the routers that connect autonomous systems (here, *gateway* means "router"). Intranets used by corporations and institutions generally employ an IGP such as OSPF for the exchange of routing information within their networks. Customers connect to ISPs, and ISPs use an EGP such as the Border Gateway Protocol (BGP) to exchange customer and ISP routes.

In general terms, the functions of a router include

- Interface physically to the inbound and outbound network channel by supporting the appropriate Layer 1 and Layer 2 protocols (e.g., SONET, ATM, Gigabit Ethernet, etc.)
- Receive packets
- Buffer packets
- Process IP-level packet headers
- Forward packets to appropriate destination, as specified in IP-level packet header over appropriate outbound interface
- Manage queues so that packets can be forwarded based on the priority specified in the IP-level packet header (when priorities are specified)
- Process topology information via the appropriate internal/external gateway protocol
- Maintain routing tables
- Perform firewall function (optional)

Most, if not all, of the NEs described in previous sections handle the incoming traffic in a fairly direct and transparent manner. While routers (also) do not deal with the content of the information, they do perform a significant amount of processing related to either protocol management (several dozen protocols are typically supported by a high-end router), queue management, or media management, or combinations of all three.

At a broad level, routers include the following components: network interfaces engines (specifically, line cards) attached to the incoming and outgoing

Table 2.5 Basic Routing Terminology

Border Gateway Protocol (BGP)	A routing protocol used for exchanging routing information between gateway hosts in an Autonomous System (AS) network; it is an inter-autonomous system routing protocol. An AS is a network or group of networks under a common administration and with common routing policies. BGP is used for exchanging routing information between gateway hosts (each with its own router) in a network of ASs. It is also used to exchange routing information for the Internet and is the protocol used between Internet Service Providers (ISPs); that is, BGP is often the protocol used between gateway hosts on the Internet. Intranets used by corporations and institutions generally employ an IGP such as OSPF for the exchange of routing information within their networks. Customers connect to ISPs, and ISPs use BGP to exchange customer and ISP routes. When BGP is used between ASs, the protocol is referred to as External BGP (EBGP). If a service provider is using BGP to exchange routes within an AS, then the protocol is referred to as Interior BGP (IBGP).
Enhanced Interior Gateway Routing Protocol (EIGRP)	A vendor-specific (Cisco) network protocol (IGP) that lets routers exchange information more efficiently than with earlier network protocols. EIGRP evolved from the Interior Gateway Routing Protocol (IGRP).
Exterior Gateway Protocol (EGP)	A protocol for distribution of routing information to the routers that connect autonomous systems (here, *gateway* means *router*).
External Border Gateway Protocol (eBGP)	An EGP used to perform interdomain routing in Transmission Control Protocol/Internet Protocol (TCP/IP) networks. A BGP router needs to establish a connection to each of its BGP peers before BGP updates can be exchanged. The BGP session between two BGP peers is said to be an eBGP session if the BGP peers are in different ASs.
Interior Gateway Protocol (IGP)	A protocol for exchanging routing information between gateways (hosts with routers) within an autonomous network. Three commonly used IGPs are the RIP, the OSPF Protocol, and Cisco's EIGRP.

Continued

Table 2.5 Basic Routing Terminology (Continued)

Interior Gateway Routing Protocol (IGRP)	The IGRP is a routing protocol (an IGP) developed in the mid-1980s by Cisco Systems.
Internal Border Gateway Protocol (iBGP)	A BGP router needs to establish a connection to each of its BGP peers before BGP updates can be exchanged. A BGP session between two BGP peers is said to be an internal BGP (iBGP) session if the BGP peers are in the same autonomous systems.
Open Shortest-Path First (OSPF)	A link-state routing protocol; an IGP defined in RFC 1583 and RFC 1793. The multicast version, MOSPF, is defined in RFC 1584 (some routing protocols are distance-vector types of protocols.)
Route	In general, a "route" is the n-tuple <prefix, nexthop, [other routing or non-routing protocol attributes]>. A route is not end-to-end, but is defined with respect to a specific next hop that should take packets on the next step toward their destination as defined by the prefix. In this usage, a route is the basic unit of information about a target destination distilled from routing protocols. This term refers to the concept of a route common to all routing protocols. With reference to the definition above, typical non-routing-protocol attributes would be associated with diffserv or traffic engineering [BER200501].
Route change events	A route can be changed implicitly by replacing it with another route or explicitly by withdrawal followed by the introduction of a new route. In either case, the change may be an actual change, no change, or a duplicate [BER200501].
Route flap	A change of state (withdrawal, announcement, attribute change) for a route.
Route mixture	The demographics of a set of routes.
Route packing	The number of route prefixes accommodated in a single Routing Protocol UPDATE Message, either as updates (additions or modifications) or as withdrawals.
Route reflector	A network element owned by a Service Provider (SP) that is used to distribute BGP routes to the SP's BGP-enabled routers [AND200501, NAG200401].

Table 2.5   Basic Routing Terminology (Continued)

Router	(aka gateways in original parlance, this term now has limited use.) A relaying device that operates at the network layer of the protocol model. Node that can forward datagrams not specifically addressed to it. An interconnection device that is similar to a bridge but serves packets or frames containing certain protocols. Routers interconnect logical subnets at the network layer. A computer that is a gateway between two networks at Layer 3 of the OSIRM and that relays and directs data packets through that internetwork. The most common form of router operates on IP packets [SHI200001].
	Internet usage: In the context of the Internet protocol suite, a networked computer that forwards Internet Protocol packets that are not addressed to the computer itself [IPV200601].
Router advertisement	Neighbor discovery message sent by a router in a pseudo-periodic way or as a router solicitation message response.
Router discovery	The process by which a host discovers the local routers on an attached link and automatically configures a default router. In IPv4, this is equivalent to using Internet Control Message Protocol v4 (ICMPv4) router discovery to configure a default gateway [NAR199801].

telecommunication links (each link can be of a different type/technology), processing module(s), buffering module(s), and an internal interconnection module (or switch fabric). Datagrams are received at an inbound network interface card; they are processed by the processing module and stored, even if transiently, in the buffering module. The kind of processing is based on the protocol envelope they carry: The header is examined and understood in the context of the protocol syntax; the protocol type is derived from a field contained in the lower layer (typically Ethernet frame) (in Ethernet-encapsulated datagrams, the field is the EtherType). Queue management is critical to the operation of a router, particularly if it is intended to support multimedia and QoS. Many queuing algorithms have been developed over the years. One of the most common models is the Weighted Fair Queuing algorithm; other algorithms include First-In, First-Out (FIFO), and class-based queuing. Datagrams are then forwarded through the internal interconnection unit to the outbound interface engines (line cards) that transmit them on the next hop. The destination is reached by traversing a series of hops; the sequence of hops selected is based on routing tables, which are maintained through the routing protocols. The aggregate packet stream of all incoming interfaces needs to be processed, buffered, and relayed; this implies that the processing and memory modules need sufficient power. Often, this is accomplished by replicating processing functions either fully or partially on the interfaces to allow for parallel operation.

Router vendors are now beginning to look into the possibility of adding "job functions" to a router, such as network-based computing. With what some vendors call "application-oriented networking" or "application-aware networking," functions such as, but not limited to, proxying, content-level inspection, and advanced voice-data convergence are being advocated. In-network processing and in-network data analysis have been proposed for large sensor networks; however, commercial deployment of sensor networks has, by and large, not yet implemented these in-network features. Some now refer to this approach as "Cloud Computing" (see Chapter 8).

Proponents view this trend for in-network computing as the ultimate form of convergence; here, the network function is more holistically integrated with the application, particularly where quality-of-service requirements exist. Others see this trend as a mechanism to move to a Service-Oriented Architecture (SOA) where not only the application is expressed in terms of functional catalogues, but also the networking functions are seen as functional blocks that are addressable and executable; Web Services support this view. In principle, "Cloud Computing" supports green network processes (greening).

These recent forays into network-computing hybridization are reminiscent of the efforts by the telephone carriers in the late 1980s to build an Advanced Intelligent Network (AIN) that would indeed provide value-added capabilities in the transport fabric. They also remind us of the tug-of-war that has existed between two camps: the traditional carriers that would like to maximize the intelligence and value proposition of the network and the IP data people who would like the network to have very thin functionality and to be general, simple, and able to carry any type of traffic. In trying to make the network intelligent, some inherent generality is lost as functions specific to targeted tasks and applications are overlaid. Nonetheless, it is interesting to note that the very same people who once argued against network intelligence, now that box-level revenues may experience erosion, advocate this once-discarded idea. It should be noted that the AIN efforts of the late 1980s and 1990s have generally failed to gain traction. The idea of a network-resident Service Management System (SMS) has not materialized.

2.2 Networking Equipment

This section briefly discusses the internal architecture of some of the NEs discussed in Section 2.1. As seen in the previous section, there is an abundance of network devices. Hence, only a very small subset is discussed here; even then, some of the devices are specific to the information-handling approach (e.g., circuit mode), the information media (e.g., voice), the transmission class (e.g., wireline), the transmission channel technology (e.g., optical), and the information-handling protocols, to list a few. It follows that only some specific instances are examined. Finally, there generally is no canonical form or architecture, even for the simplest NEs; vendor-based approaches are generally the norm. Just focusing on the simplest NEs, such as DCSs or traditional voice switches, one finds that there are multiple internal and/or logical architectures; this plethora of architectures is generally even more pronounced for the other NEs.

2.2.1 Regeneration Equipment

As implied earlier in the chapter, regeneration equipment is available for various types of transmission channels; it is also available for wired and wireless LANs (here, covering the PHY and a portion of the datalink layer). Figure 2.10 depicts the internal architecture of an optical regenerator that includes the re-amplification, re-timing, and re-shaping (3R) functionality (as noted earlier, a regenerator can also be of the 1R or 2R type).

2.2.2 Grooming Equipment

As discussed in the previous section, grooming entails optimizing physical-level information routing and traffic engineering functions in a network. As in the case of the regenerator, there are various types of grooming NEs based on the transmission channel and other factors. Figure 2.11 and Figure 2.12 provide a view into the architecture and function of a simple and more complex DCS, which is the most basic and typical grooming NE in a traditional telco (carrier) network.

2.2.3 Multiplexing Equipment

As implied earlier, multiplexing entails combining multiple streams into a composite stream that operates at higher speeds and with higher transport capacities. As noted, a number of approaches exist, based on the underlying channel (e.g., twisted-pair/coaxial cable, radio, fiber-optic); the technology (e.g., space-division multiplexing, frequency division multiplexing, wavelength division multiplexing, time division multiplexing, code division multiplexing, demand-assignment multiple access, random access); the discipline (e.g., deterministic multiplexing, statistical multiplexing); and the protocol layer (e.g., physical layer, datalink layer, packet layer). If one were to take just these factors, one would already have $3 \times 7 \times 2 \times 3 = 126$ different multiplexers, each with one or more internal architectures. Figures 2.13 and 2.14 depict two illustrative examples. Figure 2.13 shows a basic TDM that is very commonly used in a traditional telephone plant: multiplexes DS1s to DS3s; this multiplexer is also known as the M13 multiplexer. Figure 2.14 shows a basic SONET multiplexer, which is also very commonly used in a traditional telephone plant. These are just illustrative examples; many more could be provided.

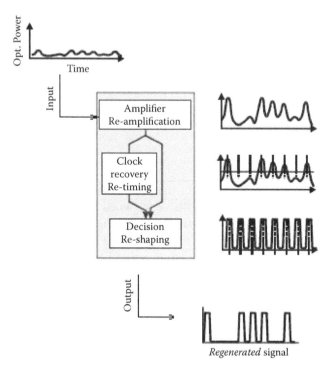

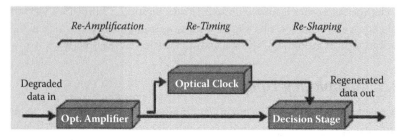

Figure 2.10   Basic architecture of a 3R regenerator.

2.2.4 Switching Equipment

There is quite a plethora of switching equipment; this is related to the fact that switches are one of the most fundamental types of NEs, being that they drive the economics of a network. Without switches, every source and sink would have to be connected either directly or via a series of links (the latter is undesirable in many instances because of the accumulation of delay).

The most common types of switches are

■ Traditional Layer 1 voice switch (originally analog switching, but now exclusively digital)

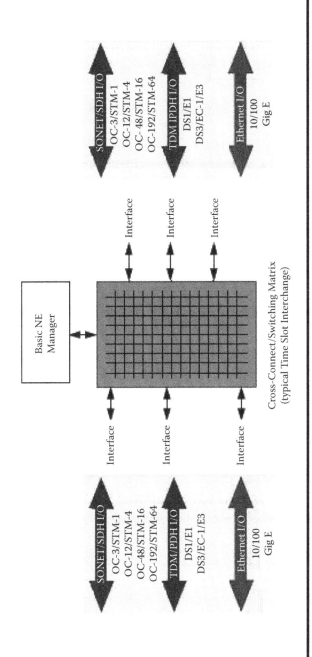

Figure 2.11   Basic digital cross-connect system.

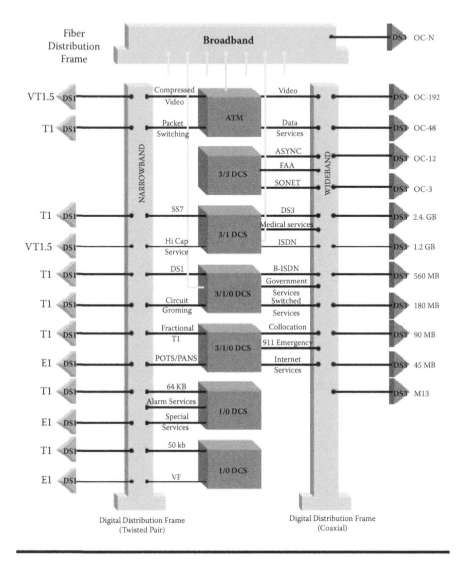

Figure 2.12   Digital cross-connect system arrangements (example).

- Ethernet/VLAN Layer 2 switches
- ATM/Frame Relay Layer 2 switches
- Layer 3 IP switches

Traditional voice switches can be of the single-stage kind or of the multi-stage kind. Single-stage switches typically are found in smaller and nonblocking applications (say, less than 10,000 lines); multi-stage switches are found in large-population environments (say, 50,000 lines)—here, there may be a level of concentration and ensuing blocking. In nonblocking environments, any inlet can always be connected

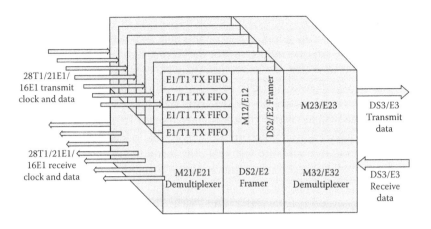

Figure 2.13   Basic TDM multiplexer (multiplexes DS1s to DS3s, aka M13).

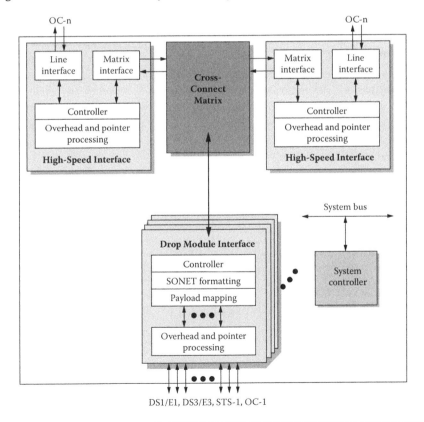

Figure 2.14   Basic SONET multiplexer.

to any specific unused outlet; in blocking environments, an inlet may or may not be able to connect to a specific outlet at a given instance even if the outlet is unused; this is because of the internal architecture of the switch. Blocking, however, is usually designed to be, at a very minimum, less than 1% of the cases (call attempts). Both the multi-stage feature and the possibly resulting blocking are utilized to decrease the cost of the switch fabric. The basic switching mechanism for traditional digital voice switches is the Time Slot Interchange (TSI).

Figure 2.15 depicts the architecture of a basic traditional voice switch. The switch has incoming line cards (to terminate the incoming loop, which could be a discrete loop or a T1 access line—either in copper form or even in a fiber form); outgoing line cards or trunks to terminate the outgoing loop or trunk, which could be a discrete loop or a T1/T3/SONET line—either in copper form or even in a fiber form); and a call control mechanism to set up and terminate end-to-end connections. Note that outgoing line cards are just loop interfaces if the call needed to reach a local subscriber, or multiplexed trunks if the switch were to connect to a remote switch, as shown in Figure 2.16.

Next we look at Asynchronous Transfer Mode (ATM). ATM, a connection-oriented, medium- to high-speed service that was standardized by the ITU in the late 1980s and early 1990s, is based on cell forwarding. A cell is a small (53 octet) Layer 2 PDU. With the use of interface cards, an ATM switch accepts multiple streams of data in many different formats, segments them into cells, and statistically multiplexes them for delivery at the remote end, where they are reassembled into the original higher-layer datagram. For example, ATM uses encapsulation of input IP packets before they are segmented into cells; after transiting the network, the cells are reassembled into an IP packet for delivery (typically to a router, or a traditional or IP voice switch.)

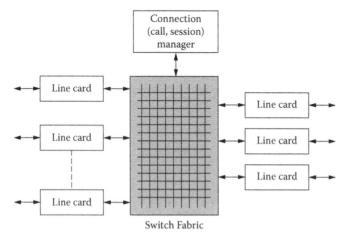

Figure 2.15 Basic voice switch architecture.

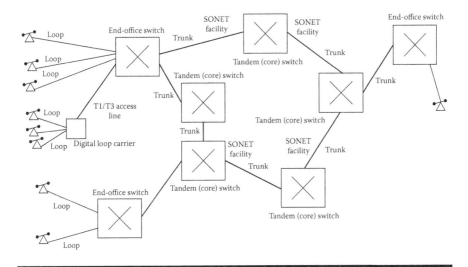

Figure 2.16   A typical carrier voice network (traditional).

Sophisticated traffic management techniques are used to achieve a tightly defined QoS level (usually a small set of QoS levels are supported). ATM can support multiple media (voice, video, and data). Access links range from 1.544 Mbps (DS1/T1), to 45 Mbps (DS3/T3), 155 Mbps (SONET), as well as, in some cases, OC-12 and OC-48 links (although these rates are much less common). ATM cells are typically transported via SONET- or SDH-based facilities. Mapping and delineation of cells into a SONET/SDH frame is done by the PHY layer's Telecommunication Convergence (TC) sublayer. The streams of information are managed using Virtual Circuits (VCs). VCs are permanent or semi-permanent logical ("soft") connections set up from source to destination and are utilized as long as they are needed and then torn down.

Cell forwarding is based on label switching. Here, cells with a common destination are assigned a label that the ATM switch uses to index a routing table to determine the outgoing port on which the cells need to be transferred. Note that the label is not the address of the destination; it is just a shorter pointer into a stateful table. The switch maintains "state" information on each active connection that indicates which outgoing port is to be used in order to reach another switch along the way to the destination, or the destination itself. The switching label only has local (not end-to-end) significance. As part of the switching process, the ATM switch may assign new labels to new cells that are going to other switches, a technique referred to as label swapping.

Types of ATM switches typically include (see Figure 2.17) carrier backbone (the Tier 1 core of a public ATM Network); carrier edge, located in a carrier's locations closer to the actual information sources or sinks (typically a local Central Office); and enterprise backbone switches, although of late enterprise networks have transitioned to high-speed core IP/MPLS routers for this application.

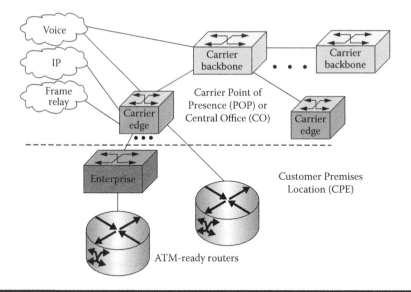

Figure 2.17   Typical ATM network environment.

Some of an ATM switch key's attributes include the following:

- Blocking behavior (e.g., blocking, virtually non-blocking, or non-blocking)
- Switch fabric architecture (e.g., single bus, multiple bus, self-routing, augmented self-routing)
- Buffering method (e.g., internal queuing, input queuing, output queuing, output queuing/shared buffer)

Ethernet Layer 2 switching has become very prevalent in enterprise networking applications. Basically, an Ethernet switch supports multiple Ethernet ports on the access side (e.g., 24, 96, 386, etc.) and a number of trunks on the uplink side. In addition, Ethernet switches almost invariably now support VLAN management (subswitching based on the VLAN ID contained in IEEE 802.1p/1Q-conformant frames.) Table 2.6 lists typical functions of an Ethernet switch; of course, frame forwarding is its basic function. Figure 2.18 depicts a typical enterprise environment, and Figure 2.19 shows one example of an Ethernet switch.

2.2.5 Routing Equipment

Routers are very common NEs in IP networks such as Internet backbone/access networks, VoIP networks, IPTV networks, and enterprise networks. Routers perform two main functions: path control functions and data path control (switching) functions. As noted earlier, these functions include

Table 2.6 Typical Functions of an Ethernet Switch

Function	Definition
802.1D Spanning Tree Protocol (STP)	Layer 2 feature that enables switches to find the shortest path between two points and eliminates loops from the topology
802.1p packet prioritization	Layer 2 feature that supports Quality-of-Service (QoS) queuing to reserve bandwidth for delay-sensitive applications (e.g., VoIP)
802.1Q and port-based VLAN	Layer 2 feature that isolates traffic and enables communications to flow more efficiently within specified groups
802.3 frame forwarding	Layer 2 feature that enables Ethernet frames to be forwarded on the uplink of the switch
802.3ad link aggregation	Layer 2 feature that allows multiple network links to be combined, forming a single high-speed channel
802.3x Flow Control	Layer 2 feature implemented in hardware to eliminate broadcast storms DVMRP (Distance Vector Multicast Routing Protocol) Layer 3 feature for routing multicast datagrams
GMRP (GARP Multicast Registration Protocol)	Layer 2 feature that determines which VLAN ports are listening to which multicast addresses to reduce unnecessary traffic through the switch
GVRP (GARP VLAN Registration Protocol)	Layer 2 mechanism for dynamically managing port memberships
IGMP (Internet Group Management Protocol) snooping	Layer 2 feature that allows switch to "listen in" on the IGMP communications between hosts and routers to automatically add port entries
Jumbo frames support	Layer 2 feature that enables packets up to 9K to transmit unfragmented across the network for lower overhead and higher throughput
OSPF (Open Shortest Path First)	Layer 3 feature that supports the calculation of a shortest path tree and maintains a routing table to reduce the amount of hops (and latency) it takes to reach the destination
RIP (Routing Information Protocol)	Layer 3 feature that supports the determination of a route based on the smallest hop count between source and destination

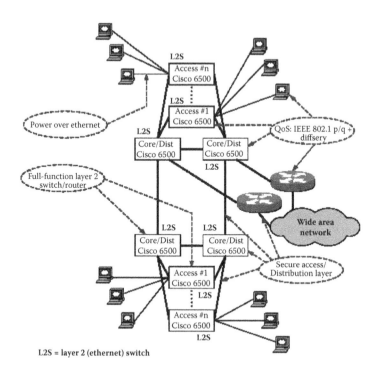

Figure 2.18   Typical switched Ethernet environment.

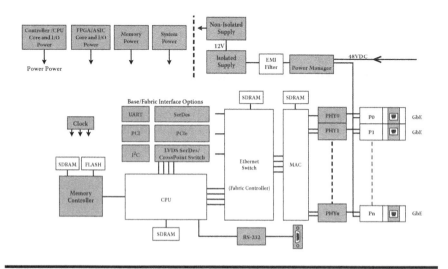

Figure 2.19   Example of an Ethernet switch. (Courtesy of Znyz Networks.)

- *Classification of packets for handling/queuing/filtering:* Specifically, compare packets to classification lists and perform control via a variety of sophisticated queue management techniques.
- *Packet switching:* Layer 3 switching based on routing information and QoS markings. This includes generating outbound Layer 2 encapsulation; performing Layer 3 checksum; managing Time-To-Live (TTL)/hop count update.
- *Packet transmission:* Access outbound transmission channels.
- *Packet processing (manipulation):* Change contents of packet (e.g., compression, encryption).
- *Packet consumption:* Maintain/manipulate routing information (track updates/update neighbors, e.g., absorb routing protocol updates; issue services advertisements/routing protocol packets).
- *Support management functions:* Interface statistics, queue statistics, Telnet, SNMP alerts, ping, trace route).

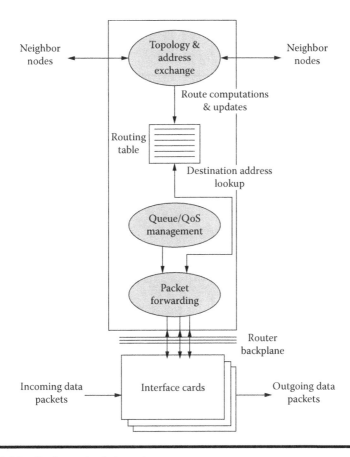

Figure 2.20   Basic router internals.

How some of these functions as supported is illustrated in Figures 2.20 through 2.23. Figure 2.20 and Figure 2.21 depict some basic router internals, Figure 2.22 shows the typical hardware modules of a router, and Figure 2.23 illustrates some vendor- and model-specific architectures.

Figure 2.24 shows a typical carrier-provided MPLS network. MPLS networks are based on Layer 3/Layer 2 routing/switching functionality and are generally replacing the carrier-based ATM networks of the late 1990s and early 2000s.

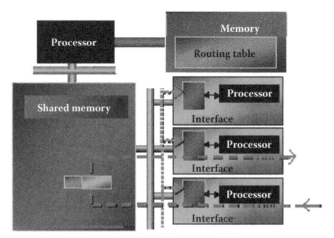

Shared Memory Distributed Processors Architecture

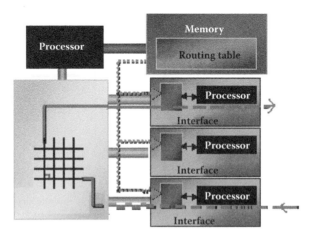

Crossbar Architecture

Figure 2.21   Basic router internals (another view).

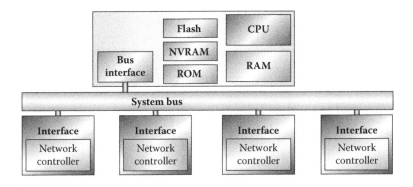

Figure 2.22   Typical router hardware.

2.3 Conclusion

This chapter looked at the basic components of carrier and/or enterprise networks. Networks are basically comprised of nodes and transmission channels. NEs support a variety of communication functions, including multiplexing, switching, and routing. NEs invariably use a considerable amount of power to operate. The same is

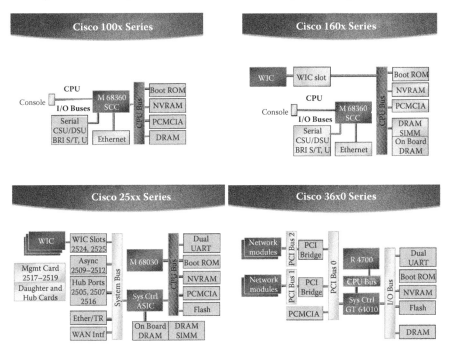

Figure 2.23   Examples of router architecture.

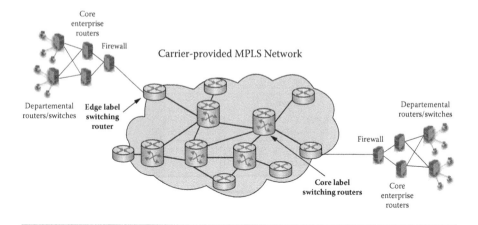

Figure 2.24 Typical MPLS network environment.

true for transmission channels, particularly for regeneration. Therefore, any greening effort will have the benefit of reducing the overall consumption of electricity, both to operate and to cool these devices. Of course, networks exist to connect users (or other nodes) to data processing centers, whether these are corporate data centers, web server farms, content-storage farms, or search-engine farms. For enterprise networks, the data processing centers probably consume much more power than the corporate network itself. However, when looking at the aggregate of all voice, video, data, multimedia, Internet, wireless, and satellite communication network infrastructure, such underlying consumption is nontrivial and, therefore, a greening initiative may prove important and deserving of consideration.

References

[AND200501]   L. Andersson and T. Madsen, Provider Provisioned Virtual Private Network (VPN) Terminology, IETF RFC 4026, March 2005.
[ATI200902]   ATIS Report on Environmental Sustainability, March 2009, A Report by the ATIS Exploratory Group on Green, ATIS, 1200 G Street, NW, Suite 500, Washington, DC 20005.
[BAR200101]   R. Barr and R. Patterson, Grooming telecommunications networks, *Optical Networks Magazine,* 2(3): 20–23, May/June 2001.
[BAR200501]   R.S. Barr, M.S. Kingsley, and R.A. Patterson, Grooming Telecommunications Networks: Optimization Models and Methods, 22 June 2005, Technical Report 05-emis-03.
[BER200501]   H. Berkowitz, E. Davies, S. Hares, P. Krishnaswamy, and M. Lepp, Terminology for Benchmarking BGP Device Convergence in the Control Plane, RFC 4098, June 2005.

[CLI200301]   T. Cinkler, Traffic and λ grooming, *IEEE Network,* 17(2): 16–21, March–April 2003.

[DUT200001]   R. Dutta and G.N. Rouskas, Traffic grooming in WDM networks: Past and future, *IEEE Network,* 16(6): 46–56, November–December 2002.

[ESC200501]   E. Escalona, S. Figuerola, S. Spadaro, and G. Junyent, Implementation of a Management System for the ASON/GMPLS CARISMA Network, White Paper, 2005, CARISMA project, http://carisma.ccaba.upc.es.

[ETNO200801]   Energy Task Team, European Telecommunications Network Operators' Association, The Hague, London, April 2008. [IPV200601]   IPv6   Portal,   http://www.ipv6tf.org/meet/faqs.php, 2006.

[IPV200601]   IPv6 Portal, http://www.ipv6tf.org/meet/faqs.php, 2006.

[NAG200401]   A. Nagarajan, Generic Requirements for Provider Provisioned Virtual Private Networks (PPVPN), RFC 3809, June 2004.

[NAR199801]   T. Narten, E. Nordmark, and W. Simpson, RFC 2461, Neighbor Discovery for IP Version 6 (IPv6), December 1998.

[SHI200001]   R. Shirey, Internet Security Glossary, RFC 2828, May 2000, Copyright © The Internet Society (2000). All Rights Reserved. This document and translations of it may be copied and furnished to others, and derivative works that comment on or otherwise explain it or assist in its implementation may be prepared, copied, published, and distributed, in whole or in part, without restriction of any kind, provided that the above copyright notice and this paragraph are included on all such copies and derivative works.

[WES200201]   J. Weston-Dawkes and S. Baroni, Mesh network grooming and restoration optimized for optical bypass, *Technical Proceedings of the National Fiber Optic Engineers Conference (NDOEF 2002),* 2002, pp. 1438–1449.

[ZHU200301]   K. Zhu, H. Zang, and B. Mukherjee, A comprehensive study on next-generation optical grooming switches, *IEEE Journal on Selected Areas in Communications,* 21(7): 1173–1186, September 2003.

[ZHU200302]   K. Zhu and B. Mukherjee, A review of traffic grooming in WDM optical networks: Architectures and challenges, *Optical Networks Magazine,* 4(2): 55–64, March/April 2003.

Chapter 3

Concepts and Analytical Measures for Green Operations

This chapter looks at a number of analytical measures and metrics that can be employed to measure the "greenness" of business operations in general and IT/network operations in particular. Technical standards or performance benchmarks to consistently "quantify" conformity to, or levels of being, "green" are needed. A planner may seek to establish the entire "carbon footprint" of an operation (say, of the IT department assets of the firm, or of the networking assets of the firm), or may seek to determine the usage/consumption/efficiency of an individual IT asset. In addition, some basic concepts covered in greater detail in the chapters that follow are introduced briefly in this overview chapter. Metrics are discussed in the second part of the chapter. As noted in Chapter 1, the focus of a greening initiative relating to IT data centers and networks (whether intranets or carrier infrastructure) covers two main areas:

1. *System Load:* This relates to the consumption efficiency of the equipment in the data center or telecom node (e.g., IT equipment such as servers, storage, and network elements such as switches, routers, and repeaters). It represents the IT/networking work capacity available for the given IT/networking power consumption. *System Load* should be defined as a function of the utilization of said capacity.

2. *Facilities Load:* This relates to the mechanical and electrical systems that support the IT/networking electrical load, such as cooling systems (e.g., chiller plant, fans, and pumps), air-conditioning units, Uninterruptible Power Supplies (UPS), Power Distribution Units (PDUs), and so on.

Organizations are increasingly interested in conducting GHG (Greenhouse Gases)/Carbon Inventory/Footprint analyses. Clearly, GHG/carbon measurement methodologies and metrics are required to obtain meaningful results. A manager is not able to control an environment or process that cannot be measured; therefore, it is important to be able to (1) describe data center/telco/networking room efficiency, (2) measure data center/telco/networking room efficiency, (3) specify data center/telco/networking room efficiency, (4) analyze data center/telco/networking room efficiency, and (5) benchmark data center/telco/networking room efficiency.

In general terms, *efficiency* is the ratio of output (usable) power/work to the input power. It is the amount of usable energy produced by a machine or device, divided by the amount of energy supplied to it. The *coefficient of performance* is the ratio of the amount of work obtained from a machine to the amount of energy supplied. *Energy efficiency* refers to designing systems (typically buildings, but in this context networks, network elements, telecom rooms, and data centers) to use less energy for the same or higher performance as conventional systems. All subsystems (e.g., electrical power; Heating, Ventilation and Air Conditioning (HVAC); building structure) can contribute to higher energy efficiency. As noted in Chapter 1, data center power and cooling consumes 50% to 70% of the electrical power, with 45% in HVAC, 25% in UPSs and PDUs, and 30% in the actual IT and networking equipment. The theme of the text is clearly energy efficiency.

A number of organizations have offered metrics for energy efficiency at the device or system level. For example, the ATIS Network Interface, Power, and Protection (NIPP) Committee has published standards for telecom and Information Technology (IT) network equipment; both the Consumer Electronics Association (CEA) and EPA have specifications for consumer equipment (namely, ENERGY STAR®); and the EPA and Green Grid are looking at system-level efficiencies relative to data centers. A uniform approach to network-level measurements, however, is still missing [ATIS200902]. As highlighted in Chapter 1, it is important to include power management requirements in network specifications as part of a network/IT equipment procurement process. Some of these approaches are discussed in this chapter. To put some context to this discussion, one can make note of some recommended Best Practices. Table 3.1 identifies, for illustrative purposes, some Best Practices for power, cooling, and energy savings that are of possible interest to network operators as defined by The European Telecommunications Network Operators' (ETNO) Association [ETNO200801]. Many of these practices require definitions of efficiency and related measurements, which is the topic of this chapter.

Table 3.1 Best Practices for Network Operators—These Best Practices Require Definitions of Efficiency and Related Measurements

Best Practice on Power
☐ Availability and reliability at interface A in accordance with ETSI guidelines
☐ Design and resilience of the concept
☐ Back-up time for the No-Break facilities
☐ Redundancy of the No-Break facilities
☐ Generator tests & fuel reserve generator set
☐ Mobile generator sets for emergency purpose
☐ Protection against electro-static discharge (ESD) in equipment rooms
Best practice on Cooling
☐ Minimum requirements regarding availability & reliability
☐ Redundancy of the cooling facilities
☐ Hardware concept, air distribution, flow pattern an cable management
☐ Set points on room temperature within full ETSI-climate range
☐ Mobile cooling/ventilation units for emergency purpose
Best practice Energy Savings
☐ Use of variable fan speed for low heat density rooms
☐ Energy Management Systems
☐ Coefficient of Performance (COP)
☐ Use of free-air cooling by low and medium heat density
☐ Raised floors and floor management
☐ Filtering
☐ Use of rest heat from equipment rooms
☐ Good maintenance on cleaning condensation units

3.1 Electric Energy and Applicable Metrics

A key factor in the design of green networks/green IT (or any green system for that matter) is energy consumption in general and electrical energy consumption in particular. This section starts the basic discussion of this topic and continues in the chapters that follow. We discuss electrical energy distribution and billing mechanisms; more technical aspects of electricity are discussed in Chapter 4.

Two key units in this context are volts and watts. Voltage is the difference in electrical potential between two points; 1 watt (W) is the unit of electrical power, equal to the flow of 1 amp (A) at a potential difference of 1 volt (V); also equal to 1 Joule per second (J/s) and to 1/746 of 1 horsepower. Distribution is the process of delivering electric energy from points on the transmission system (usually a substation) to institutional and end-user consumers. Distribution uses a grid of conductors and equipment that distributes, transports, and delivers electricity to customers. Electric energy is carried at high voltages along transmission lines; electric energy is, in turn, reduced in voltage at a substation and delivered over primary distribution lines. The distribution line is comprised of one (or more) circuit(s) of a distribution system that is (are) direct-buried, placed in conduit, or strung on poles, or a combination thereof. The costs to support, operate, and maintain the local delivery system of wires and equipment that carries electric energy from the transmission system to the customer's premises are included in the rates and are usually priced in cents per kilowatt-hour for energy-only customers and in dollars per kilowatt for demand-billed customers. This is also known as Distribution Service.

The Electricity Service is the network of generating plants, wires, and equipment needed to produce or purchase electricity (generation) and to deliver it to the local distribution system (transmission). This service is priced in cents per kilowatt-hour for energy-only customers, and in dollars (euros, etc.) per kilowatt and in cents per kilowatt-hour for demand-billed customers.

Some related nomenclature follows [MGE200901]:

Energy Charge: That part of the charge for electric service based upon the electric energy (kWh) consumed or billed.

Energy Costs: Costs, such as fuel, related to and varying with energy production or consumption.

Energy, Electric: As commonly used in the electric utility industry, it means kilowatt-hours. Energy, Off-Peak is the energy supplied during periods of relatively low system demand as specified by the supplier. Energy, On-Peak is the energy supplied during periods of relatively high system demand as specified by the supplier.

Fuel Cost Adjustments: A provision in a rate schedule that provides for an adjustment to the customer's bill if the cost of fuel at the supplier's generating stations varies from a specified unit cost.

Power, Firm: Power or power-producing capacity intended to be available at all times during the period covered by a commitment, even under adverse conditions.

Power, Interruptible: Power made available under agreements that permit curtailment or cessation of delivery by the supplier.

Power, Nonfirm: Power or power-producing capacity supplied or available under an arrangement that does not have the guaranteed continuous availability feature of firm power. Power supplied based on the availability of a generating unit is one type of such power.

Primary Voltage: The voltage of the circuit supplying power to a transformer is called the primary voltage, as opposed to the output voltage or load-supply voltage, which is called *secondary voltage.* In power supply practice, the primary voltage is almost always the high-voltage side, and the secondary is the low-voltage side of a transformer, except at generating stations.

Service Area: Territory in which a utility system is required or has the right to supply electric service to ultimate consumers.

Service, Customer's: That portion of conductors usually between the last pole or manhole and the premises of the customer served.

Service Drop: The overhead conductors between the electric supply, such as the last pole, and the building or structure being served.

Service Entrance: The equipment installed between the utility's service drop, or lateral, and the customer's conductors. Typically consists of the meter used for billing, switches and/or circuit breakers and/or fuses, and a metal housing.

Service Lateral: The underground service conductors between the street main and the first point of connection to the service entrance conductors.

Single-phase service: Service where the facility has two energized wires coming into it. Typically serves smaller needs of 120 V/240 V. Requires less and simpler equipment and infrastructure to support than three-phase service.

Step-down: To change electricity from a higher to a lower voltage.

Step-up: To change electricity from a lower to a higher voltage.

Substation: A collection of equipment for the purposes of switching and/or changing or regulating the voltage of electricity. Service equipment, line transformer installations, or minor distribution and transmission equipment are not classified as substations.

Three-Phase Service: Service where the facility has three energized wires coming into it. Typically serves larger power needs of greater than 120 V/240 V.

Transformer: An electromagnetic device for changing the voltage level of alternating-current electricity.

Transmission: The act or process of transporting electric energy in bulk from a source or sources of supply to other principal parts of the system or to other utility systems.

Data center and telecom providers are interested in service reliability. In this context, *reliability* is the guarantee of system performance at all times and under all reasonable conditions to ensure constancy, quality, adequacy, and economy of electricity. Reliability includes the assurance of the continuous supply of electricity for customers at the proper voltage and frequency.

In most jurisdictions in the United States, public utilities operate as controlled monopolies but are subject to regulation. They are obligated to charge fair, nondiscriminatory rates and to provide reliable service to the public. In return, they are generally free from direct competition and are permitted to get a fair return on investment. The Federal Energy Regulatory Commission (FERC) is an agency within the U.S. Department of Energy that has broad regulatory authority. Nearly every facet of electric (and natural gas) production, transmission and sales conducted utilities, corporations, or public marketing agencies is under FERC control.

Demand represents the rate at which electric energy is delivered to (or by) a system; it is typically expressed in kilowatts at a given instant in time or averaged over any specified period of time. *Average demand* is the demand on an electric system over any interval of time (this is derived by dividing the total number of kilowatt-hours by the number of units of time in the interval.) *Maximum annual demand* is the largest demand that occurred during a (15-minute) interval in a year.

Consumption of electrical energy is typically measured in kilo-, mega-, or giga-watt-hour: a kilowatt-hour (kWh) is the basic unit of electric energy equal to 1 kilowatt (1,000 watts) of power supplied to or taken steadily from an electric circuit for 1 hour. One kilowatt-hour equals 1,000 watt-hours. Clearly, 1 megawatt-hour (MWh) equals 1 million (1,000,000) watt-hours; 1 gigawatt-hour (GWh) is 1 billion (1,000,000,000) watt-hours, or 1 million (1,000,000) kilowatt-hours, or 1 thousand (1,000) megawatt-hours. Also see Appendix 3B for some related parameters.

Load management, defined next, may be advantageous for the user. *Load management* is the reduction in electric energy demand during a utility's peak generating periods typically resulting in lower overall costs to the user. Load-management strategies are designed to either reduce or shift demand from on-peak to off-peak times (by way of contrast, conservation strategies may primarily reduce usage over the entire 24-hour period). Actions may take the form of normal or emergency procedures. Many utilities encourage load management by offering customers a choice of service options with various price incentives. Related concepts include the following:

Load Curve: A plot showing power (kilowatts) supplied versus time of occurrence that captures the varying magnitude of the load during the period covered.

Load Factor: The ratio of the average load in kilowatts supplied during a given time period to the peak or maximum load in kilowatts occurring in that period.*

Load Shifting: Entails moving load from on-peak to off-peak periods to take advantage of time-of-use or other special rates; examples include use of storage water heating and cool storage.

---

* Load factor, in percent, is derived by multiplying the kilowatt-hours (kWh) in the period by 100 and dividing by the product of the maximum demand in kilowatts and the number of hours in the period.

The following concepts cover some of the typical billing-related factors related to electrical energy, based in part on reference [MGE200901]:

Base Rate: That part of the total electric rate covering the general costs of doing business unrelated to fuel expenses.

Conjunctive Billing: The combination of the quantities of energy, demand, or other items of two or more meters or services into respective single quantities for the purpose of billing, as if the bill were for a single meter or service.

Connection Charge: An amount to be paid by a customer in a lump sum or in installments for connecting the customer's facilities to the supplier's facilities.

Customer Charge: An amount to be paid periodically by a customer for electric service based upon costs incurred for metering, meter reading, billings, etc., exclusive of demand or energy consumption.

Demand, Billing: The demand upon which billing to a customer is based, as specified in a rate schedule or contract. It may be based on the contract year, a contract minimum, or a previous maximum and therefore does not necessarily coincide with the actual measured demand of the billing period.

Demand Charge: That part of the charge for electric service based upon the electric capacity (kW) consumed and billed on the basis of billing demand under an applicable rate schedule.

Demand Costs: Costs that are related to and vary with power demand (kW), such as fixed production costs, transmission costs, and a part of distribution costs.

Demand, Customer Maximum 15-Minute: The greatest rate at which electrical energy has been used during any period of 15 consecutive minutes in the current or preceding 11 billing months.

Demand, Instantaneous Peak: The demand at the instant of greatest load, usually determined from the readings of indicating or graphic meters.

Demand Interval: The period of time during which the electric energy flow is averaged in determining demand, such as 60-minute, 30-minute, 15-minute, or instantaneous.

Demand, Maximum: The greatest demand that occurred during a specified period of time, such as a billing period.

Firm Obligation: A commitment to supply electric energy or to make capacity available at any time specified during the period covered by the commitment.

Fixed Costs: Costs that do not change or vary with usage, output, or production.

Minimum Charge: A provision in a rate schedule stating that a customer's bill cannot fall below a specified level.

Primary Discount: A discount provision that is available to customers who can take delivery of electrical energy at primary distribution voltage levels. The transformer equipment discount is also available to customers taking primary voltage service who own their own transformers and transformer equipment.

Rate Case: The process in which a utility appears before its regulatory authority to determine the rates that can be charged to customers.

Rate Class: A group of customers identified as a class subject to a rate different from the rates of other groups.

Rate Level: The electric price a utility is authorized to collect.

Rate Structure: The design and organization of billing charges to customers.

Rates, Block: A certain specified price per unit is charged for all or any part of a block of such units, and reduced/increased prices per unit are charged for all or any part of succeeding blocks of such units, each such reduced/increased price per unit applying only to a particular block or portion thereof.

Rates, Demand: Any method of charging for electric service that is based upon, or is a function of, the rate of use, or size, of the customer's installation or maximum demand (expressed in kilowatts) during a given period of time such as a billing period.

Rates, Flat: The price charged per unit is constant, does not vary due to an increase or decrease in the number of units.

Rates, Seasonal: Rates vary depending upon the time of year. Charges are generally higher during the summer months when greater demand levels push up costs for generating electricity. Typically there are summer and winter seasonal rates. Summer rates are effective from June 1 through September 30. During all other times of the year, winter rates are effective.

Rates, Step: A certain specified price per unit is charged for the entire consumption, the rate or price depending on the particular step within which the total consumption falls.

Rates, Time-of-Use: Prices for electricity that vary depending upon what time of day or night a customer uses it. Time-of-use rates are designed to reflect the different costs an electric company incurs in providing electricity during peak periods when electricity demand is high and off-peak periods when electricity demand is low.

Retail Wheeling: An arrangement in which retail customers can purchase electricity from any supplier as opposed to their local utility. The local utility would be required to allow the outside generating company to wheel the power over the local lines to the customer.

Summer Peak: The greatest load on an electric system during any prescribed demand interval in the summer (or cooling) season.

Tariff: A schedule of prices or fees.

Tariff Schedule: A document filed with the regulatory authority(ies) specifying lawful rates, charges, rules, and conditions under which the utility provides service to the public.

Unbundling: Itemizing some of the different services a customer actually receives and charging for these services separately.

Variable Costs: Costs that change or vary with usage, output, or production. Example: fuel costs.

These concepts can all be applied to the analysis of the efficiency of a NE or data center component. A typical datasheet or hardware guide may include a number of energy metrics, such as [ECR200801]

Component-based Consumption Estimate (watts): This metric allows a customer to estimate a power draw in a "customized" configuration by adding together the configured parts (components) with known power ratings. This can be a fairly precise estimate, assuming the vendor has published accurate power numbers for all base, optional, and physical interface modules in the current board revisions. The availability and usability of such data is vendor dependent and requires a thorough knowledge of system structure and operation.

Maximum Power Consumption (watts): This metric can also be used as an upper boundary estimate for the power draw. However, this metric tends to penalize modular systems designed for the optional high-power components. For example, an Ethernet switch designed with Power over Ethernet (PoE) modules in mind will have much higher maximum power draw than a fixed copper-port model, yet both systems may yield identical consumption in a pure 1000 BaseT mode of operation. In addition, a maximum power consumption estimate can change without notice when new modules are introduced and old modules are withdrawn from production.

Power System Rating (amps or watts), aka *Power Supply Rating*: This metric reflects the site preparation requirements recommended by the vendor. It can potentially be used as an estimate for consumption, but with a possible error margin. In some cases, vendors outfit their platforms with high-capacity power supplies in planning for future system upgrades; the actual system consumption may be a fraction of what the power supplies can deliver.

Typical (Average) Power Draw (watts): This metric tends to underestimate the power consumption range. Motivated to demonstrate the low current draw, vendors are free to report this metric with underpowered configurations, components, or load profiles that yield the best results; omission of a published test methodology typically signifies these and similar issues. In the lack of a public disclosure on measurement conditions, "typical" or "average" power draw cannot be reasonably used to rate the device against any other platform.

Another metric of interest is the *power availability* metric. As implied earlier, availability refers to the percentage of time that a system is available, on-line, and capable of doing productive work. It is typically described as an annual percentage, or number of "nines." A system with "three nines" availability is 99.9% available, which translates into 8.8 hours of downtime annually. "Fives nines" availability—the standard many data centers aspire to—translates into less than 6 minutes of downtime annually. With power availability, an additional factor should be considered: the availability of conditioned power [LIE200901]. The key question is: Is it acceptable for systems being protected to operate on unconditioned utility power for short periods of time? Answering this question requires trading off the increased risk of operating on unconditioned power as a function of the added cost of UPS redundancy, which is required to minimize or eliminate the time protected equipment is exposed to utility power. Clearly, if the cost of downtime is low, the investment in redundant systems may not have a sufficiently high ROI (Return on Investment). If the cost of downtime is high, omitting redundancy could result in significant losses.

3.2 Air Conditioning

Another major energy-consumption-related issue deals with cooling. Energy in a system is lost through undesired heat loss or gain, noise, friction, and other phenomena. Cooling is achieved using air conditioners. For data centers and telco rooms, these are also called Computer Room Air Conditioners (CRACs) and may have features specific to the application at hand (data center cooling).

An air-conditioner typically consists of one or more factory-made assemblies that normally include an evaporator or cooling coil(s), compressor(s), and condenser(s). Air conditioners provide the function of air-cooling, and may include the functions of air circulation, air cleaning, dehumidifying, or humidifying. A split system is a system with components located both inside and outside a building. A single-package-unit is a system that has all components completely contained in one unit [ENE200901].

A heat pump model consists of one or more factory-made assemblies that normally include an indoor conditioning coil(s), compressor(s), and outdoor coil(s), including means to provide a heating function. Heat pumps provide the function of air heating with controlled temperature, and may include the functions of air-cooling, air circulation, air cleaning, dehumidifying, or humidifying.

Basically, a heat pump or AC unit "moves" heat from the conditioned space to the unconditioned space. A compressor is used in both a heat pump and an AC unit. The compressor "squeezes" heat out of the conditioned air using a refrigerant and a coil, thus "moving" the heat from where it is not wanted to someplace more acceptable, typically outside. In the heating mode, a heat pump still "moves" heat, but now it is taking it from the unconditioned space (outside) and delivering it to the conditioned space (inside) [CRI200001].*

The cooling capacity is the quantity of heat in British thermal units (Btu, BTU) that an air conditioner or heat pump is able to remove from an enclosed space during a 1-hour period. One British thermal unit is the amount of heat required to raise the temperature of 1 pound of water 1 degree Fahrenheit at 60 degrees Fahrenheit. A therm (thm) is the quantity of heat energy that is equivalent to 100,000 Btu.

---

* There is a limit to how cold the outside temperature can be for a heat pump to operate; this is the reason why heat pumps require the addition of electrical resistance heat or natural gas in cold temperatures, typically below 30°F

BTUH (BTU per hour) represents the thermal energy requirement per hour to heat or cool a specific volume of air. Ton is also a measure of cooling; 1 ton is 12,000 BTUH. A ton is the amount of heat removed by an air-conditioning system that would melt 1 ton of ice in 24 hours. Outside the United States, cooling is usually measured in kW. In summary,

■ Therm: A unit of heat energy equal to 100,000 Btu.
■ Ton (refrigeration): The amount of heat absorbed by melting 1 ton of ice in 24 hours. It is equal to 288,000 Btu per day, 12,000 Btu per hour or 200 Btu per minute.

Table 3.2 provides some conversion factors. See Appendix 3B for a more inclusive conversion table.

Some general air-conditioning-related parameters include the following [ENE200901]:

Coefficient of Performance (COP): COP is a measure of efficiency in the heating mode that represents the ratio of total heating capacity (Btu) to electrical input (also in Btu).

Table 3.2 Conversion Factors of Interest in the Context of Air Conditioning Systems

To Convert From	To	Multiply By
Energy & Power & Capacity		
British Thermal Unit (Btu)	Kcalorie (Kcal)	.252
British Thermal Unit Per Hr. (BTUH)	Kilowatt (kW)	.000293
Tons (refrig. effect)	Kilowatt (refrig. effect)	3.516
Tons (refrig. effect)	Kilocalories per hour (Kcal/hr)	3024
Horsepower	Kilowatt (kW)	.7457
Pressure		
Feet of Water (ftH$_2$O)	Pascals (PA)	2890
Inches of Water (inH$_2$O)	Pascals (PA)	249
Pounds per Square Inch (PSI)	Pascals (PA)	6895
PSI	Bar or KG/CM^2	6.895× 10^2

Energy Efficiency Ratio (EER): EER is a measure of efficiency in the cooling mode that represents the ratio of total cooling capacity (Btu/hour) to electrical energy input (watts).

Heating Seasonal Performance Factor (HSPF): HSPF is a measure of a heat pump's energy efficiency over one heating season. It represents the total heating output of a heat pump (including supplementary electric heat) during the normal heating season (in Btu) as compared to the total electricity consumed (in watt-hours) during the same period.

Integrated Energy Efficiency Ratio (IEER): IEER is a measure that expresses cooling part-load EER efficiency for commercial unitary air-conditioning and heat pump equipment on the basis of weighted operation at various load capacities.

Seasonal Energy Efficiency Ratio (SEER): SEER is a measure of equipment energy efficiency over the cooling season. It represents the total cooling of a central air conditioner or heat pump (in Btu) during the normal cooling season as compared to the total electric energy input (in watt-hours) consumed during the same period.

COP is essentially the ratio of electricity used to heat moved. An efficient (residential) device typically has a COP in the range of 5 to 6; higher is more efficient. A low-end AC unit (e.g., a residential window air conditioner) has a SEER around 10, while larger central air systems can have a SEER of 17 or 18; higher is better. A unit with a SEER of 18 costs half as much to operate as one with a SEER of 9.

Outdoor weather conditions affect HVAC efficiency and, in turn, data center/telco room efficiency. These outdoor conditions vary with the time of day and with the seasons. Factors such as ambient temperature, sunlight, humidity, and wind speed impact efficiency, with temperature being the most important factor. Note that modern CRACs can operate in an "economizer mode" when the outside temperature is low (say below 12°C), thereby improving system efficiency. The efficiency of the HVAC (and thus the data center/telco room) declines as the temperature increases because heat rejection systems consume more power when handling data center heat and because outdoor heat infiltration into the data center/telco room represents an additional heat load that requires handling [RUS200801].

An *air economizer* is a ducting arrangement and automatic control system that allows a cooling supply fan system to supply outdoor (outside) air to reduce or eliminate the need for mechanical refrigeration during mild or cold weather. A *water economizer* is a system by which the supply air of a cooling system is cooled directly, indirectly, or both by evaporation of water or by other appropriate fluid (in order to reduce or eliminate the need for mechanical refrigeration) [GRE201001].

Another measure of interest is Air Conditioning Airflow Efficiency (ACAE), which is defined as the amount of heat removed (watts heat) per standard cubic foot of airflow per minute.

When considering commercial building HVACs, the indoor comfort characteristics of a building are defined by two main categories of elements: (1) active elements (part of the building equipment), and (2) passive elements (part of the building envelope). The HVAC systems are active elements that are dimensioned according to the characteristics of the building envelope and according to climatic conditions. The efficiency of HVAC systems depends on (1) the quality and efficiency of the passive elements of the building envelope, and (2) the quality and efficiency of the HVAC equipment and its concepts. Table 3.3 depicts key elements of a commercial building HVAC "ecosystem" [ICA199701]. Chapter 5 revisits this topic.

Table 3.3　Key Elements of a Commercial Building HVAC 'ecosystem'

Active Elements	Thermal systems	**Production of heat and cold** Compression Absorption **Production of heat** Solar thermal Boilers Cogeneration
	Distribution systems	**Individual unit systems** Compact Split **Centralized systems** All-air All-water Air-water Refrigerant **Terminal units** Fan-coils Inductors Radiators Diffusers
Main Passive Elements	Insulation	Roofs Walls Floors
	Protection and utilization of solar radiation	External shading IR-reflecting glazing Building orientation
	Passive ventilation	Chimney effect Cross ventilation Control of air movement Ventilated window

Next we focus on two areas of direct relevance to the data center/telco room: air management and power density.

In a data center/telco room the goal of air management is (a) to minimize recirculation of hot air back from a hot air aisle to a cold aisle, and (b) to minimize bypass of cold air from a cold aisle over the rack rather than through the rack (to cool the equipment in the rack).

Successfully implemented, both measures result in energy savings and better thermal conditions. Metrics play an important role in providing a measure of the performance of air management systems. Metrics condense multidimensional information into understandable, objective, and standardized numbers. Air management metrics include the Rack Cooling Index (RCI), the Return Temperature Index (RTI), and the Supply Heat Index (SHI) [DOE200901].

■ The RCI is a measure of how effectively the equipment is cooled and maintained within an intake temperature specification. The index helps evaluate equipment cooling effectiveness. Interpretation: 100% is ideal; <90% is often considered poor.

■ The RTI is a measure of the level of by-pass air or recirculation air in data centers. Both phenomena are detrimental to the thermal and energy performance of the facility. Interpretation: 100% is the target; >100% implies recirculation air; <100% implies by-pass air.

■ The SHI is a dimensionless measure of the recirculation of hot air into the cold aisles. SHI not only provides a tool to understand convective heat transfer in the equipment room but also suggests means to improve the energy efficiency. Interpretation: SHI is a number between 0 and 1, the lower the better. SHI is typically <0.40.

We then note that there is a lack of industry standards for calculating data center power and heat density. This makes it difficult to obtain consistent measures and comparisons of Watts Per Square Foot (WPSF). For example, two data centers may have identical equipment, but if one occupies a 1,000-square-foot (ft²) room and the other a 1,300-ft² room, the first will have a higher WPSF. (Note: At press time the average data center density was 100–200 WPSF.) The following three measures have been offered to accurately and consistently compare WPSF [GAR200601]:

1. *WPSF within a work cell:* This standard measure is used to normalize power and heat density on a per-cabinet basis. A work cell is the area dedicated to one cabinet. For example, if a cabinet is 2 ft wide and 4 ft deep, it occupies 8 ft². The cold aisle in front is 4 ft, shared between two rows of cabinets, so the portion dedicated to each cabinet is 8 divided by 2, or 4 ft². The same calculation applies to the hot aisle behind each cabinet. In this example, the work cell dedicated to each cabinet is 16 ft². Some data centers are laid out this way, with an 8-ft distance, or pitch, between the center of the cold aisle

and the center of the hot aisle to accommodate cabinets up to 4 ft deep. Another advantage of the 8-ft pitch is that it allows us to coordinate the drop ceiling grid with the floor grid so that standard 2 × 4 ft light fixtures and fire sprinklers can be centered above the hot and cold aisles.

2. *WPSF over the Raised Metal Floor (RMF)/server room area:* One defines the RMF area (or server room area in data centers lacking RMF) as the total area occupied by work cells plus the perimeter egress aisles.

3. *WPSF over the total data center area:* To obtain a true measure of the efficiency of data center layout and design, one includes the area occupied by air handlers, whether or not they are located on the RMF. Therefore, when calculating WPSF for the total data center area, one should include the CRAC/ recirculation air handlers (RAHs) area and egress aisles, but not exterior ramps or staging areas.

An industry rule of thumb for an efficient layout is 25 cabinets per 1,000 ft² of RMF and air handling equipment area.

3.3 IT Assets

Typical corporate IT assets, which of course require electric energy, include, but are not limited to, the following (see Appendix 3A for a more inclusive but not exhaustive list):

■ Desktops PCs and laptops
■ Mobile devices and wireless networks (e.g., PDAs, Wi-Fi/Bluetooth devices)
■ Application servers, mainframes
■ Mail servers
■ Web servers
■ Database servers (data warehouses, storage) as well as the entire universe of corporate data, records, memos, reports, etc.
■ Network elements (switches, routers, firewalls, intrusion detection systems, load sharing devices, appliances, etc.)
■ PBXs, IP-PBXs, VRUs, ACDs, voicemail systems, etc.
■ Mobility (support) systems (Virtual Private Network (VPN) nodes, wireless e-mail servers, etc.)
■ Power sources
■ Systems deployed in remote/branch locations (including international locations)
■ Key organizational business processes (e.g., order processing, billing, procurement, customer relationship management, etc.)

As discussed in the previous chapter, typical service provider communications assets include Network Elements (NEs) to support the basic communication functions: regenerating, multiplexing, grooming, switching, and routing; information transfer (transmission). NEs include

- Regenerators
- Multiplexers
- SONET add/drop multiplexers
- Digital cross-connects
- Edge and core voice switches
- Edge and core routers and data switches

3.4 Discrete Metrics

Focusing on networks and IT operations, resources are used at the following "resource-consumption-points":

- End-user computation nodes, such as but not limited to computational servers, file/web servers, and database servers
- End-user storage nodes, such as but not limited to data warehousing, video/content storage
- End-user switching/routing/firewalling nodes for voice, data, and video media
- Transmission systems, such as but not limited to fiber-optic links, satellite links, microwave links, cellular links, 3G/4G links
- Transmission-support systems, such as but not limited to network switches, routers, multiplexers/digital cross-connect systems, (optical) regenerators/repeaters (shelter nodes)
- Network computing nodes, such as but not limited to Cloud Computing, Grid Computing, Web Services nodes (including search engines)
- Network-based content nodes, including IPTV, music sites, YouTube, social network sites, network storage, etc.
- Support sites, such as support worker depots, warehouses, offices, monitoring sites, etc.

Resources needed include

- Energy use
- Ozone-depleting substances
- Paper and toner
- Media
- Solid waste
- Business travel (and "truck rolls")
- Electronic equipment and batteries discarded

Table 3.4 shows where resources are being consumed as a function of the topology described above. Table 3.5 provides a view as to the types and metrics that are applicable.

Table 3.4 Resource Use Matrix

	End-user computation nodes	End-user storage nodes	End-user switching/routing	Transmission systems	Transmission-support systems	Network computing nodes	Network content nodes	Support sites
Energy Use	Medium	Medium	Medium	High	High	High	High	Medium-High
Ozone-depleting substances	Medium	Medium	Medium	Medium	Medium	Medium	Medium	Medium-High
Paper and Toner	Medium	Medium	Medium	Medium	Medium	Medium	Medium	Medium-High
Media	Medium	Medium	Medium	High	High	High	High	Medium-High
Solid Waste	Medium	Medium	Medium	High	High	High	High	Medium-High
Business Travel	Medium	Medium	Medium	High	High	High	High	Medium-High
Electronic discarded	Medium	Medium	Medium	Medium	Medium	Medium	Medium	Medium-High

Table 3.5   Measures of Use of Discrete Resource

Resource	Definition
Energy Use	For all resource-consumption-points, electricity purchased from energy suppliers (in kWh), for node operation, HVAC, routing/switching/transmitting; also the amount of each type of fuel consumed at the sites (e.g., natural gas, fuel oil, diesel, kerosene) (in kWh or kg) per year
Ozone-depleting substances	For all resource-consumption-points, the number of refrigerant systems that use ozone-depleting substances (e.g., air conditioning units), the type used (e.g., R22, R407C) and the amount (kg) of refrigerant leakage that occurred over the year
Paper and Toner	For all resource consumption points, the total amount of paper consumed (kg) and the amount of toner (all colors) consumed per year
Media	For all resource consumption points, the total amount of magnetic tapes and optical disks (eventually) discarded (kg of each), per year
Solid Waste	For all resource consumption points, the total amount of waste (kg) produced at the site (excluding electronic equipment and batteries), and the amount sent to landfill, incineration, or for recycling by material type (plastic, cardboard, aluminum cans, etc.) per year
Water and Sewer	For all resource consumption points, the total amount of water consumed at the location in cubic meters (m^3) and the amount of liquid sewer (gallons) per year
Business Travel (and "truck rolls")	For all resource consumption points, the amount of business travel made by employees based at the site, including fleet/hire car use, train, bus and air travel. This includes 'truck rolls' to handle remote installations. Data can be expressed in terms of distance (km), fuel consumed (gallons) and/or cost per year
Electronic Equipment and batteries discarded	For all resource consumption points, the total amount of electronic equipment discarded (without being donated) (kg) and batteries used for business purposes (kg) discarded per year

ATIS-0600015.2009 notes that accurate measurement of energy consumed by a given piece of equipment is critical in determining the overall energy efficiency for that product. Insufficient sample intervals and measurement durations can lead to power measurement errors. Use of a power analyzer (or equipment with equivalent capability and accuracy) is required for measuring the energy consumption of a given piece of equipment. Also, environmental conditions must be controlled. For example,

- The equipment should be evaluated at a temperature of 25°C ± 3°C (77 ± 5°F).
- The equipment should be evaluated at a relative humidity of 30% to 75%.
- The equipment should be evaluated at a barometric pressure between 1,020 and 812 mbar. This corresponds to typical barometric pressure between an altitude of 60 m (197 ft) below sea level to 1,829 m (6,000 ft) above sea level.
- For DC powered equipment, the equipment should be evaluated at a DC voltage of –53 V ±1 V.
- For AC powered equipment, the equipment should be evaluated with a source providing the following conditions:
    - Total Harmonic Distortion ≤2% up to and including the 13th harmonic
    - At either of the following:
        - 115 VAC ± 1%, 60 Hz ±1%
        - 230 VAC ± 1%, 50 or 60 Hz ± 1%

3.5 Aggregate Metrics

This section discusses some efficiency metrics (measures). In general, efficiency is difficult to measure due to variable workloads (percentage utilization) of the NE or IT devices; vendors make claims about efficiency of their products, but there is often no reference to the workload for which these efficiencies actually make sense, especially for servers.

3.5.1 Power Usage Efficiency (PUE)

Power equipment efficiency is usually defined as a percentage value. As noted in Chapter 1, The Green Grid defines two metrics for measuring data center infrastructure overhead as follows: Power Usage Effectiveness (PUE*) and Data Center Infrastructure Efficiency (DCiE) [GRE200701]. DCiE is compliant with the recommendations of the 2007 EPA report to Congress on data center efficiency. The definitions are as follows:

PUE = Total Facility Power / IT Equipment Power
PUE is a ratio. It should be less than 2; the closer to 1, the better.

DCiE = IT Equipment Power × 100 / Total Facility Power
DCiE is a percentage. The larger the number, the better.

---

* PUE is also called Site Infrastructure Power Overhead Multiplier (SI-POM) by some.

Hence, DCiE is the ratio of the IT lower power to the total data center input power. PUE is the inverse of DCiE; namely, DCiE = 1/PUE. The PUE for traditional data centers is generally between 1.8 and 3.0; the goal is clearly to design data and telecom centers that have much lower PUEs, say around 1.1.

DCiE describes the electrical efficiency of a data center as the fraction of the total electrical power supplied to the data center that is, in final terms, delivered to the IT/networking load, as depicted pictorially in Figure 3.1. Total Facility Power is defined as the power measured at the utility meter. If all the power supplied to the data center would reach the IT data and networking loads, the data center would be 100% efficient. In reality, energy is also consumed by transformers, UPSs, ACs, pumps, humidifiers, lighting, security systems, and so on. For the purpose of the calculation of the DCiE metric, the only power that "counts" is the power actually delivered to the IT/networking loads. Note that typically the UPS and transformers are in series with the IT/networking loads, while cooling and lighting are in parallel with the loads. The elements in series provide a power path to the IT loads; the elements in parallel provide protection and support to the loads. Figure 3.2 (modeled in part after [RUS200801]) depicts a logical view of power demand/consumption in a data center/telco room. Note that PUE and/or DCiE comparisons are more meaningful when comparing facilities located in geographic areas that have similar geographic weather patterns. Figure 3.3 shows the typical (but unique) power usage distribution of a data center. Figure 3.4 shows the typical (but not unique) power usage distribution of a carrier's central office.

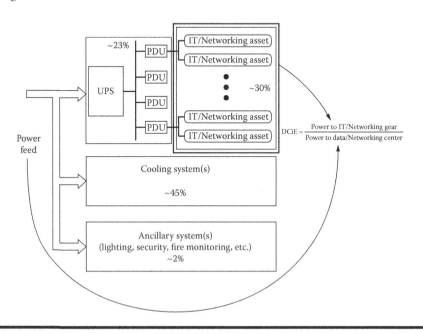

Figure 3.1   Pictorial view of DCiE.

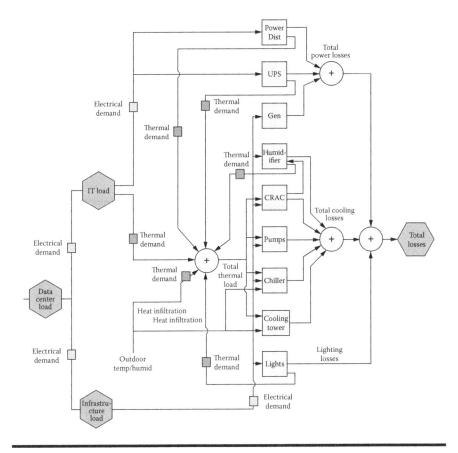

Figure 3.2 Logical view of power demand/consumption in a data center/telco room.

Both the power that is part of the numerator in the DCiE as well as the power that is not part of the numerator in the DCiE can be optimized/reduced to improve the efficiency. Consider the before and three after cases in the example that follows:

■ Before:
  Total IT load power: 40
  Total power: 100
  DCiE = 40%

■ After 1: (reducing the "other power")
  Total IT load power: 40
  Total power: 90
  DCiE = 44%

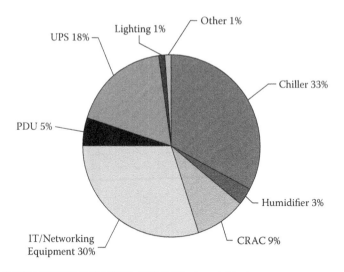

Figure 3.3   Typical power distribution in a data center.

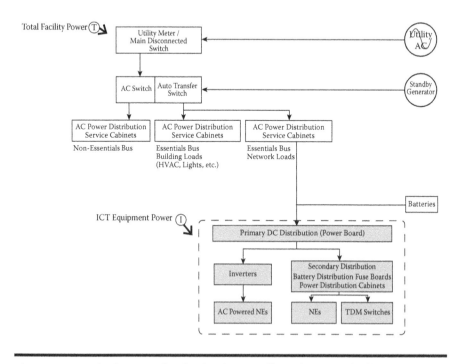

Figure 3.4   Typical power distribution in a small carrier central office. (Courtesy of ATIS.)

■ After 2: (reducing the "IT power")
Total IT load power: 30
Total power: 90
DCiE = 33%

■ After 3: (reducing both)
Total IT load power: 30
Total power: 80
DCiE = 38%

Most—if not virtually all—of the power in a data center/telco room ends up in heat.

For networking equipment, the Equipment Power is calculated utilizing the Busy Hour Drain (the actual average current drawn by a circuit or group of circuits during the busy hour of the season) captured by the DC power plant controller/monitor, multiplied by the DC power plant float voltage, multiplied by the number of hours in the given time period. If monthly Busy Hour Drain data is available, then sum for each month (using appropriate hours per month). If the Busy Hour Drain is not available, then an instantaneous drain reading can be used [ATI200901]. An example of a PUE calculation for a small telco shelter is shown below.

Month	Year	Total Facility Energy (kWh)	NE Busy Hour Drain (A)	Float Voltage (V)	NE Eqpt Energy (KWh)	PUE	DCiE (%)
Nov	2008	895	16.7	52.08	605	1.48	67.6
Dec	2008	1051	16.7	52.08	730	1.44	69.5
Jan	2009	875	16.7	52.08	626	1.40	71.6
Feb	2009	903	16.7	52.08	647	1.40	71.6
Mar	2009	899	16.7	52.08	626	1.44	69.6
Apr	2009	945	17.3	52.08	627	1.51	66.4
May	2009	1128	17.3	52.08	627	1.80	55.6
Jun	2009	1145	17.3	52.08	670	1.71	58.5
Jul	2009	1117	17.3	52.08	670	1.67	60.0
Aug	2009	1250	16.8	52.08	672	1.86	53.7
Sep	2009	1022	16.8	52.08	588	1.74	57.5
Oct	2009	1037	16.8	52.08	630	1.65	60.7
		12273			7720	1.59	62.9

As previously discussed for air conditioners, the measures of interest are the COP (the ratio of heat removed to the electrical power consumed) for chillers and the EER for rooftop systems (one can convert from one to the other by formula). Note that these metrics are calculated by the manufacturers at a single point for humidity and temperature; therefore, they do not provide a complete view of the efficiency. The actual efficiency of the cooling equipment depends on the System Load and the actual environmental conditions (e.g., the variations in outdoor air conditions that occur during the year).

It should be noted that most cooling devices, such as humidifiers, fans (dry cooler fans and fresh ventilation fans), and many types of CRACs and chillers, cycle on and off over the course of the day. This causes the total data center input power consumption to vary over time (even at instantaneous moments). Therefore, the DCiE measure will vary over the day, even if the IT load remains constant. Figure 3.5 depicts an actual example. It follows that when computing the DCiE metric, one should use the average consumption. This is also shown in Figure 3.5. It turns out that most data center devices have cycling periods in the 10-minute range; this implies that averaging data over a 1-hour interval is reasonable. However, given certain seasonal variances, daily, weekly, or monthly averaging may be appropriate in some instances. In some cases it is best to calculate the ratio using energy with a time period of 1 year, where the Total Facility Power includes a rolling 12-month cumulative kilowatt-hour figure (e.g., obtained from commercial AC utility bills)—the 12-month cumulative total allows one to capture energy usage fluctuations due to seasons, weather variation, in-year business cycles, and so on. In summary, keep in mind that DCiE calculated over a very short period of time (an "instantaneous DciE") is not all that meaningful, and it will likely differ from a DCiE calculated over a day, week, month, or year. Single evaluations of data center efficiency are not reliable benchmarking mechanisms for decision making.

Another factor that should be taken into account when computing the efficiency (specifically, the DCiE) is the load of the IT infrastructure. In a number of instances, the consumption of the IT/telecom equipment depends on the load of the equipment. For example, power management features in newer IT/telecom equipment cause the load to vary over time (again, see Figure 3.5); this may be an instantaneous process or it may be a shift-level (day, night) process. The implication is that that the numerator of the DCiE equation changes as a function of the load, and so does the DCiE, as shown graphically in Figure 3.6. This means that one needs to be careful as to how one compares two installations. Figure 3.7 illustrates this point. In this example, Company Alpha has a more efficient data center/telco room environment than Company Beta; but because the calculations can be carried out at two different loads, the results may be misleading.

As discussed in a previous section, the efficiency of a CRAC depends on the outdoor temperature. In turn, this can impact the efficiency of the data center/telco room, as depicted graphically in Figure 3.8.

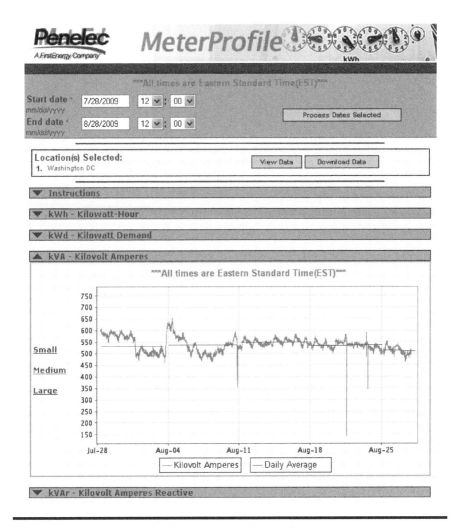

Figure 3.5   Actual example of power usage for a telco room/center.

3.5.2 *Energy Consumption Rating (ECR)*

In the context of networking, we noted in Chapter 1 that the Energy Consumption Rating (ECR) Initiative is a framework for measuring the energy efficiency of network and telecom devices. ECR is a "performance-per-energy unit" rating that can be reported as a peak (scalar) or synthetic (weighted) metric that takes dynamic power management capabilities into account [ECR200901]. One primary and one secondary metric are calculated. The primary is a peak ECR metric, which is calculated according to the following formula:

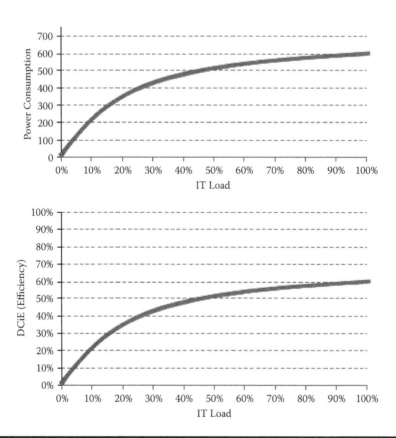

Figure 3.6   DCiE as a function of the load.

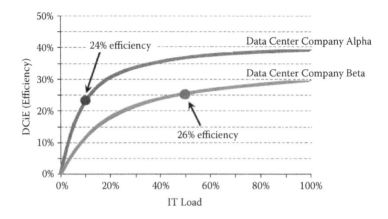

Figure 3.7   Care required when comparing two environments.

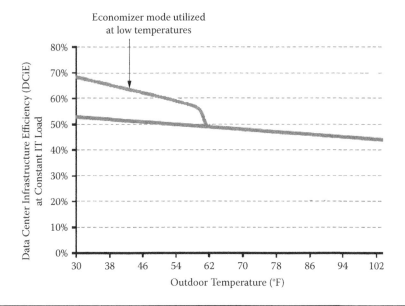

Figure 3.8 Typical data center/telco room efficiency as a function of outside temperature.

$$ECR = E_f / T_f \text{ (expressed in watts per 10 Gbps)}$$

where T_f is the maximum throughput (Gbps) achieved in the measurement, E_f is the energy consumption (watts) measured during running test T_f, and ECR is normalized to watts/10 Gbps and has a physical meaning of the energy consumption to move 10 Gbits worth of user data per second. This reflects the best possible platform performance for a fully equipped system within a chosen application and relates to the commonly used interface speed. Sometimes, for convenience purposes, energy efficiency is also reported in Gbps/watt under the name of EER; EER = 1 / (ECR).

The second metric is a weighted (synthetic) metric (ECRW) that takes the idle mode into account. It is used in addition to the primary metric to estimate the power management capabilities of the device.

$$ECRW = ((\alpha \times E_f) + (\beta \times E_h) + (\gamma \times E_i)) / T_f \text{ (dimensionless)}$$

where T_f is the maximum throughput (Gbps) achieved in the measurement; E_f is the energy consumption (watts) measured during running test T_f; E_h is the energy consumption (watts) measured during half-load test; E_i is the energy consumption (watts) measured during idle test; α, β, γ are the weight coefficients to reflect the mixed mode of operation (ECR specifies $\alpha = 0.35$, $\beta = 0.4$, and $\gamma = 0.25$); and

ECRW reflects the dynamic power management capabilities of the device, which match energy consumption to the actual work accomplished. An ideal system following the Barroso principle of energy-proportional computing should yield an ECRW = $|0.55 \times ECR|$ or better result. A system with no power management should yield ECRW = $|ECR|$.

Results can be variably reported based upon a class definition or a combination of application and packet size, such as: ECR (class A) = Y, or ECRx (B) = Z, where A = equipment class, B = payload type, x = packet size, and Y, Z = calculated efficiency. For example, ECR (Class 1) = 12 W/Gbps; ECR256 (IPv6) = 5.2 W/Gbps.

For comparison purposes, the data can be collected in tables to reflect a head-to-head competitive situation typical of RFP qualification; for example:

	Product A	*Product B*	*Product C*	*Product D*
Product Class	Core	Core	Core	Core
	Product A	*Product B*	*Product C*	*Product D*
Nominal capacity	640 G	1.28 T	1.6 T	3.2 T
ECR (Class 1)	10 W/Gbps	12 W/Gbps	9 W/Gbps	12 W/Gbps
ECRW	9.1 W/Gbps	11 W/Gbps	7 W/Gbps	10 W/Gbps

The measurement procedure is discussed next [ECR200901]. The actual measurement cycle is designed to be simple, fast, and inexpensive to run. It can be fully automated and is designed to reflect the utilization profile and conditions frequently experienced in the field. Note: There is no System Under Test (SUT) configuration change allowed any time during the test. All energy savings adjustments (if done) by the SUT should be automatic.

The procedure consists of four main steps:

System Under Test Preparation: SUT is configured according to class requirements and offered load defined in the class requirements.

Router tester equipment is used to simulate the load and collect the results.

Step 1. First run determines the maximum system throughput T_f (full duplex, measured in Gbps) with methodology similar to RFC 2544 within a selected application (at zero packet loss, full-mesh port configuration). There is no time limit for this run. The run is complete after T_f is determined.

Step 2 (full load): Second run offers the load T_f (identified at Step 1) to SUT for a period of 1,200 seconds. Energy consumption is being sampled for the entire period, and average consumption E_f calculated.

Step 3 (half load): Third run reduces the load T_f twice (T_h = 0.5 × T_f) and runs for another 1,200 seconds. Energy consumption is being measured for the entire period, and average consumption E_h calculated. Load reduction is achieved by reducing the packet rate on all configured ports. Packet loss during second or third run (if seen) invalidates the measurement and resets testing to first run to provide a better T_h estimate.

Step 4 (idle load): Idle run removes the load and runs for another 1,200 seconds. Energy consumption is being measured for the entire period, and average consumption E_i calculated. Load reduction is achieved by idling packet rate on all configured ports, or disabling ports on packet tester side, at vendor discretion.

For the purposes of public testing, all platforms should be tested with publicly available (shipping) software images, publicly available (shipping) board hardware revisions, and fully documented and supported configurations. See Table 3.6 for measurement conditions. Table 3.7 depicts equipment classes covered by the metric.

Table 3.6   Measurement Conditions

Temperature	The equipment shall be evaluated at an ambient temperature of 25°C ± 3°C. The System Under Test (SUT) itself should stay offline or operate at this air temperature for no less than 3 hours prior to the test. No ambient temperature changes are allowed until the test is complete.
Humidity	The equipment shall be evaluated at a relative humidity of 30% to 75%.
Air pressure	The equipment shall be evaluated at site pressure between 860 and 1,060 mbar.
DC voltage	The input to the SUT (all active feeds) shall be at a nominal DC voltage of ±0.5%.
AC voltage	The input to the SUT (all active feeds) shall be the specified voltage ±1% and the specified frequency ±1%.
Metrology requirements	Every active power feed should have the power (amp) meter installed in-line, with desired accuracy no less than ±1% of the actual power consumption.
Sampling frequency	E_f, E_h, and E_i calculations are based on averaging multiple readings over the course of measurements. Power meter(s) should be able to produce no less than 100 evenly spaced readings in every 1,200-second test cycle.

Table 3.7 Product Classes & Test Applications

Equipment Class	Description	Qualification	Test Application	Interface Types	Other Notes
(1) Core routers	Core routing platforms are systems with Terabit (half-duplex) or higher capacity. They are designed to provide line-rate performance in network cores with minimum functions (packet lookup and forwarding/ switching). Core routing platforms come in various form factors, in standalone and multi-chassis enclosures.	1 Tbps or better forwarding capacity	IPv4, IPv6 or MPLS forwarding at discretion of the vendor; packet size: 64B; forwarding over any type of forwarding entries (static, connected, IGP, EGP) — no less than active 16K routes.	10GE or 100GE as designated by the vendor, SR optics	For the purposes of testing, all redundant components (fabric, routing engines, power supplies, etc.) should be present in the system. Metric awarded: ECR64 (MPLS, IPv4, IPv6)

(1) Edge Routers	Edge routing platforms	200 Gbps or better forwarding capacity	IP VPN, PWE, or VPLS forwarding at discretion of the vendor; packet size: 256B; forwarding over any types of forwarding entries across all VPN instances, no less than 2K VPN instances active (PWE circuits, VPLS instances, IP VPN VRFs)	At vendor discretion	For the purposes of testing, all redundant components (fabric, routing engines, power supplies, etc.) should be present in the system. Metric awarded: ECR256 (PWE, VPLS, IP VPN)
(1) Multipurpose routers	Routing platforms of variable purposes (enterprise, edge, etc.)	L3 forwarding	IPv4 or IPv6 forwarding at vendor discretion. packet size: 576B; forwarding over any types of forwarding entries, no less than 16K active routes	Electrical or optical at vendor discretion	For the purposes of testing, redundant component may be removed. Metric awarded: ECR576 (IPv4, IPv6)

Continued

Table 3.7   Product Classes & Test Applications (Continued)

Equipment Class	Description	Qualification	Test Application	Interface Types	Other Notes
(2) WAN/ Broadband Aggregation Device: BRAS devices	Legacy broadband aggregation devices	Any capacity, PPPoE, PPPoA, PPP, per-subscriber QoS	PPPoE, PPPoA, PPP forwarding at discretion of the vendor; packet size: 256B; forwarding over any type of per-subscriber entries, no less than 64K subscribers with no less than four (4) queues assigned to each.	SR optical at vendor discretion	For the purposes of testing, all redundant components (fabric, routing engines, power supplies, etc.) should be present in the system. Metric awarded: ECR256 (PPP, PPPoE, PPPoA)
(2) WAN/ Broadband Aggregation Device: BSR/ Common Edge devices	Broadband aggregation devices, Ethernet-oriented	Any capacity, PPPoE, PPP, IP DHCP, per-subscriber QoS	IP/DHCP, PPPoE, PPP forwarding at discretion of the vendor; packet size: 256B; forwarding over any type of per-subscriber entries, no less than 64K subscribers with no less than four (4) queues assigned to each	SR optical at vendor discretion	For the purposes of testing, all redundant components (fabric, routing engines, power supplies, etc.) should be present in the system. Metric awarded: ECR256 (PPP, IP DHCP, PPPoE)

(3) Ethernet L2/L3 Switches — Carrier Ethernet Platforms	Carrier-grade ethernet switching platforms	L2 (Ethernet) forwarding, MPLS forwarding, IPv4, or IPv6 forwarding	Ethernet or MPLS forwarding at vendor discretion. Payload packet size: 256B frames; forwarding over any type of forwarding entries and encap type.	SR optical (10/100/1000/10GE) at vendor discretion	For the purposes of testing, redundant components must be present. Metric awarded: ECR256 (Ethernet, MPLS, IPv4, IPv6)
(3) Ethernet L2/L3 Switches — Generic Ethernet Platforms	Ethernet switching platforms	L2 (Ethernet) forwarding, MPLS forwarding, IPv4, or IPv6 forwarding	Ethernet or MPLS forwarding at vendor discretion. Payload packet size: 1500B frames; forwarding over any type of forwarding entries and encapsulation type.	Copper or SR optical (10/100/1000/10GE) at vendor discretion	For the purposes of testing, redundant components may be removed. Metric awarded: ECR1500 (Ethernet, MPLS, IPv4, IPv6)
(4) Experimental					

Continued

Table 3.7 Product Classes & Test Applications (Continued)

Equipment Class	Description	Qualification	Test Application	Interface Types	Other Notes
(5) Security appliances (DPI, Firewalls, VPN Gateways, etc.)	Security platforms of variable purposes (IP Sec VPN, HTTPS, DPI, IDS, etc.)	L3 forwarding, security features	IPSec or HTTPS, minimum number of firewall or DPI forwarding rules at vendor discretion; 512B payload packets	at vendor discretion	For the purposes of testing, redundant component may be removed. Metric awarded: ECR512 (IPSec DES, IPSec 3DES, HTTPS, DPI, IDS, etc)
(6) Application Gateways (Layer 5–7 accelerators, load balancers, etc.)	Application platforms of variable purposes (SLB, accelerators, compressors)	Application-specific features	User traffic at vendor discretion (need more qualification for setup); 512B payload packets	at vendor discretion	For the purposes of testing, redundant components may be removed. Metric awarded: ECR512 (SLB, TCP acceleration, compression, etc.)

3.5.3 Telecommunications Energy Efficiency Ratio (TEER)

The Telecommunications Energy Efficiency Ratio (TEER) is a measure of network-element efficiency defined by the Alliance for Telecommunications Industry Solutions (ATIS). The standards provide a comprehensive methodology for measuring and reporting energy consumption, and uniformly quantify a network component's ratio of "work performed" to energy consumed. The efficiency standards are specific to equipment type, network location, and classification. Normalizing these ratings by functionality enables "apples-to-apples" equipment comparison. This systemized assessment results in repeatable and comparable energy consumption measurement.

ATIS-0600015.2009 is the base standard for determining telecommunications energy efficiency. The latter two documents (server requirements, and transport system or network configuration requirements) are part of an ongoing series to define the telecommunications energy efficiency of various telecommunications components.

In general, each TEER follows the formula:

$$\text{TEER} = \frac{\text{Useful Work}}{\text{Power}}$$

where Useful Work is defined in the supplemental standard based on the equipment function. Examples could be, but are not limited to, data rate, throughput, processes per second, and so on. Power is the power in watts (dependent on the equipment measurement).

3.5.4 Telecommunications Equipment Energy Efficiency Rating (TEEER)

Verizon* has published a Telecommunications Equipment Energy Efficiency Rating (TEEER) methodology for procuring equipment assemblies used in Verizon's telecommunications networks [VER200901]. TEEER is a calculated value representing the energy efficiency rating of a specific product. The rating applies, as a minimum, to the set of NEs and other equipment shown in Table 3.8. Table 3.9 provides some key terms.

The following list identifies some specific conditions and approaches to be used in deriving TEEER:

■ *Temperature:* The equipment shall be evaluated at a temperature of 25°C ± 3°C.
■ *Humidity:* The equipment shall be evaluated at a relative humidity of 30% to 75%.
■ *Pressure:* The equipment shall be evaluated at site pressure between 860 and 1060 mbar.

---

* This section is based, in its entirety, on Verizon materials.

Table 3.8 NEs and Other Equipment for TEEER

Transport	Optical transport system
	Video transport system
	Point-to-point microwave transport
Switch/Router	Digital switch
	Soft switch
	Enterprise router
	Core router
	Edge router
	Backbone router
	Feature application router
Gateways	Media gateway
Access	Digital Subscriber Line Access Multiplexer (DSLAM)
	Optical Line Terminal (OLT)
Power	Rectifiers
	Converters
	Inverters
	Uninterruptable Power Supply (UPS)
Data center equipment	Servers
Customer Premises Equipment (CPE)	External power adapters
	Set-top boxes
	ONT power supplies
Wireless	Power amplifiers

- *Test equipment and set-up:* Power measurements shall be made with a suitably calibrated voltmeter and ammeter, or power analyzer. The power measurement instrument shall have a resolution of 0.1 W or better for active power. Power measurements shall be taken immediately adjacent to the powered product being evaluated. Support equipment shall be provided to verify proper operation of the equipment under test.
- *Test voltage, DC powered equipment:* The input to the EUT shall be at a DC voltage of –53 ± 0.25 V. Equipment using voltages other than –48 V DC shall be evaluated at ±0.25 of its nominal voltage.

Table 3.9 Definition of Key TEEER Terms

Term	Definition
Forwarding capacity	The number of bits per second (bps) that a device can be observed to transmit successfully to the correct egress interface
Nameplate Output Power (P_{no})	The manufacturer stated output power for AC–DC or AC–AC adapters provided on the power adapter nameplate
P_{max}	The measured input power with the Equipment Under Test (EUT) operating at maximum load
P_{sleep}	The measured input power of the EUT while operating in a sleep/no activity mode
P_{Total}	The weighted total input power to be used in the formation of the TEEER value
P_{50}	The measured input power of the EUT while operating at 50% of maximum load
SPECpower_ssj2008TM	SPEC benchmark that evaluates the power and performance characteristics of volume server class computers
Throughput	The number of bits passing through the data communication system expressed in bits per second (bps)

■ *Test voltage, AC powered equipment:* The input to the Equipment Under Test (EUT) shall be the specified voltage ±1% and the specified frequency ±1%.

■ *Utilization conditions:* The equipment shall be fully loaded with all card slots populated with functioning modules and all redundancies in place. The equipment shall have all cables installed as in a typical deployment. All system functions or features that increase power consumption shall be activated during testing. If the equipment has any energy saving features that are controlled by internal software, then they should be enabled for testing. The EUT shall be tested at the following utilization conditions:

Utilization Condition 1	100%
Utilization Condition 2	50%
Utilization Condition 3	0%

Testing sequence: With the equipment configured as stated above, the EUT shall be operated at 100% utilization for at least 15 minutes prior to conducting power measurements.

> After the 15 minute initialization period, the EUT input power shall be monitored to assess the stability of the EUT. If the power level does not drift by more than 5% from the maximum value observed, the EUT can be considered stable and the measurements can begin.
>
> With the equipment operating under normal maximum power conditions, record the average input power to the equipment under test over a 15 minute time period for Utilization Condition 1. This value shall be recorded as P_{max}.
>
> Repeat power input measurements for Utilization Condition 2 and Utilization Condition 3, and record these values at P_{50} and P_{sleep}, respectively. The total power consumption for the EUT shall be represented by the weighting formula

$$P_{Total} = (0.35 \times P_{max}) + (0.4 \times P50) + (0.25 \times P_{sleep})$$

> where P_{max} is the average power measured during Utilization Condition 1, P_{50} is the average power measured during Utilization Condition 2, and P_{sleep} is the average power measured during Utilization Condition 3.

■ *Weighting values:* Verizon assigns weighting values to accommodate for the variable utilization of equipment in each duty cycle.
■ *Power equipment:* Equipment is to be configured as would be for a typical Verizon deployment. Input power measurements shall be taken immediately adjacent to the input terminals. Output measurements shall be taken from the main output distribution bus. Breakers and fuses shall not be included for the calculations and measurements.
■ *Power utilization levels:* Power equipment testing will be performed at utilization levels of 100% and 50%.

$$P_{Out\ Total} = (P_{Out\ max} + P_{Out\ 50})/2$$

$$P_{In\ Total} = (P_{In\ max} + P_{In\ 50})/2$$

With these definitions in mind, TEEER formulas for telecom equipment follow. The total average power, Ptotal (calculated from above), is used in calculating the TEEER. Using the type of equipment that most closely resembles the equipment tested, calculate the TEEER for the given system.

Telecom Equipment Type	TEEER Formula
Transport	$-\log(P_{Total}$ / Throughput)
Switch/Router	$-\log(P_{Total}$ / Forwarding Capacity)
Media gateway	$-\log(_{PTotal}$ / Throughput)
Access	(Access Lines / P_{Total}) + 1
Power	$(P_{Out\ Total}$ / $P_{In\ Total})$ × 10
Power amplifiers (wireless)	(Total RF Output Power / Total Input Power) × 10

Examples of TEEER calculations follow.

■ Transport:
Throughput = 40 Gbps
P_{max} = 1,000 W
P_{50} = 950 W
P_{sleep} = 900 W
P_{Total} = (0.35 × 1,000) + (0.4 × 950) + (0.25 × 900) = 955 W
TEEER = $-\log$ (PTotal / Throughput)
= $-\log$ (955 / 40,000,000,000)
= $-\log$ (0.000000023875)
= 7.62

■ Switch/Router:
Forwarding Capacity = 160 Gbps
P_{max} = 4,320 W
P_{50} = 3,000 W
P_{sleep} = 1,500 W
P_{Total} = (0.35 × 4,320) + (0.4 × 3,000) + (0.25 × 1,500) = 3,087 W
TEEER = $-\log$ (P_{Total} / Forwarding Capacity)
= $-\log$ (3087 / 160,000,000,000)
= $-\log$ (0.00000001929375)
= 7.71

■ Access:
Access Lines = 284
P_{max} = 120 W
P_{50} = 80 W
P_{sleep} = 40 W
P_{Total} = (0.35 × 120) + (0.4 × 80) + (0.25 × 40) = 84 W
TEEER = (Access Lines / PTotal) + 1

$= (284 / 84) + 1$

$= 4.38$

■ Power:

$P_{Out\ max} = 800\ W$

$P_{Out\ 50} = 400\ W$

$P_{In\ max} = 844\ W$

$P_{In\ 50} = 462\ W$

$P_{Total\ Out} = (800 + 400)/2 = 600$

$P_{Total\ In} = (838 + 462)/2 = 650$

$TEEER = (P_{Total\ Out} / P_{Total\ In}) \times 10$

$= (600 / 650) \times 10$

$= 9.23$

■ Power Amplifiers (Wireless):

Sectors = 3

Carriers = 8

RF Output Power/Carrier, measured at the input of the Antenna P1 = 20.0 W

Input Power/Watt of output power P2 = 11.425 W

Total Input Power for 3 Sectors, 8 Carriers Amplification = Sectors × Carriers
  × P1 × P2

$= 3 \times 8 \times 20 \times 11.425\ W$

$= 5,484\ W$

Total RF Output Power for 3 Sectors, 8 Carriers = Sectors × Carriers × P1

$= 3 \times 8 \times 20\ W$

$= 480\ W$

$TEEER = (Total\ RF\ Output\ Power\ /\ Total\ Input\ Power) \times 10$

$= 480 / 5,458 \times 10$

$= 0.875$

With previous definitions in mind, TEEER formulas for Data Center Equipment follow. Server-type equipment is tested using the methods and procedures as defined in the SPECpower_ssj2008 specification (see Section 3.5.5). The value derived from SPECpower_ssj2008 is used in the formula provided below to arrive at a TEEER value for that server. To calculate the SPECpower_ssj2008 value, this benchmark program must be obtained from the Web site www.spec.org. SPECpower_ssj2008 is the first industry-standard SPEC benchmark that evaluates the power and performance characteristics of volume server class computers. The initial benchmark addresses the performance of server-side Java.

Data Center Equipment Type	TEEER Formula
Server	(SPECpower_ssj2008) / 100

The TEEER value calculated from above shall meet the minimum TEEER value allowable (e.g., as specified by Verizon).

With the previous definitions in mind, TEEER formulas for CPE can be developed for

■ External Power Adapters
■ Set-Top Boxes
■ ONT Power Supply Load Levels

External Power Adapters

Customer premises equipment that is supplied with either an AC-to-AC adapter or AC-to-DC adapter is expected to follow the methods and procedures of the most current version of the ENERGY STAR® requirements for external power adapters. ENERGY STAR requirements for external power supplies can be found at www. energystar.gov. An external power supply model must meet or exceed a minimum average efficiency for Active Mode, which varies based on the model's nameplate output power.

Calculate the model's single average Active Mode efficiency value by testing at 100%, 75%, 50%, and 25% of rated current output and then computing the simple arithmetic average of these four values as specified in the ENERGY STAR Test Method.

Based on the model's nameplate output power, select the appropriate equation from Table 3.10 and calculate the minimum average efficiency.

Compare the model's actual average efficiency; if greater than or equal to the minimum average efficiency, the model has satisfied the requirement.

Set-Top Boxes

Set-Top Box equipment efficiency is expected to follow the methods and procedures of the most current version of the ENERGY STAR requirements for Set-Top Boxes. ENERGY STAR requirements for Set-Top Boxes can be found at www. energystar.gov. A Set-Top Box model must meet or exceed a minimum average efficiency for base functionality plus allowances for specific, additional functionalities (see Table 3.11) present across a duty cycle. To calculate the allowance for a given device, the sum of the base functionality allowance and all applicable additional functionalities allowances are added. This value is compared to the measured values following the procedures as stated in the ENERGY STAR Set-Top Box test procedures to determine compliance.

Table 3.10   Energy Efficiency Criteria

Energy Efficiency Criteria for AC–AC and AC–DC External Power Supplies in Active Mode: Standard Models	
Nameplate Output Power (P$_{no}$) (watt)	*Minimum Average Efficiency in Active Mode (expressed as a decimal)*
0 to ≤ 1	≥ 0.480 × P$_{no}$ + 0.140
> 1 to ≤ 49	≥ [0.0626 × Ln (P$_{no}$)] + 0.622
> 49	≥ 0.870
Energy-Efficiency Criteria for AC–AC and AC–DC External Power Supplies in Active Mode: Low-Voltage Models	
Nameplate Output Power (P$_{no}$) (watt)	*Minimum Average Efficiency in Active Mode (expressed as a decimal)*
0 to ≤ 1	≥ 0.497 × P$_{no}$ + 0.067
> 1 to ≤ 49	≥ [0.0750 × Ln (P$_{no}$)] + 0.561
> 49	≥ 0.860

Energy Consumption Criteria for No Load		
Nameplate Output Power (P$_{no}$) (watt)	*Maximum Power in No Load (watt, W)*	
	AC–AC EPS	*AC–DC EPS*
0 to < 50	≤ 0.5	≤ 0.3
≥ 50 to ≤ 250	≤ 0.5	≤ 0.5

Example: High-Definition, Cable Set-Top Box with DVR

- ■ Annual Energy Allowance (kWh/year) = Base Functionality + Additional Functionalities
- ■ Annual Energy Allowance (kWh/year) = 70 + 60 + 35
- ■ Annual Energy Allowance (kWh/year) = 165
- ■ Optical Network Terminal (ONT) Power Supplies

Equipment is to be configured as it would be for a typical Verizon deployment. Input power measurements are taken immediately adjacent to the input terminals of the Optical Network Terminal Power Supply Unit (OPSU). Output measurements will be taken immediately adjacent to the output terminal of the Battery

Table 3.11 Annual Energy Allowance

Base Functionality	Tier 1 Annual Energy Allowance (kWh/year)
Cable	70
Satellite	88
IP	45
Terrestrial	27
Thin-client/remote	27
Additional Functionalities	Tier 1 Annual Energy Allowance (kWh/year)
Additional tuners	53
Additional tuners — terrestrial/IP	14
Advanced video processing	18
DVR	60
High definition	35
Removable media player	12
Removable media player/recorder	23
Multi-room	44
CableCard	15
Home network interface	20

Backup Unit (BBU). A resistive load may used to represent the ONT. Testing will be performed with the battery in a fully charged state.

ONT Power Supply testing will be performed at load levels of 100% and 50%:

$$P_{Out\ Total} = (P_{Out\ max} + P_{Out\ 50})/2$$

$$P_{In\ Total} = (P_{In\ max} + P_{In\ 50})/2$$

The ONT Power Supply TEEER Formation is as follows:

Equipment Type	TEEER Formula
ONT Power Supply	$(P_{Out\ Total} / P_{In\ Total}) \times 10$

3.5.5 Standard Performance Evaluation Corporation (SPEC)

SPEC* is a non-profit organization that establishes, maintains, and endorses standardized benchmarks to evaluate performance for the newest generation of computing systems. Its membership comprises more than 80 leading computer hardware and software vendors, educational institutions, research organizations, and government agencies worldwide.

SPECpower_ssj2008 is the first industry-standard SPEC benchmark that evaluates the power and performance characteristics of volume server class and multinode class computers. With SPECpower_ssj2008, SPEC is defining server power measurement standards in the same way it has done for performance. The drive to create the power and performance benchmark comes from the recognition that the IT industry, computer manufacturers, and governments are increasingly concerned with the energy use of servers. Currently, many vendors report some energy efficiency figures, but these are often not directly comparable due to differences in workload, configuration, test environment, etc. Development of this benchmark provides a means to measure power (at the AC input) in conjunction with a performance metric. This should help IT managers to consider power characteristics along with other selection criteria to increase the efficiency of data centers.

The purpose of any general benchmark is to provide a comparison point between offerings in a specific environment. In performance benchmarks, it is important to select a benchmark that relates well to the desired environment for a computer installation. Different solutions perform differently in various benchmarks, and a single benchmark cannot be considered a direct indicator of how a system will perform in all environments.

The same statement is true for a metric that combines performance and power. Because performance is a very significant part of this equation, it is important to consider the workload being performed by the benchmark and to ensure that it is relevant. Power metrics add an additional dimension to this caution. Power characteristics depend heavily on both workload and configuration. Thus, when examining power metrics between two systems, it is important to know whether the configurations of the systems being compared are relevant to an environment that is important to the firm.

The use of a performance benchmark or of a power-and-performance benchmark should focus on a comparison of solutions, and not specific planning purposes. Unless the workload and configurations of the benchmarked solution match the planned solution, it could be very misleading to assume that a benchmark result will equate to reality in a production data center. Benchmark designers can help prevent customers and analysts from drawing inappropriate conclusions by creating power-related metrics that allow easy comparisons, but do not readily lead to predictions of real power characteristics in a customer environment.

---

* This section is based on SPEC documentation [SPE200901].

The initial SPEC benchmark, SPECpower_ssj2008, addressed only one subset of server workloads: the performance of server-side Java. In 2009, SPEC released SPECpower_ssj2008 V1.10, the latest version of the only industry-standard benchmark that measures power consumption in relation to performance for server-class computers. SPECpower_ssj2008 v1.10 adds several new capabilities, including support for measuring multi-node (blade) servers, further automation of power measurement, support for multiple power analyzers, and a visual activity monitor that provides real-time graphic display of data collected while running the benchmark. The benchmark workload represents typical server-side Java business applications. The workload is scalable, multi-threaded, portable across a wide range of operating environments, and economical to run. It exercises the CPUs, caches, memory hierarchy, and the scalability of Shared Memory Processors (SMPs) as well as the implementations of the JVM (Java Virtual Machine), JIT (Just-In-Time) compiler, garbage collection, threads, and some aspects of the operating system. The benchmark runs on a wide variety of operating systems and hardware architectures, and should not require extensive client or storage infrastructure.

The major new feature in SPECpower_ssj2008 V1.10 is support for measuring multi-node servers. Other new features further automate power measurement, including support for multiple power analyzers. With this release, SPEC is responding to market demand to measure power and performance for blade servers, thereby further extending the benchmark as a tool that assists IT managers in deploying energy-efficient data centers. SPECpower_ssj2008 is both a benchmark and a toolset that generates customized loads allowing site-specific, in-depth analysis of server efficiency. Results from the latest version of the benchmark are comparable to V1.00 and V1.01, although the previous versions are limited to measuring one server at a time. SPEC member companies active in developing SPECpower_ssj2008 V1.10 include AMD, Dell, Fujitsu, HP, Intel, IBM, and Sun Microsystems.

SPECpower_ssj2008 reports power consumption for servers at different performance levels—from 100% to idle in 10% segments—over a set period of time. The graduated workload recognizes the fact that processing loads and power consumption on servers vary substantially over the course of days or weeks. To compute a power-performance metric across all levels, measured transaction throughputs for each segment are added together, then divided by the sum of the average power consumed for each segment. The result is a figure of merit called "overall ssj_ops/watt."

3.5.6 Standardization Efforts

Table 3.12 depicts the worldwide state of network efficiency standardization as of the end of 2008 [ALI200801]. Chapter 6 provides additional information on some of these efforts. In addition to the infrastructure and load efficiency metrics, some are developing other metrics, such as the following [EUC200801]:

Table 3.12 Network Efficiency Standardization Work

	High-Level Definition[a]	Metric Definition		SUT Test Procedures	
		Peak	Weighted	Methodology	Load/ Profiles
ATIS draft NIPP-TEE-2008-031R4	Yes[d]	TEER[d]	Not defined	Not defined	Not defined
METI "Top Runner"	Yes[d]	Not defined	Yes[e]	Yes[f]	Yes[f]
ITU-T[b]	Not defined	Not defined	Not defined	Not defined	Not defined
EC	Not defined	Not defined	Not defined	Not defined	Not defined
BWF[c]	Not defined	Not defined	Not defined	Not defined	Not defined
Green Grid	Yes[d]	ATIS	ATIS	ATIS	ATIS
ECR	Yes	Yes	Yes	Yes	Yes

[a] Definition of general ICT energy efficiency as "payload per energy unit."
[b] ECR draft submitted as formal contribution to Climate Change ICT FG.
[c] ECT draft submitted to Marketing and Test & Interoperability/Green OC.
[d] Identical to ECR.
[e] Currently incompatible with ECR.
[f] Compatible with ECR.

■ *IT productivity metric:* An advanced metric providing an indicator of how efficiently the IT equipment provides useful IT services.
■ *Total energy productivity metric:* Similar to the IT metrics but relating the useful IT services to the total energy consumption of facility.

3.6 Metrics on Carbon Footprints

The benefits of the "green agenda" are frequently discussed in terms of reduction in power consumption (energy efficiency) or decrease in Greenhouse Gases (GHG) emissions. At the macro level, it is estimated that there are now about 390 CO_2 molecules in the atmosphere for every million molecules of air; environmental activists claim that the number should not exceed 350 parts per million. Quantifying the relationship between parameters in these two domains can be helpful in planning and assessing performance. While power metrics discussed in the sections above are useful, once a power consumption-based measure is known, estimates of the equivalent GHG emission is possible using tools such as the EPA Carbon-Equivalence Calculator. This tool provides the estimated GHG equivalents based on a U.S. "national average" (using the statistics on energy generation technologies across the country). An equivalency calculator for specific regions, considering the current power generation sources in various regions of the United States, is also available from the EPA [EPA200901].

"Carbon footprint" is a term used to describe a calculation of total carbon emissions of some system, namely, to describe energy consumption. It is the amount of GHG produced to support a given activity (or piece of equipment), typically expressed in equivalent tons of carbon dioxide (CO_2). Carbon footprints are normally calculated for a period of 1 year. One way to calculate carbon footprints is to use a Life-Cycle Assessment; this takes into account the emissions required to produce, use, and dispose the object. Another way is to analyze the emissions resulting from the use of fossil fuels. Several carbon footprint calculators are available. A well-known calculator is offered by the EPA: It is a personal emissions calculator that allows individuals to estimate their own carbon footprint by asking the individual to supply various fields relating to transportation, home energy, and waste; associated with each question, the calculator shows how many pounds of carbon dioxide are formed per year. Safe Climate is another carbon footprint calculator; to determine the footprint, this calculator requires information regarding energy use in gallons, liters, therms, thousand cubic feet, and kilowatt-hours per month (or year). See Appendix 3C for some examples.

At a more macro level, recent discussions have addressed global-level metrics such as carbon intensity and carbon productivity; while these do not apply directly to green IT/green networks, they are worth mentioning.

- *Carbon intensity*, also called per capita annual emissions, is a measure of how many carbon equivalents (CO_2e) are emitted per capita of GDP.
- *Carbon productivity* is the inverse of carbon intensity; it is the amount of GDP product per unit of carbon equivalent.

Stabilization of the amount of GHG should be achieved while maintaining economic growth. The climate change community targets emissions at 20 gigatons per year. To achieve this, carbon productivity will have to increase from the $740 GDP per ton of CO_2e to $7,300 GDP per ton of CO_2e by 2050. In 2008, the carbon intensities (tons of CO_2e per $1,000) of the United States and Russia were 21.5 and 15.9, respectively, compared to values of 5.7 and 1.9 for China and India, respectively. According to proponents, the carbon intensity average should be around 2.2 [PRA200901].

Carbon neutral refers to calculating total carbon emissions for an operation and/or firm, reducing them where possible, and balancing remaining emissions with the purchase of carbon offsets [GAL200801].

3.7 Other Metrics

Other metrics are described, as needed, in the chapters that follow. For example, a number of terms related to power are discussed in Chapter 4, and terms related to CRACs/HVACs are discussed in Chapter 5.

Appendix 3A: Common Information Systems Assets

This appendix, drawn from *Microsoft Solutions for Security and Compliance and Microsoft Security Center of Excellence, The Security Risk Management Guide* (2006), lists information system assets commonly found in organizations of various types [MIC200601]. It is not intended to be comprehensive, and it is unlikely that this list will represent all the assets present in an organization's unique environment. Therefore, it is important that an organization customizes the list during the Risk Assessment phase of the project. Table 3A.1 provides common information systems assets.

Table 3A.1　Common Information Systems Assets

Asset Class	*Overall IT Environment*	*Asset Name*	*Asset Rating*
	Highest Level Description of Asset	**Next-Level Definition (if needed)**	**Asset Value Rating**
Tangible	Physical infrastructure	Data centers	5
Tangible	Physical infrastructure	Servers	3
Tangible	Physical infrastructure	Desktop computers	1
Tangible	Physical infrastructure	Mobile computers	3
Tangible	Physical infrastructure	PDAs	1
Tangible	Physical infrastructure	Cell phones	1
Tangible	Physical infrastructure	Server application software	1

Table 3A.1 Common Information Systems Assets (Continued)

Asset Class	Overall IT Environment	Asset Name	Asset Rating
	Highest Level Description of Asset	**Next-Level Definition (if needed)**	**Asset Value Rating**
Tangible	Physical infrastructure	End-user application software	1
Tangible	Physical infrastructure	Development tools	3
Tangible	Physical infrastructure	Routers	3
Tangible	Physical infrastructure	Network switches	3
Tangible	Physical infrastructure	Fax machines	1
Tangible	Physical infrastructure	PBXs	3
Tangible	Physical infrastructure	Removable media (tapes, floppy disks, CD-ROMs, DVDs, portable hard drives, PC card storage devices, USB storage devices, etc.)	1
Tangible	Physical infrastructure	Power supplies	3
Tangible	Physical infrastructure	Uninterruptible power supplies	3
Tangible	Physical infrastructure	Fire suppression systems	3
Tangible	Physical infrastructure	Air-conditioning systems	3

Continued

Table 3A.1   Common Information Systems Assets (Continued)

Asset Class	Overall IT Environment	Asset Name	Asset Rating
	Highest Level Description of Asset	**Next-Level Definition (if needed)**	**Asset Value Rating**
Tangible	Physical infrastructure	Air filtration systems	1
Tangible	Physical infrastructure	Other environmental control systems	3
Tangible	Intranet data	Source code	5
Tangible	Intranet data	Human resources data	5
Tangible	Intranet data	Financial data	5
Tangible	Intranet data	Marketing data	5
Tangible	Intranet data	Employee passwords	5
Tangible	Intranet data	Employee private cryptographic keys	5
Tangible	Intranet data	Computer system cryptographic keys	5
Tangible	Intranet data	Smart cards	5
Tangible	Intranet data	Intellectual property	5
Tangible	Intranet data	Data for regulatory requirements (GLBA, HIPAA, CA SB1386, EU Data Protection Directive, etc.)	5

Table 3A.1   Common Information Systems Assets (Continued)

Asset Class	Overall IT Environment	Asset Name	Asset Rating
	Highest Level Description of Asset	**Next-Level Definition (if needed)**	**Asset Value Rating**
Tangible	Intranet data	U.S. employee Social Security numbers	5
Tangible	Intranet data	Employee drivers' license numbers	5
Tangible	Intranet data	Strategic plans	3
Tangible	Intranet data	Customer consumer credit reports	5
Tangible	Intranet data	Customer medical records	5
Tangible	Intranet data	Employee biometric identifiers	5
Tangible	Intranet data	Employee business contact data	1
Tangible	Intranet data	Employee personal contact data	3
Tangible	Intranet data	Purchase order data	5
Tangible	Intranet data	Network infrastructure design	3
Tangible	Intranet data	Internal web sites	3
Tangible	Intranet data	Employee ethno-graphic data	3

Continued

Table 3A.1   Common Information Systems Assets (Continued)

Asset Class	Overall IT Environment	Asset Name	Asset Rating
	Highest Level Description of Asset	**Next-Level Definition (if needed)**	**Asset Value Rating**
Tangible	Extranet data	Partner contract data	5
Tangible	Extranet data	Partner financial data	5
Tangible	Extranet data	Partner contact data	3
Tangible	Extranet data	Partner collaboration application	3
Tangible	Extranet data	Partner cryptographic keys	5
Tangible	Extranet data	Partner credit reports	3
Tangible	Extranet data	Partner purchase order data	3
Tangible	Extranet data	Supplier contract data	5
Tangible	Extranet data	Supplier financial data	5
Tangible	Extranet data	Supplier contact data	3
Tangible	Extranet data	Supplier collaboration application	3
Tangible	Extranet data	Supplier cryptographic keys	5
Tangible	Extranet data	Supplier credit reports	3

Table 3A.1   Common Information Systems Assets (Continued)

Asset Class	Overall IT Environment	Asset Name	Asset Rating
	Highest Level Description of Asset	**Next-Level Definition (if needed)**	**Asset Value Rating**
Tangible	Extranet data	Supplier purchase order data	3
Tangible	Internet data	Web site sales application	5
Tangible	Internet data	Web site marketing data	3
Tangible	Internet data	Customer credit card data	5
Tangible	Internet data	Customer contact data	3
Tangible	Internet data	Public cryptographic keys	1
Tangible	Internet data	Press releases	1
Tangible	Internet data	White papers	1
Tangible	Internet data	Product documentation	1
Tangible	Internet data	Training materials	3
Intangible		Reputation	5
Intangible		Goodwill	3
Intangible		Employee morale	3
Intangible		Employee productivity	3
IT Services	Messaging	E-mail/scheduling (e.g., Microsoft Exchange)	3
IT Services	Messaging	Instant messaging	1

Continued

Table 3A.1 Common Information Systems Assets (Continued)

Asset Class	Overall IT Environment	Asset Name	Asset Rating
	Highest Level Description of Asset	**Next-Level Definition (if needed)**	**Asset Value Rating**
IT Services	Messaging	Microsoft Outlook® Web Access (OWA)	1
IT Services	Core infrastructure	Active Directory® directory service	3
IT Services	Core infrastructure	Domain Name System (DNS)	3
IT Services	Core infrastructure	Dynamic Host Configuration Protocol (DHCP)	3
IT Services	Core infrastructure	Enterprise management tools	3
IT Services	Core infrastructure	File sharing	3
IT Services	Core infrastructure	Storage	3
IT Services	Core infrastructure	Dial-up remote access	3
IT Services	Core infrastructure	Telephony	3
IT Services	Core infrastructure	Virtual Private Networking (VPN) access	3
IT Services	Core infrastructure	Microsoft Windows® Internet Naming Service (WINS)	1
IT Services	Other infrastructure	Collaboration services (e.g., Microsoft SharePoint®)	

Appendix 3B: Conversion of Units

Table 3B.1 depicts some common cross-conversion of units of interest.

Table 3B.1 Conversion Units

Mass			
1 pound (lb)	453.6 grams (g)	0.4536 kilograms (kg)	0.0004536 metric tons (tonne)
1 kilogram (kg)	2.205 pounds (lb)		
1 short ton (ton)	2,000 pounds (lb)	907.2 kilograms (kg)	
1 metric ton (tonne)	2,205 pounds (lb)	1,000 kilograms (kg)	
Volume			
1 cubic foot (ft³)	7.4805 U.S. gallons (gal)	0.1781 barrel (bbl)	
1 cubic foot (ft³)	28.32 liters (L)	0.02832 cubic meters (m³)	
1 U.S. gallon (gal)	0.0238 barrel (bbl)	3.785 liters (L)	0.003785 cubic meters (m³)
1 barrel (bbl)	42 U.S. gallons (gal)	158.99 liters (L)	0.1589 cubic meters (m³)
1 liter (L)	0.001 cubic meters (m³)	0.2642 U.S. gallons (gal)	
1 cubic meter (m³)	6.2897 barrels (bbl)	264.2 U.S. gallons (gal)	1,000 liters (L)
Energy			
1 kilowatt hour (kWh)	3,412 Btu (BTU)	3,600 kilojoules (kJ)	
1 megajoule (MJ)	0.001 gigajoules (GJ)		
1 gigajoule (GJ)	0.9478 million Btu (million Btu)	277.8 kilowatt-hours (kWh)	
1 Btu (Btu)	1,055 joules (J)		

Continued

Table 3B.1   Conversion Units (Continued)

Energy			
1 million Btu (million Btu)	1.055 gigajoules (GJ)	293 kilowatt-hours (kWh)	
1 therm (thm)	100,000 Btu	0.1055 gigajoules (GJ)	29.3 kilowatt-hours (kWh)
1 hundred cubic feet of natural gas (CCF)	1.03 therm (therm)		
Other			
1 land mile	1.609 land kilometers (km)		
1 nautical mile	1.15 land miles		
1 metric ton carbon	3.664 metric tons CO_2		

Appendix 3C: Example Carbon Footprint Calculations

This appendix provides some general examples of carbon footprint calculations. The information that follows comes directly from EPA sources (www.epa.gov).

Electricity Use (kilowatt-hours)

The Clean Energy Equivalencies Calculator uses the Emissions & Generation Resource Integrated Database (eGRID) U.S. annual non-baseload CO_2 output emission rate when converting reductions in kilowatt hours to avoided units of CO_2 emissions.

$$7.18 \times 10^{-4} \text{ metric tons } CO_2 \text{ / kWh}$$

(eGRID2007 Version 1.1, U.S. annual non-baseload CO_2 output emission rate, year 2005 data)

Coal-Fired Power Plant for A Year

In 2005 there were 2,134,520,641 tons of CO_2 emitted from power plants in the United States whose primary source of fuel was coal. In 2005 there was a total of 417 power plants whose primary source of fuel was coal. CO_2 emissions per power plant were calculated by dividing the number of power plants by the total emissions from power plants whose primary source of fuel was coal. The quotient was then converted from tons to metric tons:

$$2,134,520,641 \text{ tons } CO_2 \times 1/417 \text{ power plants} \times 0.9072 \text{ metric tons } /$$
$$1 \text{ short ton} = 4,643,734 \text{ metric tons } CO_2/\text{power plant}$$

Gallons of Gasoline Consumed

The average heat content of conventional motor gasoline is 5.22 million Btu (MMBtu) per barrel. The average carbon coefficient of motor gasoline is 19.33 kilograms (kg) carbon per million Btu. The fraction oxidized to CO_2 is 100%. Carbon dioxide emissions per barrel of gasoline were determined by multiplying heat content times the carbon coefficient time the fraction oxidized times the ratio of the molecular weight ratio of carbon dioxide to carbon (44/12). A barrel equals 42 gallons (gal).

$$5.22 \text{ MMBtu/barrel} \times 19.33 \text{ kg C/MMBtu} \times 1 \text{ barrel/42 gal} \times 44 \text{ g } CO_2/12 \text{ g C} \times$$
$$1 \text{ metric ton/1,000 kg} = 8.81 \times 10^{-3} \text{ metric tons } CO_2/\text{gal}$$

Passenger Vehicles per Year

Passenger vehicles are defined as two-axle, four-tire vehicles, including passenger cars, vans, pickup trucks, and sport/utility vehicles. In 2005 the weighted average combined fuel economy of cars and light trucks combined was 19.7 miles/gal. The average vehicle miles traveled (VMT) in 2005 was 11,856 miles per year. In 2005, the ratio of CO_2 emissions to total emissions including carbon dioxide (CO_2), methane (CH_4), and nitrous oxide (N_2O), all expressed as carbon dioxide equivalents) for passenger vehicles was 0.971. The amount of CO_2 emitted per gallon of motor gasoline burned is 8.81×10^{-3} metric tons. Thus,

$$8.81 \times 10^{-3} \text{ metric tons } CO_2/\text{gal gasoline} \times 11,856 \text{ VMT car/truck average} \times$$
$$1/19.7 \text{ miles/gal car/truck average} \times 1 \text{ } CO_2, CH_4, \text{ and } N_2O/0.971 \text{ } CO_2$$
$$= 5.46 \text{ metric tons } CO_2E \text{ /vehicle/year}$$

Therms of Natural Gas

The average heat content of natural gas is 0.1 MMBtu/therm. The average carbon coefficient of natural gas is 14.47 kg carbon/MMBtu. The fraction oxidized to CO_2 is 100%. Carbon dioxide emissions per therm were determined by multiplying heat content times the carbon coefficient times the fraction oxidized times the ratio of the molecular weight ratio of carbon dioxide to carbon (44/12).

$$0.1 \text{ MMBtu/1 therm} \times 14.47 \text{ kg C/MMBtu} \times 44 \text{ g CO}_2/12 \text{ g C} \times$$
$$1 \text{ metric ton/1,000 kg} = 0.005 \text{ metric tons CO}_2/\text{therm}$$

Barrels of Oil Consumed

The average heat content of crude oil is 5.80 MMBtu per barrel. The average carbon coefficient of crude oil is 20.33 kg carbon/MMBtu. Fraction oxidized is 100%. Carbon dioxide emissions per barrel of crude oil are determined by multiplying heat content times the carbon coefficient times the fraction oxidized times the ratio of the molecular weight of carbon dioxide to that of carbon (44/12).

$$5.80 \text{ MMBtu/barrel} \times 20.33 \text{ kg C/MMBtu} \times 44 \text{ g CO}_2/12 \text{ g C} \times$$
$$1 \text{ metric ton/1,000 kg} = 0.43 \text{ metric tons CO}_2/\text{barrel}$$

Home Electricity Use

In 2001, there were 107 million homes in the United States; of those, 73.7 million were single-family homes. On average, each single-family home consumed 11,965 kilowatt hours (kWh) of delivered electricity. The national average carbon dioxide output rate for electricity in 2005 was 1,329 pounds (lb) CO_2 per megawatt hour. Annual single-family home electricity consumption was multiplied by the carbon dioxide emission rate (per unit of electricity delivered) to determine annual carbon dioxide emissions per home:

$$11,965 \text{ kWh/home} \times 1,329.35 \text{ lb CO}_2/\text{MWh delivered} \times 1 \text{ MWh/1000 kWh} \times$$
$$1 \text{ metric ton/2,204.6 lb} = 7.21 \text{ metric tons CO}_2/\text{home.}$$

Home Energy Use

In 2001, there were 107 million homes in the United States; of those, 73.7 million were single-family homes. On average, each single-family home consumed 11,965 kWh of delivered electricity, 52,429 cubic feet (ft³) of natural gas, 57.3 gal of fuel oil, 46.6 gal of liquid petroleum gas, and 2.6 gal of kerosene. The national average CO_2 output rate for electricity in 2005 was 1,329 lb CO_2 per megawatt hour (MWh). The average carbon dioxide coefficient of natural gas is 0.0546 kg CO_2/ft^3. The fraction oxidized to CO_2 is 100%. The average carbon dioxide coefficient of distillate fuel oil is 462.1 kg CO_2 per 42 gallon barrel; the fraction oxidized to CO_2 is 100%. The average carbon dioxide coefficient of liquefied petroleum gases is 231.9 kg CO_2 per 42 gallon barrel; the fraction oxidized is 100%. The average carbon dioxide coefficient of kerosene is 410.0 kg CO_2 per 42 gallon barrel; the fraction oxidized to CO_2 is 100%. Total single-family home electricity, natural gas, kerosene distillate fuel oil, and liquefied petroleum gas consumption figures were converted from their various units to metric tons of CO_2 and added together to obtain total CO_2 emissions per home:

1. Delivered electricity: 11,965 kWh/home × 1,329.35 lb CO_2/MWh delivered × 1 MWh/1,000 kWh × 1 metric ton/2,204.6 lb = 7.21 metric tons CO_2/home

2. Natural gas: 52,429 ft³/home × 0.0546 kg CO_2/ft³ × 1/1,000 kg/metric ton = 2.86 metric tons CO_2/home

3. Fuel oil: 57.3 gal/home × 1/42 barrels/gal × 462.1 kg CO_2/barrel = 0.63 metric tons CO_2/home

4. Liquid petroleum gas: 46.6 gal/home × 1/42 barrels/gal × 231.9 kg CO_2/barrel = 0.26 metric tons CO_2/home

5. Kerosene: 2.6 gal/home × 1/42 barrels/gal × 410 kg CO_2/barrel = 0.03 metric tons CO_2/home

Therefore, the total CO_2 emissions for energy use per single-family home:

7.21 metric tons CO_2 for electricity + 2.86 metric tons CO_2 for natural gas + 0.63 metric tons CO_2 for fuel oil + 0.26 metric tons CO_2 for liquid petroleum gas + 0.03 metric tons CO_2 for kerosene = 10.99 metric tons CO_2 per home per year.

Propane Cylinders Used for Home Barbeques

The average heat content of liquefied petroleum gas is 21,591 Btu/lb. The average carbon coefficient of liquefied petroleum gases is 16.99 kg C/MMBtu. The fraction oxidized is 100%. Carbon dioxide emissions per pound of propane are determined by multiplying heat content times the carbon coefficient times the fraction oxidized times the ratio of the molecular weight of carbon dioxide to that of carbon (44/12). Propane cylinders vary with respect to size; for the purpose of this equivalency calculation, a typical cylinder for home use was assumed to contain 18 lb propane.

21,591 Btu/lb × 1 MMBtu/106 Btu × 16.99 kg C/MMBtu × 44 g CO_2/12 g C × 18 lb/1 canister × 1 metric ton/1,000 kg = 0.024 metric tons CO_2/canister

One Google Search

It has been stated in the press that one Google search from a desktop computer generates 7 grams of CO_2.

References

[ALI200801]   A. Alimian, B. Nordman, and D. Kharitonov, Network and Telecom Equipment — Energy and Performance Assessment. Test Procedure and Measurement Methodology. Draft 1.0.4, 10 November 2008.

[ATI200901]   ATIS Report on Environmental Sustainability, March 2009, A Report by the ATIS Exploratory Group on Green, ATIS, 1200 G Street, NW, Suite 500, Washington, DC 20005.

[CRI200001]   Criterium Engineers, What Does It All Mean?, White paper. 2000. 22 Monument Square – Suite 600, Portland, ME 04101.

[DOE200901] U.S. Department of Energy. Energy Efficiency and Renewable Energy, Data Center Energy Efficiency Training Materials, 5/19/2009.

[ECR200801]   ECR, Energy Efficiency for Network Equipment: Two Steps Beyond Greenwashing, Technical Report, August 10 2008, Revision 1.0.2, 10-08-2008.

[ECR200901]   The Energy Consumption Rating (ECR) Initiative, IXIA, Juniper Networks, http://www.ecrinitiative.org.

[ENE200901]   ENERGY STAR® Program Requirements for Light Commercial HVAC Partner Commitments, Version 2.0 — Draft 1, 5 January 2009.

[EPA200901]   Environmental Protection Agency, On-line Footprint Calculator, www.epa.gov, retrieved 11/11/2009.

[ETNO200801]   Energy Task Team, European Telecommunications Network Operators' Association, The Hague, London, April 2008.

[EUC200801]   EU Code of Conduct on Data Centers Energy Efficiency — Version 1.0, 30 October 2008.

[GAL200801]   Galley Eco Capital LLC, Glossary, 2008, San Francisco, CA.

[GAR200601]   D. Garday and D. Costello, Air-Cooled High-Performance Data Centers: Case Studies and Best Methods, White paper, Intel Corporation, November 2006.

[GRE200701]   The Green Grid, The Green Grid Data Center Power Efficiency Metrics: PUE and DCiE, White paper, 2007.

[GRE201001]   The Green Grid. Glossary and Other Reference Materials. 2010.

[ICA199701]   HVAC Systems for Buildings, ICAEN, REHVA, and Partex, THERMIE Programme, EUROPEAN Commission, 1997.

[LIE200901]   Liebert, Five Questions to Ask before Selecting Power Protection for Critical Systems, A Guide for It and Data Center Managers, 2003. Liebert Corporation, 1050 Dearborn Drive, P.O. Box 29186, Columbus, OH 43229.

[MGE200901]   Madison Gas and Electric, *Electric Glossary,* Madison Gas and Electric, Madison, WI, 2009.

[MIC200601]   Microsoft Corporation, Microsoft Solutions for Security and Compliance and Microsoft Security Center of Excellence, The Security Risk Management Guide, 2006, Redmond, WA.

[PRA200901]   V. Prakash, India One of the Least Carbon Intensive Countries in the World, Reuters, May 25, 2009.

[RUS200801]   N. Rasmussen, Electrical Efficiency Measurements for Data Centers, White Paper #154, American Power Conversion/Schneider Electric, 2008.

[SPE200901]   Standard Performance Evaluation Corp. (SPEC), Warrenton, VA, 15 April 2009, www.spec.org.

[VER200901]   Verizon, Verizon NEBSTM Compliance: Energy Efficiency Requirements for Telecommunications Equipment, Verizon Technical Purchasing Requirements, VZ.TPR.9205, Issue 4, August 2009.

Chapter 4

Power Management Basics

Data center and networking professionals need to gain an understanding of key power usage and power management concepts. These concepts clearly apply to enterprises with large network operations, Web hosting operations, "Cloud Computing/Grid Computing" operations, carriers, and service providers; while smaller telecom nodes as well as floor-level (closet) installations typically have less stringent/formal requirements, the basic ideas are still applicable.

Industry studies show, at least as seen from the recent past, that data center energy usage and costs double every 5 years. For example, in the mid-1990s, a fully populated server rack could house 14 single-corded servers operating at 120 volts (V); this rack consumed approximately 4 kilowatts (kW) of power. By 2001, a fully populated rack could house 42 servers, which were likely to be dual-corded (84 receptacles) and operating at 208 V, single phase; this rack consumed almost 20 kW. More recently, the emergence of blade servers has driven even more demand: A standard rack can now house six dual-corded blade chassis operating at 208 V, single phase, with a power consumption of 24 kW [LIE200701]. Press time requirements were reaching 28 kW per rack in some applications [MIT201001]. If one assumes that the trend continues, data center energy usage will have increased significantly between 2005 and 2025. While the underlying cost of energy may appear to stabilize during economic downturns, an increased political emphasis on "cap and trade" in an effort to foster a transition to renewable energy sources will invariably lead to additional increases in the cost of power—at least in the short term (say the decade of the 2010s). It follows that understanding power issues and finding ways of managing usage are critical for IT management in general, and networking professionals in particular.

The chapter reviews some of the basic concepts related to power and traditional data center/networking node use. We review current types, some key metrics for usage, and Backup Power Systems, specifically Power Protection Systems and Uninterruptible Power Supplies (UPS).

4.1 Current Types

Current used in data centers and in telco/network nodes can be of the Direct Current (DC), Single-Phase Alternating Current (AC), and Three-Phase AC types. In DC, the amplitude of the voltage is constant over time. Voltage is a measure of the work needed to move an electric charge; such movement gives rise to a current. Single-phase AC is an alternating current with a sinusoidal waveform for the time–amplitude relationship of the voltage. Three-phase AC is an alternating current with three sinusoidal waveforms for the time–amplitude relationship of the voltage that follow each other in precise timing, typically 120° out-of-phase. See Figure 4.1. Table 4.1 identifies some key electrical terms.

Some simple examples of Ohm's law follow:

■ Example 1: A motor running on a single-phase 240 V electric circuit is drawing 50 A. How many watts (W) does it use?
$$P = E \times I$$
Watts = Volts × Amps
Watts = 240 V × 50 A = 12,000 W = 12 kW

■ Example 2: A 10,000 W resistance coil heater is connected to a 240 V supply. How many amps (A) does it draw?
$$I = P / E$$
Amps = Watts / Volts
Amps = 10,000 W / 240 V = 41.6 A

Residential and small commercial power is delivered as "Single Phase." Large commercial and industrial establishments use "Three-Phase" service. Figure 4.2 depicts the typical U.S. arrangement. Three-phase power can be delivered in a "Delta" (Δ) or "Wye" (Y) arrangement, with Y being more prevalent. In the Y arrangement, one can derive a full-value three-phase service, say, at 208 V or 480 V (e.g., for medium versus large building service) or a single-phase, scaled-down service with 120 V or 277 V. (The lower values are obtained as follows: 120 V = 208 /√3; 277 V = 480 /√3). Appendix 4A provides a short primer on three-phase power.

One of the issues of key concern for data centers and network nodes is the phenomenon of "transients." Transients are short but typically high-amplitude bursts of electrical energy (transient voltage) that can travel across AC power, telephone, or data lines. Studies show that about 35% of all transients originate outside the facility from lightning, utility grid switching, electrical accidents, and other sources, while the remaining 65% come from inside the facility, often when large

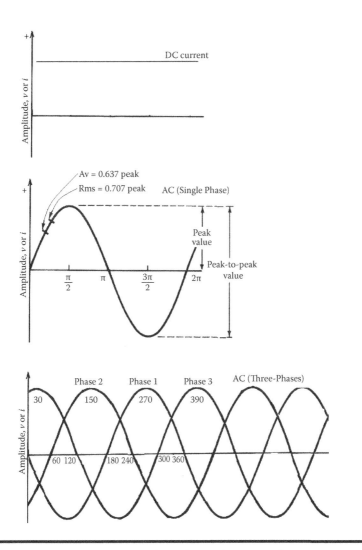

Figure 4.1   DC, AC single-phase, and AC three-phase waveforms.

power-consuming systems (such as motors or building air-conditioning systems) are switched on [LIE200301]. Note that circuit breakers are designed for overcurrent protection and do not provide transient voltage protection.

4.2 Power Measurements

The load on an electrical installation, or the load of an item of equipment, can be expressed either in amperes (A) or in kilovoltamperes (kVA). In the case of a SoHo

Table 4.1 Basic Electrical Terms

Amperage (Amp)	The *rate* or intensity of electric flow through a conductor. When the charge moves at the rate of 6.25 × 10^{18} electrons flowing past a given point per second, the value of the current is 1 ampere.
Circuits	An arrangement where one has a power source, a conductor, and a load. An "open" circuit may be caused by a switch or conductor failure. When circuits are wired in parallel, the voltage is constant at any point in the circuit for any load.
Direct Current (DC)	A current where the amplitude is constant over time.
Horsepower (HP, single phase)	Another measure of power defined as: HP = (Volts × Amps × Efficiency × Power Factor) / 746 HP = kW / 0.746 -or- 1 HP = 746 W = 0.746 kW (with assumed 100% efficiency and no Power Factor penalty)
Ohm	The *resistance* to the flow of electricity. A resistance that develops 0.24 calories of heat when 1 ampere of current flows through it for 1 second has 1 ohm of opposition.
Ohm's law	Ohm's law equation: V = R × I Power law equation: P = V × I See graphics below for additional combinations V measured in volts (V) I measured in amps (A) R measured in ohms (Ω) P measured watts (W)
Amperage (Amp)	The *rate* or intensity of electric flow through a conductor. When the charge moves at the rate of 6.25 × 10^{18} electrons flowing past a given point per second, the value of the current is 1 ampere.
Circuits	An arrangement where one has a power source, a conductor, and a load. An "open" circuit may be caused by a switch or conductor failure. When circuits are wired in parallel, the voltage is constant at any point in the circuit for any load (an excessive much load will drop the voltage, but the drop is constant).
Direct Current (DC)	A current where the amplitude is constant over time.

Table 4.1   Basic Electrical Terms (Continued)

Horsepower (HP, single phase)	Another measure of power defined as: HP = Volts × Amps × Efficiency × Power Factor) / 746 HP = kW / 0.746 -or- 1 HP = 746 W = 0.746 kW (with assumed 100% efficiency and no Power Factor penalty)
Ohm	The *resistance* to the flow of electricity. A resistance that develops 0.24 calories of heat when 1 ampere of current flows through it for 1 second has 1 ohm of opposition.
Ohm's law	Ohm's law equation: V = R × I Power law equation: P = V × I. See graphics below for additional combinations. V measured in volts (V) I measured in amps (A) R measured in resistance P measured watts (W)
Ohm's law (graphic)	
Single-phase alternating current (AC)	Alternating current with a sinusoidal waveform for the time-amplitude relationship.
Three-phase alternating current (AC)	Alternating current with three sinusoidal waveforms that follow each other in precise timing, typically 120° out-of-phase.

Continued

Table 4.1   Basic Electrical Terms (Continued)

	The diagram that follows shows how a theoretical alternator generating three phases works; actual generators join some of the outgoing leads in a Y or Δ fashion so that only three or four wires are required (not six). Three-Phase Alternator All formulas for single-phase power can be used for three-phase with the additional multiplication of 1.73 when working with volts and amps. For example, Power (in watts) = Volts × Amps × 1.73
Volts	A measure of the work needed to move an electric charge. When 0.7376 foot-pounds of work is required to move 6.25×10^{18} electrons (1 coulomb) between two points, each with its own charge, the potential difference is 1 volt. Note that the definition of a volt is for a coulomb of charge (6.25×10^{18} electrons). Also, 0.7376 foot-pounds of work is equal to 1 Joule (J), which is the metric unit of work or energy. Thus, 1 volt equals 1 Joule of work per coulomb of charge.
Watts	The volume flow or *rate of usage* of electrical energy. Power = Volts × Amps $$1\,W = 1\frac{J}{s} = 1\frac{kg \times m^2}{s^3} = 1\frac{N \times m}{s} = 1V \times 1A$$ Clearly, 1 kW = 1,000 W; 1 MW = 1,000 kW = 1,000,000 W

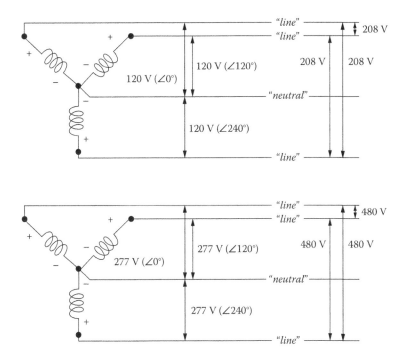

Note: In this figure the coil of wire can be an alternator or a transformer winding

Figure 4.2   Three-phase power arrangements: SoHo (upper) and commercial building (lower).

(Small office/Home office), an electricity distributor may request the maximum demand of the installation in kW, rather than kVA.

4.2.1 Single-Phase Loads: Smaller AC Operations/Nodes

Converting amperes (A) to kilovoltamperes (kVA): For a single-phase load expressed in amperes, the load in kVA can be calculated as follows:

$$kVA = \frac{\text{nominal voltage (volts)} \times \text{current (amperes)}}{1,000}$$

(4.1)

▪ Example 1: If the designer of a single-phase 230-V electrical installation has assessed the maximum demand as 100 A, the load is then equal to 23 kVA (230 V × 100 A ÷ 1,000).

Converting kVA to amperes: For a single-phase load expressed in kVA, the load in amperes can be calculated as follows:

$$\text{Current (amperes)} = \frac{\text{kVA} \times 1{,}000}{\text{nominal voltage (volts)}}$$

(4.2)

■ Example 2: If a piece of equipment of rated voltage 230 V has a nameplate rating of 20 kVA, the rated current of the equipment is then equal to 87 A (20 kVA × 1,000 ÷ 230 V).

4.2.2 Balanced Three-Phase Loads

Converting amperes to kilovoltamperes: For a balanced three-phase load expressed in amperes (A), the load in kilovoltamperes (kVA) can be calculated using either Equation (4.3), which uses the nominal phase (to neutral) voltage (U_0), or Equation (4.4), which uses the nominal line (phase-to-phase) voltage (U):

$$\text{kVA} = \frac{3 \times \text{nominal phase voltage } (U_0)\text{(volts)} \times \text{line current (amperes)}}{1{,}000}$$

(4.3)

$$\text{kVA} = \frac{\sqrt{3} \times \text{nominal line voltage } (U)\text{(volts)} \times \text{line current (amperes)}}{1{,}000}$$

(4.4)

■ Example Equation (4.3): Assume that a three-phase installation having nominal phase voltage (U_0) of 230 V has a balanced three-phase maximum demand current of 120 A, as shown in Figure 4.3 [VOL200901]. The maximum demand in kVA is then equal to 82.8 kVA (3 × 230 V × 120 A ÷ 1,000).

Converting kVA to amperes: For a balanced three-phase load expressed in kilovoltamperes, the load in amperes can be calculated using either Equation (4.5), which uses the nominal phase (to neutral) voltage ($U0$), or Equation (4.6), which uses the nominal line (phase-to-phase) voltage (U):

$$\text{Current (amperes)} = \frac{\text{kVA} \times 1{,}000}{\sqrt{3} \times \text{nominal voltage to earth } (U) \times \text{(volts)}}$$

(4.5)

$$\text{Current (amperes)} = \frac{\text{kVA} \times 1{,}000}{\sqrt{3} \times \text{nominal voltage to earth } (U) \times \text{(volts)}}$$

(4.6)

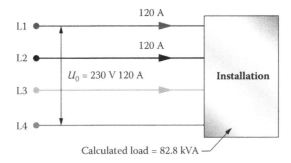

Figure 4.3   Example. A balanced three-phase load where the phase-to-neutral voltage U_0 and the load current are known.

■ Example Equation (4.6): Assume that an item of balanced three-phase, three-wire equipment having a rated line voltage (*U*) of 400 V has a nameplate rating of 70 kVA, as shown in Figure 4.4 [VOL200901]. The line current is then equal to 101 A, given by [(70 kVA × 1,000) ÷ (1.732 × 400 V)].

4.2.3 Expressing Loads in Kilowatts (kW)

Electricity distributors sometimes request values of load (maximum demand) for SoHo (and residential) installations to be expressed in kilowatts, rather than in kilovoltamperes. This is acceptable because the load in a typical SoHo/dwelling will not be significantly inductive or capacitive and therefore will have a power factor close

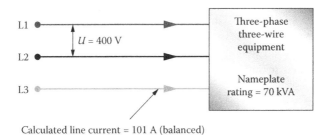

Figure 4.4   Another example. A balanced three-phase load where the phase-to-neutral voltage U_0 and the load current are known.

to unity. For a load having unity power factor, the load in kilowatts has the same numerical value as the load in kilovoltamperes, and can be calculated as follows:

$$kW = \frac{\text{nominal voltage (volts)} \times \text{current (amperes)}}{1,000}$$

(4.7)

4.2.4 Example

A significant number of data center operators reported at press time that they were in a situation where their data centers had insufficient cooling and insufficient power; about half the world's data centers were considered functionally obsolete due to insufficient power and cooling capacity to meet the demands of high-density equipment, according to some market research firms* [BOT200701]. Legacy data centers typically were built to a design specification of about 100 to 150 watts per square foot (WPSF); current designs require about 300 to 400 WPSF; and by 2011, this is expected to increase to more than 600 WSPF. A 20-A AC feed supports about 2400 W, or about 600 WPSF. If a data center has, for example, 200 racks, that would require about 480 kW (kVA).

4.3 Backup Power Systems

This section provides a basic overview of data center power backup. On a smaller scale, these techniques and approaches are applicable to network nodes. Typical causes of power interruptions include utility outages, human error in the data/ networking center, maintenance of power system components, and the failure of power system components, among others. Utility outages may be caused by lightning strikes, weather-related power line problems (e.g., ice storms, winds, hurricanes), construction projects, vehicular accidents (cars or trucks hitting poles, which occurs more commonly than often realized), and problems with power company equipment. For these reasons, backup power systems are needed.

A power protection system is configured using four basic types of equipment (a specific system may include one or all of these components):

■ *Transient Voltage Surge Suppression (TVSS) system:* As the name implies, a TVSS system is used to manage transient voltages. TVSS prevents noise and high voltages from reaching the load and possibly damaging the equipment; however, TVSS does not prevent interruptions resulting from loss of utility

---

* Gartner recently estimated that 70% of 1000 global businesses will need to significantly modify their data center facilities during the next 5 years. Few current data centers will be able to manage the next wave of high-density equipment [MOD200701].

power. TVSSs should be used at both the service entrance and the data center. This functionality can be integrated into the UPS or installed as stand-alone devices.

■ *Uninterruptible Power Supply (UPS) systems* (including batteries for power during short-term outages or on-site generator to protect against longer-term outages): A UPS provides backup power to guard against an interruption in utility power, and, often also "conditions" utility power to eliminate power disturbances such as transients. There are three types of UPS topology: passive standby (offline), line interactive, and double conversion (the kind of power conditioning provided by a UPS system is a function of its topology) [LIE200301]:

■ A passive standby, or off-line, UPS provides mainly short-term outage protection. It may include surge suppression, but does not provide true power conditioning and is typically used in noncritical desktop applications.

■ Line interactive UPS systems monitor incoming power quality and correct for major sags or surges, relying on battery power to enhance the quality of power to the load. This topology eliminates major fluctuations in power quality, but does not protect against the full range of power problems and does not isolate connected equipment from the power source.

■ An on-line double conversion system provides the highest degree of protection for critical systems. Rather than simply monitoring the power passing through it, as with the line interactive approach, a double conversion UPS system creates new, clean AC power within the UPS itself by converting incoming AC power to DC and then back to AC. This not only provides precisely regulated power, but also has the advantage of isolating the load from the upstream distribution system.

Clearly, the battery capacity of the UPS determines how long the system can provide power to the load when utility power fails. Generally, battery subsystems are sized to provide 10 to 20 minutes of power at full load in large data centers; smaller facilities or special (mission-critical) applications may be configured with additional battery capacity. Often, data center designers locate the UPS system outside the data center to avoid consuming valuable data center floor space with support systems. This approach is reasonable for a room-level approach to protection; if power protection is required on a rack-by-rack basis, the system is better positioned in the data center itself.

■ Transfer switches allow switching from a primary power source to a secondary or tertiary power source. Transfer switches are used with emergency power generators to back up power from the utility (electric) company.
  – Automatic transfer switches (ATSs) continually monitor the incoming power for any anomalies (e.g., brownouts, surges, failure); any such event will cause the internal circuitry to activate a generator and transfer the load to the generator when additional switch circuitry determines the generate has a proper voltage and frequency (when utility power returns,

the transfer switch will then transfer back to utility power and direct the generator to turn off.)

– *Static Transfer Switches (STSs):* STSs are used to implement redundancy into the power system and eliminate single points of failure. Specifically, when redundant UPS systems are used, there is a need to support switching between systems that is transparent to the load.

■ *Power Distribution Units (PDUs):* The PDU distributes power from the UPS and/or utility to the loads (systems); the configuration can include a single PDU, to a PDU/STS combination, to multiple PDUs and STSs. The PDU receives 480 V (or 600 V) power from the UPS. In the traditional approach (see Figure 4.5) to power distribution within the data center or large telecom room, the UPS feeds a number of PDUs, which then distribute power directly to equipment in the rack. This approach is reasonable when the number of servers and racks is relatively low; but at this juncture, new designs that support greater flexibility and scalability are being considered by medium-to-large data/telecom centers (e.g., two-stage power distribution).

Figure 4.6 depicts a number of arrangements that can be architected using these power-support elements.

The devices just identified can be assembled according to a number of distinct architectures. As noted elsewhere, the availability of the power system is a key measure of interest. The architecture of the data center power system determines the level of availability it can support; UPS redundancy improves the availability number and allows UPS modules to be serviced without impacting the power to downstream equipment. There are four basic system architectures, each providing a different level of protection and availability [LIE200301]:

■ *Basic protection:* Basic hardware protection of connected equipment is typically accomplished through a TVSS and PDU. This arrangement achieves around 99.9% availability, depending on the reliability of the utility power source.

■ *Operational support:* Adding a UPS to the power system provides protection against short-term interruptions in utility power and the ability to ensure a

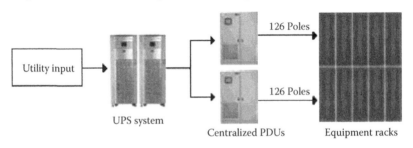

Figure 4.5 Traditional (single-stage) power distribution.

Basic Production Power System	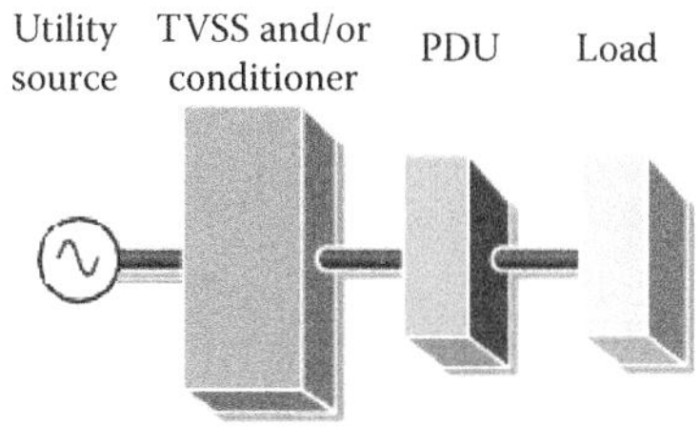 Utility source — TVSS and/or conditioner — PDU — Load	Prevents damage to electronics, but does not provide protection against interruptions in utility power
Operational Support System	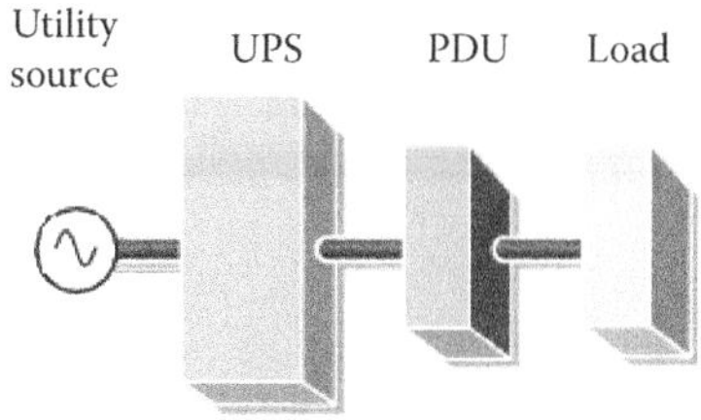 Utility source — UPS — PDU — Load	A system that mitigates brief interruptions in power and enables one to preserve data (in case that an extended outage follows).
High Availability 1+1 Redundancy	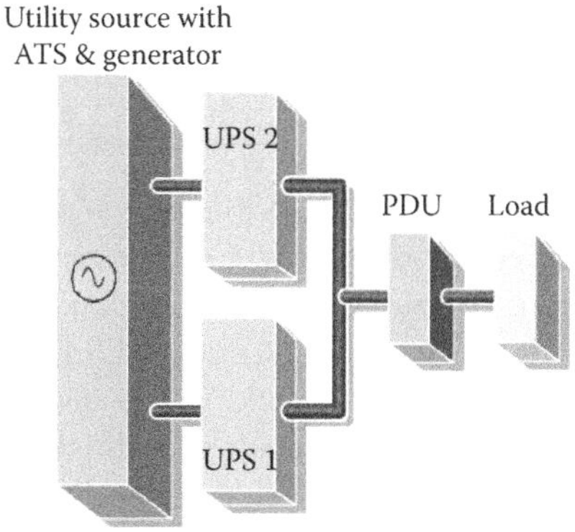 Utility source with ATS & generator — UPS 2 / UPS 1 — PDU — Load	Approach adds UPS redundancy to increase availability to 99.999 percent to 99.99999 percent.
High Availability Parallel Redundancy (N+1)	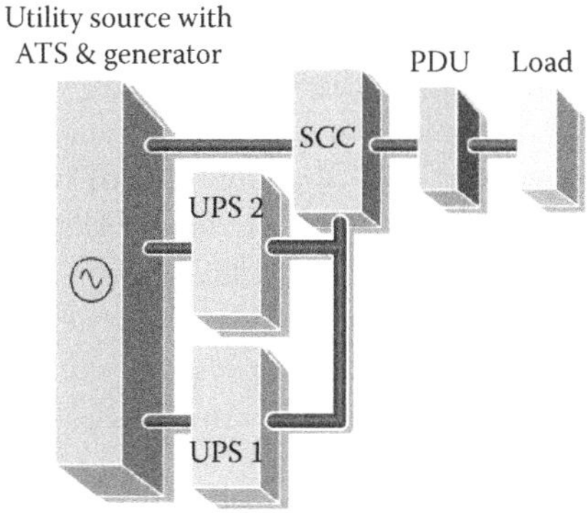 Utility source with ATS & generator — UPS 2 / UPS 1 — SCC — PDU — Load	A parallel redundant architecture allows load sharing of parallel redundant units. The sharing is coordinated by system controls, which can be located in an external cabinet (in which case it is known as system controls cabinet – SCC) or within a UPS system with such capability

Figure 4.6   Power protection approaches.

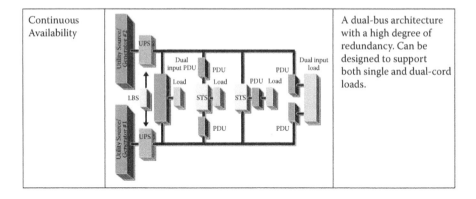

| Continuous Availability | | A dual-bus architecture with a high degree of redundancy. Can be designed to support both single and dual-cord loads. |

Figure 4.6 (Continued)

controlled shutdown in the event the outage exceeds UPS battery capacity. This increases power availability to 99.99%.* The UPS can also condition the power being delivered to the load. For critical applications, an online double-conversion UPS is recommended.

■ *High availability:* This architecture adds redundancy at the UPS level of the system to increase availability to between 99.999% (5 minutes annually) and 99.99999% (3 seconds annually). With these systems, maintenance on all devices with the exception of the PDU can be performed concurrently. This can be accomplished through parallel redundancy or 1+1 redundancy. A parallel-redundant system has two or more UPS modules connected in parallel to a common distribution network. It uses enough modules to carry the maximum projected load, plus at least one additional module for redundancy. During normal operation all modules share the load. If a module has to be taken off-line, the other modules have the capacity to carry the full load. In a 1+1 redundant system, each UPS system can carry the full load, creating redundant power paths for part of the system.

■ *Continuous availability:* This architecture utilizes a dual-bus with distributed redundancy and delivers near 100% availability. This system includes two or more independent UPS systems—each capable of carrying the entire load. Each system provides power to its own independent distribution network. No power connections exist between the two UPS systems. This allows 100% concurrent maintenance and brings power system redundancy to every piece of load equipment as close to the input terminals as possible.

The simplest approach to redundancy is a 1+1 system, in which each UPS module has the capacity to support all connected equipment; the N+1 configuration seeks to balance scalability and availability. Figure 4.7 depicts a summary of the

---

* About 1 hour of unplanned downtime annually.

System Type	Description	No. of Buses
SMS	Single module UPS system. No redundancy.	Single
1+1	2 UPS modules running in parallel.	Single
1+N	1 UPS plus N = number of UPS modules added for redundancy. Each module has its own internal bypass. (Accepted practice in Europe and gaining acceptance in US)	Single
N+1	N = number of UPS modules required, plus one additional module for redundancy. In high-availability applications, N is typically less than 3. N + 1 systems use a system-level bypass. (Accepted practice in US)	Single
Dual-bus or 2N or 2 x (N+1)	2 UPS systems feeding 2 independent output distribution systems. UPS output buses are typically in sync. Each bus often contains a redundant UPS system, which is technically a 2 x (N+1), but is sometimes also called 2N. Redundancy can also be accomplished through 1 + N configuration, which is less common but growing in acceptance.	Dual

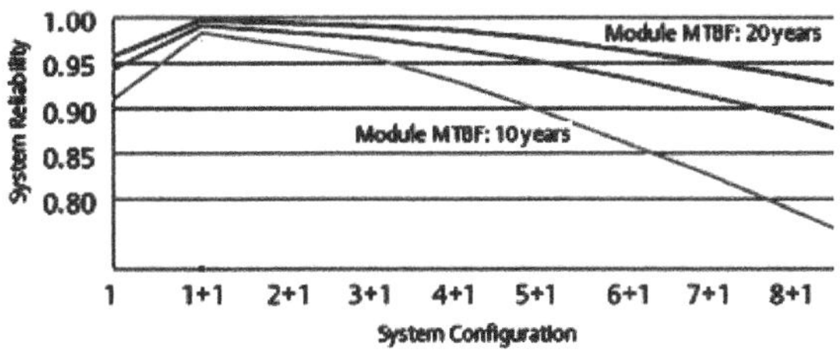

Courtesy: Liebert / Emerson Network Power

Figure 4.7   Summary of most common UPS system configurations and optimal system reliability in N+1 configurations.

most common system configurations in use and a view to the performance of the N+1 system commonly deployed in U.S. installations (the figure is parameterized on the Mean Time Between Failure (MTBF) rates of the constituent UPSs; note that reliability diminishes beyond the 3+1 configuration because the increase in the number of modules increases the system parts count, which in turn increases the probability of failure). A dependable design requires proper sizing of the UPS modules: The initial facility load is light compared to future requirements; hence, a planner may try to size UPS modules to initial requirements. However, this can give rise to reliability problems in the future if the UPS module count gets too high. If reliability is important, UPS modules should be sized no less than one third of the projected total facility load. If there is high certainty that growth will occur, start with modules sized at one half the initial load. This provides room for growth while ensuring adequate availability throughout the life of the facility. Batteries, which are one of the most expensive components of the power system, can be sized to initial load with additional capacity added as required [LIE200701].

As power requirements continue to increase in data centers and telecom rooms, the need for downstream management of and visibility into the power consumption chain becomes more critical. Two developments have emerged in the recent past [LIE200701]:

■ *Two-stage power distribution to racks:* The use of two stages (see Figure 4.8) separates deliverable capacity and physical distribution capability into separate systems to eliminate the breaker space limitations of the traditional PDU. The first stage of the two-stage system provides mid-level distribution; the mid-level distribution unit includes most of the components that exist in a traditional PDU, but with an optimized mix of circuit- and branch-level distribution breakers. Instead of doing direct load-level distribution, this device feeds floor-mounted distribution cabinets via an I-Line® panelboard distribution section. The I-Line panelboard typically provides the flexibility to add up to ten plug-in output breakers of different ratings as needed. The load-level

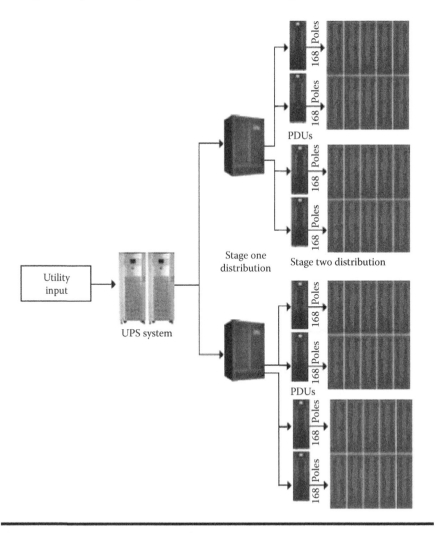

Figure 4.8   Two-stage power distribution.

distribution unit distributes power directly to the rack-mounted equipment. This approach has the benefit of enhancing the efficiency of the distribution system by enabling fewer, larger-power paths from the UPS to the mid-level distribution unit and shorter paths from the load-level units to the point of consumption. In addition to functional flexibility, the under-floor cabling is reduced, thereby improving the ability to optimize airflow.

■ *Improved in-rack power management:* In-rack power management can be accomplished with intelligent power strips. Intelligent power strips can mount either vertically or horizontally to simplify equipment changes and reduce cable clutter while providing increased visibility and control of rack power consumption. These power strips monitor the electrical attributes of an individual power strip, including real-time remote management using Simple Network Management Protocol (SNMP) and local LED display of volts, amps, watts, and phase loading or the power circuit. More advanced strips can also provide receptacle-level control of power on/off, enabling receptacles to be monitored and turned on/off locally or remotely to prevent the addition of new devices that could create an overload condition.

There are a number of commercial power and cooling management systems, but comprehensively integrated management systems that enable IT managers to map the relationships between power demand and power supply (fluctuations) in the data center have been slow to appear. Some products measure the power of their own product line but they do not support the products of other vendors. Other space- and power capacity-monitoring products provide a sense of the power being used but do not monitor or control individual server energy demand [BOT200701]. Fully integrated management tools are expected to enter the market around 2010.

Appendix 4A: Three-Phase Power

There are obvious advantages to using three-phase AC power at large data centers. This appendix reviews some of the basic concepts.

The basic equations are as follows:

$$\text{Current (single phase): } I = P / V_p \times \cos \varphi$$

$$\text{Power (single phase): } P = V_p \times I_p \times \cos \varphi$$

$$\text{Current (three phases): } I = P / \sqrt{3} \ V_l \times \cos \varphi \text{ or } I = P / 3 \ V_p \times \cos \varphi$$

$$\text{Power (three phases): } P = \sqrt{3} \ V_l \times I_l \times \cos \varphi \text{ or } P = \sqrt{3} \ V_p \times I_p \times \cos \varphi$$

Power factor $PF = |$ Real Power cos $\varphi |$ Total Apparent Power. For a purely resistive circuit, $PF = 1$

where V_1 = line voltage (volts), V_p = phase voltage (volts), I_1 = line current (amps), I_p = phase current (amps), P = power (watts), and φ = power factor angle.

The top portion of Figure 4A.1 depicts a conceptual generator (alternator) that generates three-phase AC. The issue, however, is that this arrangement would need six conductors to carry the current. To avoid this, which would make transmission of the current an expensive proposition, the same leads from each phase can be connected together to form a "Wye" (Y) (also known as "star") connection, as shown in Figure 4A.1 (middle portion). The neutral connection may or may not be present; the neutral connection is used when a single-phase load must be supplied (the single-phase voltage is available from neutral to A, neutral to B, and neutral to C as shown in Figure 4.2). Line voltage refers to the amount of voltage measured between any two line conductors

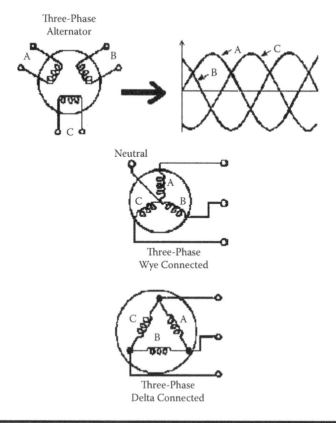

Figure 4A.1   Generation of three-phase currents.

in a balanced three-phase system; phase voltage refers to the voltage measured across any one component (source winding or load impedance) in a balanced three-phase source or load. The total voltage, or line voltage, across any two of the three line leads is the vector sum of the individual phase voltages; each line voltage is 1.73 = $\sqrt{3}$ times one of the phase voltages. The term "line current" refers to current through any one line conductor, and the term "phase current" refers to the current through any one component. Because the windings form only one path for the current flow between phases, the line and phase currents are the same (equal) [ALL200901]. Namely,

For "Y" Circuits:
$$E_{line} = \sqrt{3} = E_{phase}$$
$$I_{line} = I_{phase}$$

Based on these equations we see that Y-connected sources and loads have line voltages greater than phase voltages, and line currents equal to phase currents.

A three-phase alternator can also be configured so that the phases are connected in a Delta (Δ) arrangement, as shown at the bottom of Figure 4A.1. In this arrangement, line voltages are equal to phase voltages, but each line current is equal to 1.73 = $\sqrt{3}$ times the phase current. Namely,

For Δ ("delta") Circuits:
$$E_{line} = E_{phase}$$
$$I_{line} = \sqrt{3}\ I_{phase}$$

With both of these arrangements, only three wires come out of the alternator, thus allowing convenient connection to three-phase power distribution transformers. The symbol $\left(\!\!\sim\!\!\right)$ represents an AC source while the symbol $^+\!\!-\!\!\text{000}\!\!-^-$ represents a coil (such as the coil of a transformer.)

Three-Phase Y Configurations

In the Y arrangement, three voltage sources are connected together with a common connection point joining one side of each source, as depicted in Figure 4A.2. Notice the use of a "common" fourth wire; the common wire mitigates potential problems in case one element of a three-phase load fails (to an open state). The three conductors emanating from the voltage sources (namely, the windings) toward a load are usually called lines; the windings are usually called phases.

With a Y-connected system, a neutral wire is needed in case one of the phase loads were to fail to open in order to keep the phase voltages at the load from changing. Note that in some cases the "common/neutral" wire is not available and/or used.

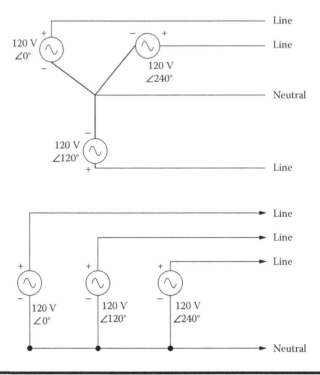

Figure 4A.2   Representations of the Y arrangement.

Three-Phase Δ Configurations

Figure 4A.3 depicts the Delta (Δ) configuration. Note that the three-phase, three-wire Δ connection has no common.

At face value it would appear that the three voltage sources create a short-circuit, with electrons flowing with nothing but the internal impedance of the windings to offer a load. However, due to the phase angles of these three voltage sources, this is not the case; in fact, the three voltages around the loop add up to zero and consequently there is no voltage available to push current around the loop.

Because each pair of line conductors is connected directly across a single winding in a Δ circuit, the line voltage is equal to the phase voltage. Also, because each line conductor attaches at a node between two windings, the line current will be the vector sum of the two joining phase currents. This is substantially more than the line currents in a Y-connected system. The greater conductor currents necessitate thicker (lower-gauge) and more expensive wire. However, although a circuit may require three lower-gauge copper conductors, the weight of the copper in this case is still less than the amount of copper of required for a single-phase system delivering the same power at the same voltage [ALL200901].

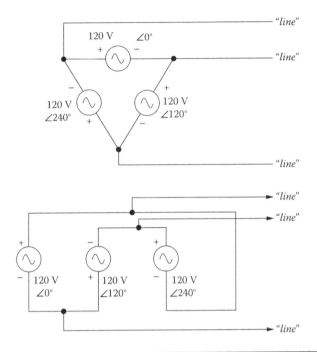

Figure 4A.3   Representations of the Δ arrangement.

One advantage of the Δ-connected source is that it is impervious to failure in the sense that it possible for one of the windings in a Δ-connected source to fail to open without affecting load voltage or current. Also, in the Δ-connected system there is no neutral wire, thus requiring less cable/copper; in a Δ-connected circuit, each load phase element is directly connected across a respective source phase winding, and the phase voltage is constant regardless of open failures in the load elements.

Figure 4A.4 depicts the failure modes. In the Δ arrangement, even with a source winding failure, the line voltage is still 120 V, and consequently the load phase voltage of a Δ-connected load is still 120 V; the only difference is the extra current in the remaining functional source windings. In a Y-connected system (source), a winding failure reduces the voltage to 50% on two loads of a Δ-connected load (two of the resistances suffer reduced voltage [e.g., to 104 V], while one remains at the original line voltage [208 V]). Open source winding of a "Y–Y" system results in halving the voltage on two loads, and loses one load completely. Hence, Δ-connected sources are preferred for reliability. However, if dual voltages are needed (e.g., 120/208) or preferred for lower line currents, Y-connected systems are used.

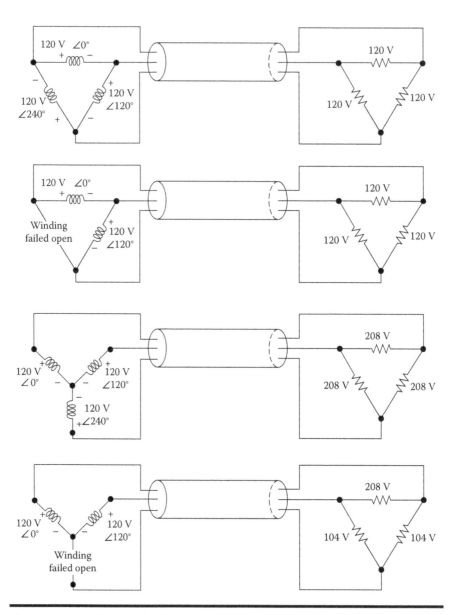

Figure 4A.4   Failure modes for Δ/Y configurations.

References

[ALL200901]   All About Circuits, Three-Phase Y and Delta Configurations, On-line resource, www.allaboutcircuits.com.

[BOT200701]   B. Botelho, Gartner Predicts Data Center Power and Cooling Crisis, SearchDataCenter.com, 14 June 2007.

[BRO200801]   J. Brodkin, Gartner in "green" data centre warning, *Network World (U.S.)*, 30 October 2008.

[LIE200301]   Liebert, Five Questions to Ask before Selecting Power Protection for Critical Systems, A Guide for IT and Data Center Managers, 2003. Liebert Corporation, 1050 Dearborn Drive, P.O. Box 29186, Columbus, OH 43229.

[LIE200701]   Liebert, *Power Management Strategies for High-Density IT Facilities and Systems, 2007,* Emerson Network Power, 1050 Dearborn Drive, P.O. Box 29186, Columbus, OH 43229.

[MIT201001]   R. L. Mitchell, Data Center Density Hits the Wall, Computerworld, Power and Cooling for the Modern Era, White paper, 2/2/2010.

[MOD200701]   A. Modine, Gartner: No Relief for Data Center Costs, Enterprise, 4 October 2007, http://www.channelregister.co.uk/2007/10/04/gartner_2011_signifi-cant_disruptions/.

[VOL200901]   Voltimum Partners, Converting amperes to kVA, Voltimum UK & Ireland Ltd., 3rd Floor Bishop's Park House, 25-29 Fulham High Street, London SW6 3JH, 2009.

Chapter 5

HVAC and CRAC Basics

This chapter reviews the basic operation of a general Heat, Ventilation and Air Conditioning (HVAC) system and then focuses on Computer Room Air Conditioners (CRACs).

5.1 General Overview of an Air-Conditioning System

Building envelopes, data centers, and telco/telecom rooms (e.g., telecommunications central switch offices) are sealed, and an HVAC system controls the air exchange and conditions the air that is delivered to the closed environment. This is called mechanical ventilation.* The HVAC system brings outdoor air into a building, humidifies or dehumidifies it, and heats or cools it to meet the ventilation and thermal comfort needs of the occupants (and/or the cooling needs of equipment, such as IT/networking equipment), and to dilute the contaminants inside the building. CRACs are optimized for data center/telecom site applications. Below we discuss some general principles about HVAC operation and the focus on CRACs.

A typical HVAC system consists of a supply air system, a return air system, and an exhaust air system. It typically also contains heating and cooling units, a humidifier, air filters, and fans. The compression and expansion of a gaseous fluid is the main physical principle used for transferring heat from the inside to the outside, and vice versa. Typically, the HVAC is a reversible heat pump providing heat and cold to a cooling/heating selector; heat and/or cold is distributed independently

---

* Natural ventilation takes place when windows or other openings in the building allow indoor and outdoor air to exchange passively. Natural ventilation is rarely used in data center application, although some hybrid systems are being studied.

to each area (room) requiring thermal management. Any substance that transfers heat from one place to another, creating a cooling effect, is called a *refrigerant*; for example, water is a refrigerant that is used in absorption machines. Figure 5.1 depicts two basic systems, one using air and one using water as an energy vector [ICA199701]. Figure 5.2 depicts the working components of a typical commercial building HVAC system [NRC200901], Figure 5.3 provides a pictorial view of the outside system, while Figure 5.4 gives a pictorial view of typical CRACs. A plethora of controls and sensors are included in the ventilation system to manage

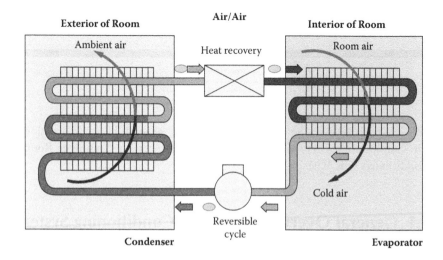

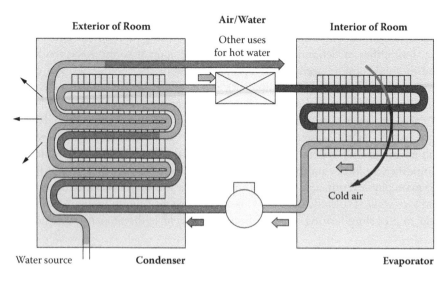

Figure 5.1 Air/air and air/water condenser/evaporator system.

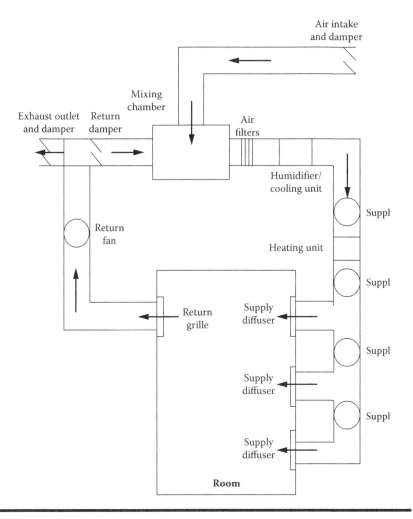

Figure 5.2   Typical HVAC system.

the air quality and thermal environment. Controls include, but are not limited to, thermostats, humidity sensors, time clocks, automatic dampers and valves, carbon monoxide and carbon dioxide sensors, air pressure sensors, and ventilation rate controls. There are five "loops" in an AC system: airside loop, chilled-water loop, refrigeration-equipment loop, heat-rejection loop, and controls loop.

A building or a space is typically partitioned into zones to support different heating, cooling, or ventilation requirements. For example, a perimeter zone may require heating or cooling due the outdoor air temperature and the solar heating, while a core inner zone may require cooling year-round due to the internal heat produced by office equipment. As another example, a telecom closet in a building may have higher cooling requirements than general office space (sometimes the telecom closet

Figure 5.3   Typical commercial rooftop system.

has its own AC unit). Clearly, a data center housed in a multi-purpose building will have different cooling requirements than the rest of the building and/or the offices of the support staff, but in this case, dedicated air conditioners (CRACs) will invariably be used. For commercial systems, ventilation rate is the amount of outdoor air that is supplied to a room with the return air. The ventilation rate is calculated based on factors such as room volume, number of occupants, types of equipment, and contaminant levels, among others. HVAC systems operate on a constant volume (CV) or variable air volume (VAV) basis, as follows [NRC200901]:

■ *CV systems* supply a constant rate of airflow to a space at reasonably high velocity and remove an equal amount of the air from the space (return air). The temperature and humidity in the space are changed by varying the temperature and humidity of the supply air. The high air velocity used in these

| Liebert
Ceiling and Wallmount | Liebert
Floor Mount | Liebert CW
Chilled Water Based |

Figure 5.4   Typical computer room air-conditioning systems.

systems creates a risk of draught for occupants who are positioned poorly in relation to the supply diffusers.

■ *VAV systems* differ in that both the supply airflow and return airflow rates vary according to the thermal demands in the space. The minimum airflow is set based on the expected number of occupants in the space and then varied to create the desired room temperature and humidity. The supply air is always slightly cooler than the room air.

VAV systems and variable air volume and temperature (VVT) provide accurate control of the air exchange rates and air temperatures. Office/commercial buildings typically use VAV, while home systems use CV air. Figures 5.5 and 5.6 depict the basic operation of these systems, based in part on reference [ICA199701]. Computer rooms use heat exchangers (compressors in the unit that enable coils to take heat from room); then the heat must be dissipated via a water/glycol line to an outside fan. Split systems are individual systems in which the two heat exchangers are separated (one outside, one inside).

The air distribution system determines the airflow pattern by controlling the location at which air is supplied. There are two common distribution systems: mixing and displacement [NRC200901]:

■ *Mixing systems* are the most common in North America, especially in office buildings. Air can be supplied either at the ceiling or at the floor, and it is exhausted at the ceiling. The airflow pattern causes room air to mix thoroughly with supply air so that contaminated air is diluted and removed.

■ *Displacement systems* supply the air at floor level at a relatively low velocity (≤ 0.5 m/s) and at a temperature typically cooler than the room temperature. The temperature and location of air supply cause the air to spread out along the floor, displacing the room air that rises to the ceiling. Contaminants rise with the air and are removed at the ceiling while the supply air fills the occupied space.

To save energy, HVAC systems in some buildings may be operated at a reduced ventilation rate or shut down during unoccupied hours. This technique, however, has only limited application to mission-critical operations (including data centers).

Some concepts that have applicability to data centers and networking (telecom/datacom) rooms follow [GRE201001]:

■ *Horizontal Displacement (HDP):* An air-distribution system used predominantly in telecommunications central offices in Europe and Asia; typically, this system introduces air horizontally from one end of a cold aisle.

■ *Horizontal Overhead (HOH):* An air-distribution system that is used by some long-distance carriers in North America. This system introduces the supply air horizontally above the cold aisles and is generally utilized in raised-floor environments where the raised floor is used for cabling.

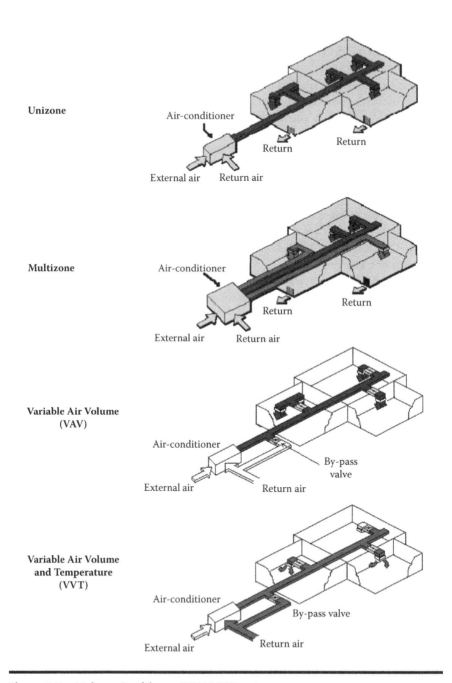

Figure 5.5 Unizone/multizone/VAV/VVT systems.

Water System	System is based on the hot or cold water production by a central unit. The water is distributed to individual heating/cooling units located in each room of the building.	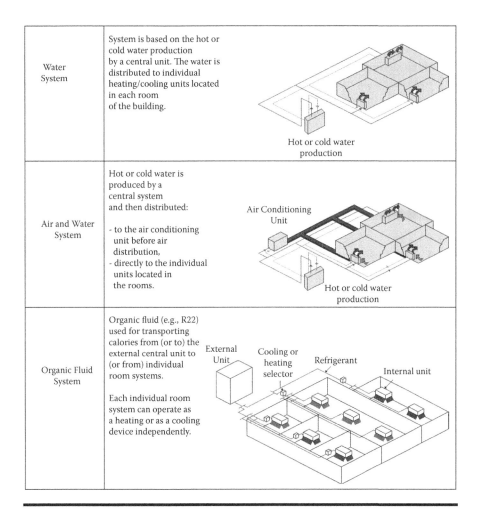 Hot or cold water production
Air and Water System	Hot or cold water is produced by a central system and then distributed: - to the air conditioning unit before air distribution, - directly to the individual units located in the rooms.	Air Conditioning Unit Hot or cold water production
Organic Fluid System	Organic fluid (e.g., R22) used for transporting calories from (or to) the external central unit to (or from) individual room systems. Each individual room system can operate as a heating or as a cooling device independently.	External Unit Cooling or heating selector Refrigerant Internal unit

Figure 5.6 AC configurations.

▪ *Hot Aisle/Cold Aisle:* A common means of providing cooling to telecom/data-com/data center rooms in which IT equipment is arranged in rows and cold supply air is supplied to the cold aisle, pulled through the inlets of the IT equipment, and exhausted to a hot aisle to minimize recirculation of the hot exhaust air with the cold supply air. Supply air is introduced into a region called the cold aisle. On each side of the cold aisle, equipment racks are placed with their intake sides facing the cold aisle. A hot aisle is the region between the backs of two rows of racks. The cooling air delivered is drawn into the intake side of the racks. This air heats up inside the racks and is exhausted from the back of the racks into the hot aisle.

■ Cold-aisle containment. Building on a hot-aisle/cold-aisle design, an enhance-ment is to use cold-aisle containment techniques; this facilitates keeping high-density server cabinets cool. This approach is applicable for cabinets exceeding about 4 kW. This approach typically involves using ducting to tar-get cold air, closing off the ends of aisles with doors, and/or installing barriers atop rows to prevent hot air from circulating over the tops of racks.

Next we describe two basic approaches to HVAC systems: DX unitary systems and chilled-water applied systems.* Direct-expansion (DX) AC is an air-condition-ing system in which the cooling effect is obtained directly from the refrigerant. It typically incorporates a compressor; almost invariably, the refrigerant undergoes a change of state in the system.

5.1.1 DX Unitary Systems: HVAC Systems

In a DX unitary system, the evaporator is in direct contact with the air stream, so the cooling coil of the airside loop is also the evaporator of the refrigeration loop. The term "direct" refers to the position of the evaporator with respect to the airside loop. The term "expansion" refers to the method used to introduce the refrigerant into the cooling coil. The liquid refrigerant passes through an expansion device (usu-ally a valve) just before entering the cooling coil (the evaporator). This expansion device reduces the pressure and temperature of the refrigerant to the point where it is colder than the air passing through the coil. See Figure 5.7 (upper). DX-type systems require Freon piping between the indoor and outdoor units. This is not an issue at free-standing, one-story buildings, but has limits on how many stories the refriger-ant lines can be run and still provide reliable year-round operation. Glycol-cooled units utilize Glycol cooling, which does not have this limitation, and additionally also allows "free cooling" in winter if electric utility costs are a factor.

The components of the DX unitary system refrigeration loop (evaporator, com-pressor, condenser, expansion device, and even some unit controls) may be pack-aged together, which provides for factory assembly and testing of all components, including the electrical wiring, the refrigerant piping, and the controls. This is called a *packaged* DX system. Alternatively, the components of the refrigeration loop may be split apart, allowing for increased flexibility in the system design. This is called a *split* DX system. Separating the elements has the advantage of providing the system design engineer with complete flexibility to match components in order to achieve the desired performance.

A common reason for selecting a DX system, especially a packaged DX system, is that, in a smaller building, it frequently has a lower installed cost than a chilled-water system because it requires less field labor and has fewer materials to install. Packaged DX systems that use air-cooled condensers can be located on the roof of a building.

* This discussion is based on educational material from Trane Corporation.

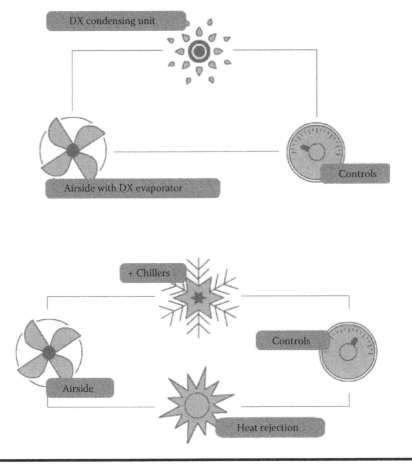

Figure 5.7 Basic HVAC systems: DX unitary system (upper), and chilled-water applied systems (lower).

5.1.2 Chilled-Water Applied Systems: HVAC Systems

A chilled-water applied system uses chilled water to transport heat energy among the airside, chillers, and the outdoors. See Figure 5.7 (lower). These systems are more commonly found in large HVAC installations, given their efficiency advantages. The components of the chiller (evaporator, compressor, an air- or water-cooled condenser, and expansion device) are often manufactured, assembled, and tested as a complete package within the factory. These packaged systems can reduce field labor, speed installation, and improve reliability. Alternatively, the components of the refrigeration loop may be selected separately. While water-cooled chillers are rarely installed as separate components, some air-cooled chillers offer the flexibility of separating the components for installation in different locations. This allows the system design engineer to position the components where they best serve the space, acoustic, and maintenance requirements of the building owner. Another benefit of a chilled-water applied system is refrigerant containment. Having the refrigeration equipment installed in a central location minimizes the potential for refrigerant leaks, simplifies refrigerant handling practices, and typically makes it easier to contain a leak if one does occur.

Table 5.1 enumerates some of the factors that relate to selecting a generic HVAC system, and Table 5.2 provides a listing of some key AC-related terms.

Table 5.1 Topics/Factors of Relevance to AC Systems

• Requirements for occupant comfort and/or equipment thermal control
• Five "loops": airside loop, chilled-water loop, refrigeration-equipment loop, heat-rejection loop, controls loop
• Chilled-water versus a DX system
• Packaged versus split systems
• Single-zone versus multiple-zone systems
• Constant-volume versus variable-air-volume systems
• Packaged terminal air conditioner (PTAC)
• Single-zone packaged DX rooftop
• DX split system
• Chilled-water terminal system (fan coils, unit ventilators, blower coils)
• Two-pipe versus four-pipe systems
• Water-source heat pump systems
• Dedicated outdoor-air systems
• Single-zone VAV system
• Multizone system
• Three-deck multizone system
• Changeover-bypass system
• Multiple-zone VAV system
• Rooftop VAV system
• Self-contained DX VAV system
• Chilled-water VAV system
• Double-duct VAV system

Table 5.2   Key Terms Related to HVACs/CRACs

Term	Definition
Absolute humidity	The total amount of moisture contained in a cubic foot of air, measured in grains per cubic foot.
Absolute pressure	The total pressure on a surface, measured in pounds per square inch. Equal to gauge pressure plus 14.7 (atmospheric pressure).
Absorbent	Any substance that can absorb another substance, changing it physically or chemically in the process. Lithium bromide is the absorbent used in absorption chillers.
Absorber	The low-pressure section of the absorption machine that has a lithium bromide solution that is cooled to attract refrigerant vapor from the evaporator.
Absorption system	Refrigeration system that uses heat to produce a cooling effect.
Air vent valve	Valves connected to the top of the water box or connecting pipe used to vent trapped air.
ASHRAE	American Society of Heating, Refrigerating and Air Conditioning Engineers.
Atmospheric pressure	The pressure exerted upon the Earth's surface by the weight of atmosphere above it.
Balancing valve	A device to balance the water flow.
Boiling point	The temperature at which the addition of any heat will begin a change of state from a liquid to a vapor.
BTU (also Btu)	Abbreviation for British Thermal Unit. The amount of heat required to raise 1 pound of water 1°F.
Bypass	A piping detour around a component.
Calibrate	To adjust a control or device in order to maintain its accuracy.
Capacity	The maximum usable output of a machine.
Charge	The refrigerant and lithium bromide contained in a sealed system; the total of refrigerant and lithium bromide required.
Coefficient of Performance (COP)	The ratio of the amount of work obtained from a machine to the amount of energy supplied.

Continued

Table 5.2   Key Terms Related to HVACs/CRACs (Continued)

Term	Definition
Compound gauge	A gauge used to measure pressures above and below atmospheric pressure (0 psig).
Condensable	A gas that can be easily converted to liquid form, usually by lowering the temperature and/or increasing the pressure.
Condensate	Water vapor that liquefies due to its temperature being lowered to the saturation point.
Condensation point	The temperature at which the removal of any heat will begin a change of state from a vapor to a liquid.
Condenser	A device in which the superheat and latent heat of condensation is removed to effect a change of state from vapor to liquid. Some sub-cooling is also usually accomplished.
Condenser water	Water that removes the heat from the lithium bromide in the absorber and from condenser vapor. The heat is rejected to the atmosphere by a cooling tower.
Constant Volume (CV) AC	ACs that supply a constant rate of airflow to a space at reasonably high velocity and remove an equal amount of air from the space (return air).
Contactor	A switch that can repeatedly cycle, making and breaking an electrical circuit: a circuit control. When sufficient current flows through the coil built into the contactor, the resulting magnetic field causes the contacts to be pulled in or closed.
Control	Any component that regulates the flow of fluid, or electricity.
Control device	Any device that changes the energy input to the chiller/heater when the building load changes, and shuts it down when the chiller is not needed.
Dew-point temperature (DPT)	The temperature at which a moist air sample at the same pressure would reach water vapor saturation. At this saturation point, water vapor begins to condense into liquid water fog or solid frost, as heat is removed.
Dry-bulb temperature (DBT)	The temperature of an air sample, as determined by an ordinary thermometer, the thermometer's bulb being dry. The SI unit is Kelvin; in the U.S., the unit is degrees Fahrenheit.
Efficiency	The amount of usable energy produced by a machine, divided by the amount of energy supplied to it.

Table 5.2 Key Terms Related to HVACs/CRACs (Continued)

Term	Definition
Equilibrium	When refrigerant (water) molecules leave the solution at the same rate that they are being absorbed, the solution is said to be in equilibrium.
Equilibrium chart	A pressure-temperature concentration chart that can be used to plot solution equilibrium at any point in the absorption cycle.
Evacuate	To remove, through the use of the vacuum pump, all noncondensables from a machine.
Evaporator	Heats and vaporizes refrigerant liquid from the condenser, using building system water.
Fahrenheit	The common scale of temperature measurement in the English system of units. It is based on the freezing point of water being 32°F and the boiling point of water being 212°F at standard pressure conditions.
Freezing point	The temperature at which the removal of any heat will begin a change of state from a liquid to a solid.
Gauge pressure	A fluid pressure scale in which the atmospheric pressure equals zero pounds and a perfect vacuum equals 30 inches of mercury.
Heat exchanger	Any device for transferring heat from one fluid to another.
Heat of condensation	The latent heat energy liberated in the transition from a gaseous to a liquid state.
Heat transfer	The three methods of heat transfer are conduction, convection, and radiation.
High temp generator	The section of a chiller where heat is applied to the lithium bromide solution to separate water vapor.
Humidity ratio	The proportion of mass of water vapor per unit mass of dry air at the given conditions (DBT, WBT, DPT, RH, etc.). The humidity ratio is also known as the moisture content or mixing ratio.
Hydrocarbon	Any of a number of compounds composed of carbon and hydrogen.

Continued

Table 5.2 Key Terms Related to HVACs/CRACs (Continued)

Term	Definition
Inches of mercury column (inHg)	A unit used in measuring pressures; 1 inch of mercury column equals a pressure of 0.491 pounds per square inch.
Inches of water column (inWC)	A unit used in measuring pressures. One inch of water column equals a pressure of 0.578 ounces per square inch. A 1-inch mercury column equals about 13.6-inch water column.
Inhibitor	A chemical in the lithium bromide solution used to protect the metal shell and tubes from corrosive attack by the absorbent solution.
Input rate	The quantity of heat or fuel supplied to an appliance, expressed in volume or heat units per unit time, such as cubic feet per hour or Btu per hour.
Latent heat	Heat that produces a change of state without a change in temperature; e.g., ice to water at 32°F, or water to steam at 212°F.
Latent heat of condensation	The amount of heat energy in BTUs that must be removed to change the state of 1 pound of vapor to 1 pound of liquid at the same temperature.
Latent heat of vaporization	The amount of heat energy in BTUs required to change the state of 1 pound of liquid to 1 pound of vapor at the same temperature.
Lithium bromide	The absorbent used in the absorption machine.
Noncondens-able gas	Air or any gas in the machine that will not liquefy under operating pressures and temperatures.
Refrigerant	Any substance that transfers heat from one place to another, creating a cooling effect. Water is a refrigerant in absorption machines.
Relative humidity (RH)	The ratio of the mole fraction of water vapor to the mole fraction of saturated moist air at the same temperature and pressure. RH is dimensionless, and is usually expressed as a percentage.
Relief valve	A valve that opens before a dangerously high pressure is reached.

Table 5.2   Key Terms Related to HVACs/CRACs (Continued)

Term	Definition
Solenoid valve	A control device that is opened and closed by an electrically energized coil.
Solution additive	Octyl alcohol added to lithium bromide to enhance operation.
Solution pump	Recirculates lithium bromide solution in the absorption cycle.
Specific enthalpy	The sum of the internal (heat) energy of the moist air in question, including the heat of the air and water vapor within. Enthalpy is given in (SI) joules per kilogram of air or BTU per pound of dry air. Specific enthalpy is also called heat content per unit mass.
Specific heat	The amount of heat necessary to change the temperature of 1 pound of a substance 1°F.
Specific volume	The volume per unit mass of the air sample. The SI units are cubic meters per kilogram of dry air; other units are cubic feet per pound of dry air. Specific volume is also called *inverse density*.
Spill point	When the evaporator pan overflows into the absorber.
Spread	Numerical difference in the percentage concentration of the concentrated and dilute solutions.
Storage	Amount of refrigerant in the unit not in solution with the lithium bromide.
Strainer	Filter that removes solid particles from the liquid passing through it.
Sub-cooling	Cooling of a liquid, at a constant pressure, below the point at which it was condensed.
Superheat	The heat added to vapor after all liquid has been vaporized.
Temperature	A measurement of heat intensity.
Therm	A unit of heat energy equal to 100,000 Btu.
Ton (refrigeration)	The amount of heat absorbed by melting 1 ton of ice in 24 hours. Equal to 288,000 Btu/day, 12,000 Btu/hour, or 200 Btu/minute.

Continued

Table 5.2   Key Terms Related to HVACs/CRACs (Continued)

Term	Definition
Transformer	A coil or set of coils that increases or decreases voltage and current by induction.
Trimming the machine	Involves automatic adjustment of the generator solution flow to deliver the correct solution concentration at any operating condition and providing the right evaporator pan water storage capacity at design conditions.
Vacuum	A pressure below atmospheric pressure. A perfect vacuum is 30 inches Hg.
Variable Air Volume (VAV) AC	ACs where both the supply airflow and return airflow rates vary according to the thermal demands in the space.
Wet-bulb temperature (WBT)	The temperature of an air sample after it has passed through a constant-pressure, ideal, adiabatic saturation process. Effectively, this is after the air has passed over a large surface of liquid water in an insulated channel. Note: WBT = DBT when the air sample is saturated with water.

Note: Most terms based on Trane, *Glossary, ABS-M-11A Operation/Maintenance Manual,* 27 April 1999. Ingersoll-Rand Company, Corporate Center, One Centennial Ave, Piscataway, NJ 08854.

5.1.3 Psychrometry

Psychrometry (also psychrometrics) defines the field of engineering concerned with the determination of physical and thermodynamic properties of gas-vapor mixtures. A psychrometric chart is a graph of the physical properties of moist air at a constant pressure (typically equated to an elevation relative to sea level). These charts are used for determining properties of moist air and analyzing air-conditioning processes. Figure 5.8 provides an example. The thermophysical properties typically found on psychrometric charts include the parameters that follow:

- *Dry-bulb temperature (DBT):* The temperature of an air sample, as determined by an ordinary thermometer, the thermometer's bulb being dry. The SI unit is Kelvin; in the United States, the unit is degrees Fahrenheit.
- *Wet-bulb temperature (WBT):* The temperature of an air sample after it has passed through a constant-pressure, ideal, adiabatic saturation process. Note: WBT = DBT when the air sample is saturated with water.
- *Dew-point temperature (DPT):* That temperature at which a moist air sample at the same pressure would reach water vapor saturation.

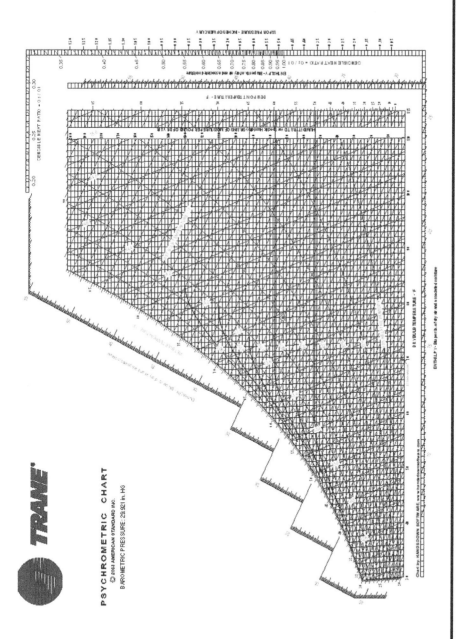

Figure 5.8 Psychrometric chart.

- *Relative humidity (RH):* The ratio of the mole fraction of water vapor to the mole fraction of saturated moist air at the same temperature and pressure.
- *Humidity ratio:* The proportion of mass of water vapor per unit mass of dry air at the given conditions (DBT, WBT, DPT, RH, etc.). The humidity ratio is also known as *moisture content* or *mixing ratio.*
- *Specific enthalpy:* The sum of the internal (heat) energy of the moist air in question, including the heat of the air and water vapor within. Enthalpy is specified in (SI) joules per kilogram of air or BTU per pound of dry air. Specific enthalpy is also called *heat content per unit mass.*
- *Specific volume:* The volume per unit mass of air sample. The SI units are cubic meters per kilogram of dry air; other units are cubic feet per pound of dry air. Specific volume is also called *inverse density.*

To use a chart, at a given elevation, at least two of the six independent properties must be known (i.e., DBT, WBT, RH, humidity ratio, specific enthalpy, and specific volume). The "ω–t" chart is often used, where the DBT appears horizontally as the abscissa (x-axis) and the humidity ratios (ω) appear as the ordinate (y-axis). (For DPT, one follows the horizontal line from the point where the line from the horizontal axis arrives at 100% RH; for WBT, one looks at the line inclined to the horizontal and intersects the saturation curve at DBT point; for RH, one looks at hyperbolic lines drawn asymptotically with respect to the saturation curve, which corresponds to 100% RH; for specific enthalpy, one looks at lines of equal-valued slope from the upper left to the lower right; and specific volume is found as an equally spaced parallel family of lines.)

5.1.4 Chiller Issues

The equipment used to produce chilled water for HVAC systems can account for up to 35% of a facility's electrical energy use. If the replacement of an existing chiller is considered, there are efficient systems on the market. The most efficient chillers currently available operate at efficiencies of 0.50 kilowatts per ton (kW/ ton), a savings of 0.15 to 0.30 kW/ton over most existing equipment. A planner should consider chiller replacement when existing equipment is more than 10 years old and the life-cycle cost analysis confirms that replacement is worthwhile. New chillers can be 30% to 40% more efficient than existing equipment. First-cost and energy performance are the major components of life-cycle costing, but refrigerant fluids may also be a factor. Older chillers using CFCs may be very expensive to recharge if a refrigerant leak occurs (and loss of refrigerant is environmentally damaging) [DOE200101]. An ideal time to consider chiller replacement is when lighting retrofits, glazing replacement, or other modifications are being done to the building that will reduce cooling loads. Conversely, when a chiller is being replaced, consider whether such energy improvements should be carried out—in some situations, those energy improvements can be essentially done for free because they will be paid for from savings achieved in downsizing the chiller.

Electric chillers use a vapor compression refrigerant cycle to transfer heat. The basic components of an electric chiller include an electric motor, refrigerant compressor, condenser, evaporator, expansion device, and controls. Electric chiller classification is based on the type of compressor used—common types include centrifugal, screw, and reciprocating. The scroll compressor is another type frequently used for smaller applications of 20 to 60 tons. Both the heat rejection system and building distribution loop can use water or air as the working fluid. Wet condensers usually incorporate one or several cooling towers. Evaporative condensers can be used in certain (generally dry) climates. Air-cooled condensers incorporate one or more fans to cool refrigerant coils and are common on smaller, packaged rooftop units. Air-cooled condensers may also be located remotely from the chillers.

The refrigerant issues currently facing facility managers arise from concerns about protection of the ozone layer and the buildup of greenhouse gases in the atmosphere. The CFC refrigerants traditionally used in most large chillers were phased out of production on 1 January 1996, to protect the ozone layer. CFC chillers still in service must be (1) serviced with stockpiled refrigerants or refrigerants recovered from retired equipment, or (2) converted to HCFC-123 (for the CFC-11 chillers) or HFC-134a (for the CFC-12 chillers), or (3) replaced with new chillers using EPA-approved refrigerants. Under current regulations, HCFC-22 will be phased out in the year 2020, and HCFC-123 will be phased out in the year 2030. Chlorine-free refrigerants, such as HFC-134a and water/lithium bromide mixtures, are not currently listed for phase-out. A chiller operating with a CFC refrigerant is not directly damaging to the ozone, provided that the refrigerant is totally contained during the chiller's operational life and that the refrigerant is recovered upon retirement [DOE200101]. See Table 5.3 for a comparison of refrigerants.

Table 5.3   Comparison of Refrigerant Alternatives

Criteria	HCFC-123	HCFC-22	HFC-134a	Ammonia
Ozone-depletion potential	0.016	0.05	0	0
Global warming potential (relative to CO_2)	85	1,500	1,200	0
Ideal kW/ton	0.46	0.50	0.52	0.48
Occupational risk	Low	Low	Low	Low
Flammable	No	No	No	No

Source: U.S. Environmental Protection Agency.

5.2 General Overview of a Computer Room Air-Conditioning System

Servers, storage devices, and communications equipment are getting smaller and more powerful. Data center and telecom designers now tend to pack more equipment into a smaller space, thus concentrating a large amount of heat. Adequate and efficient cooling equipment is needed, along with proper airflow. Traditionally the data center cooling requirement has been in the 30 to 50 watts per square foot (WPSF) range. Blade servers can generate a heat load of 15 to 25 kW per cabinet, representing a density of around 300 WPSF over the total data center area; in some cases, 500 WPSF may be encountered for the "server room" area [GAR200601]. It is typically expensive to upgrade existing data centers to support new generations of servers. Some are looking at liquid cooling as the way to solve the server density challenge. Traditional data center cabinet/WPSF loads have translated into a requirement of 1 to 1.5 T of cooling per rack.

Generally, the racks in a data center or telecom center adhere to a geometric standard: The racks are laid out in rows separated with hot and cold aisles. Figure 5.9 depicts under-floor cooling of a data center; ceiling and wall-mounted CRAC systems are also available. While the cold aisles supply cold air to the systems, hot aisles are designated to remove hot air from the systems. Cold-aisles containment techniques are also being used in very high-end data centers. The goal of a CRAC is to provide a complete environmental control package, including both precision air conditioning and humidity control. It should be noted that although racks come in standard sizes, improper layout can change the fluid mechanics inside a data center, leading to inefficient utilization of CRAC units. For example, studies reveal that minor layout changes in rack placement can lead to an imbalance in cooling loads on CRAC units by as much as 25% [PAT200201].

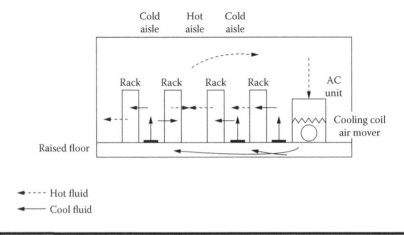

Figure 5.9   Under-floor cooling of data center.

A refrigerated or chilled-water cooling coil in the AC unit extracts the heat from the air and cools it within a range of 10°C to 17°C. A typical 3 × 1 × 2 m CRAC unit has a sensible heat removal capacity of 100 kW. The cool air is recirculated back to the racks through vented tiles in the raised under-floor plenum. The air handlers pressurize the plenum with cool air; the cool air enters the data center through vented tiles near the inlet of the racks.

There may be localized hotspots within the data center, and some servers may be ineffectively drawing in hot air from the hot aisles. Other areas can be very cold, with most cold air bypassing the hot equipment and short-circuiting right back to the CRAC units. Studies have found that some CRAC units move large quantities of air, yet their cooling coils may be only providing 10% to 25% of their rated cooling capability. A situation can arise where one is moving a lot of air and using considerable fan energy, yet receiving little benefit from most of the cooling equipment. If cooling air cannot be effectively channeled through hot servers, it is wasted. A poor design may lead to cold aisles that are not operating as a cold air delivery zone as intended; it may be a situation where servers are drawing in a mix of cold air and the hot air exiting the equipment [GAR200601]. Figure 5.10 shows a typical heat distribution for racks of equipment.

Air-cooled electronic equipment depends on the intake air temperature for effective cooling. Accordingly, at this juncture most (but not all) environmental specifications refer to the intake conditions; Table 5.4 illustrates some of the specifications available. The preferred strategy is to supply cold air as close to the equipment intakes as possible without prior mixing with ambient air and return hot exhaust air without prior mixing with ambient air, i.e., *once-through cooling*. Adequate thermal conditions (intake temperatures) are important for the reliability and longevity of electronic equipment. We mentioned in Chapter 3 that in a data center/telco room air-management techniques are used to minimize recirculation of hot air back from a hot-air aisle

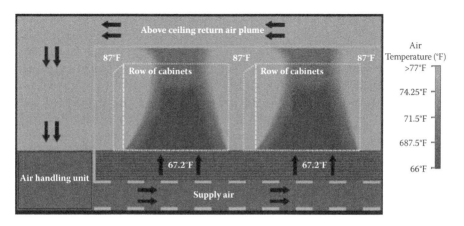

Figure 5.10   Typical heat distribution generated by racks of equipment.

Table 5.4 Rack Intake Recommendations (Various Specifications)

	Recommended (Facility)	Allowable Equipment
Temperature (@Equipment Intake) Data centers (ASHRAE)	20–25 C 18–27 C	15–32 C 5–40 C
Humidity (RH) (@Equipment Intake) Data Centers (ASHRAE) Telecom (NEBS)	4–55% -<55%	20–80% 5–85%

to a cold aisle, and to minimize *by-pass* of cold air from a cold aisle over the rack rather than through the rack (to cool the equipment in the rack). Recirculation air participates in cooling the electronic equipment multiple times and should be minimized. Recirculation may be caused by a deficit of supply air. Recirculation air leads to less control of the equipment intake conditions; the implications may be reduced reliability and longevity. Local hot spots may lead to a perceived need to increase the overall supply airflow (higher fan energy) or reduce the supply temperature (lower chiller efficiency). *By-pass air* does not participate in cooling the gear and should be minimized. By-pass air may be caused by an excess of supply air or leakage through cable cutouts. By-pass air increases the operational costs but may be a safeguard against poor thermal conditions; it requires high system airflow (higher fan energy) and leads to lower return air temperatures (lower chiller efficiency). Reducing the by-pass leads to airflow and cooling capacity regain. Air management helps reduce operating costs by enhancing economizer utilization, improving chiller efficiency, and reducing fan energy. Improved energy efficiency also results in reduced capital investments for cooling equipment, air-moving equipment, and real estate [DOE200901].

We have noted a number of times in this text that arranging the space in alternating hot and cold aisles is the first step toward once-through cooling. Cold air is supplied into the cold front aisles, the gear moves the air from the front to the rear, and the hot exhaust air is returned to the air handler from the hot rear aisles. Physical barriers can successfully be used to enhance the separation of hot and cold air. Enclosed aisles permit high supply and, in turn, return temperatures. As we have seen, data center/telco rooms can use under-floor or overhead cooling. Correctly designed, both cooling methods work well; each has its own benefits and drawbacks. Changing from one to the other is generally not an option. Tall open ceilings promote thermal stratification and the placement of the return grille is not critical. A return plenum often means a lower clear ceiling but allows placing the return grilles above the hot aisles. Open tall ceilings are simpler compared to return plenums. The cooling capacity of a raised floor depends on its *effective* height, which can be increased by removing cables and other obstructions that are not in use. The cooling capacity is generally limited by pressure variations in the plenum, which lead to erratic cooling airflow rates. Equipment aisle enclosures can increase the capacity since variations in the airflows cancel out inside the enclosure. CRAC units should be placed to promote an even pressure distribution in the floor plenum; when possible they should be centered on the hot aisles rather than on the cold aisles. This results in better cooling performance. Overhead ducted systems generally have better stability than raised-floor systems. Higher pressure losses buy high stability, which, in turn, may result in lower energy usage over time [DOE200901].

As noted, the current industry Best Practice is to employ the "hot-aisle/cold-aisle," as illustrated yet again in Figure 5.11 [ADC200601]. In a hot-aisle/cold-aisle configuration, equipment racks are arranged in alternating rows of hot and cold aisles. In the cold aisle, equipment racks are arranged face to face, while in the hot aisle, they are back to back. Perforated tiles in the raised floor of the cold aisles allow cold air to be drawn into the face of the equipment. The cold air washes over the equipment and is expelled out the back into the hot aisle (in the hot aisle there are no perforated tiles to keep the hot air from mingling with the cold air). For best results with this method, aisles should be two tiles wide, enabling the use of perforated tiles in both rows. It should be noted that while it is common for equipment cabinets to exhaust heat out the back, it is not a universal practice: Some equipment cabinets draw cold air in from the bottom and discharge the heated air out the top or sides, while other equipment brings in cold air from the sides and exhausts hot air out the top.

In summary, as noted earlier, "hot-aisle/cold-aisle" is a common means of providing cooling to data center and telecom rooms where the equipment is arranged in rows and cold supply air is supplied to the cold aisle, pulled through the inlets of the IT equipment, and exhausted to a hot aisle to minimize recirculation of the hot exhaust air with the cold supply air. Supply air is introduced into a region called the cold aisle. On each side of the cold aisle, equipment racks are placed with their intake sides

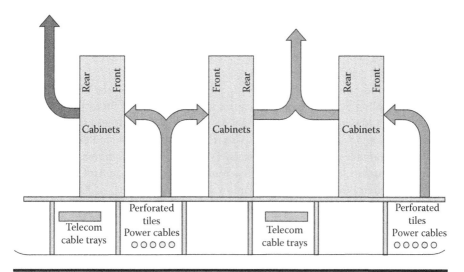

Figure 5.11   Another view of the hot-aisle/cold-aisle concept.

facing the cold aisle. A hot aisle is the region between the backs of two rows of racks. The cooling air delivered is drawn into the intake side of the racks [GRE201001].

Additional design principles include the following, among others [ADC200601]:

■ Spreading equipment out over unused portions of the raised floor if available.
■ Increasing the height of the raised floor. Doubling floor height has been shown to increase airflow by as much as 50%.
■ Using open racks instead of cabinets. If security concerns or the depth of servers makes using racks impossible, cabinets with mesh fronts and solid backs are alternatives.
■ Increasing airflow under the floor by blocking all unnecessary air escapes.
■ Replacing existing perforated tiles with ones having larger openings. Most tiles come with 25% openings, but some provide openings of 40% to 60%.
■ Planning under-floor cable pathways in such a way as to minimize cold air exit restrictions. One should place power cables low under exit tiles because they occupy less area than data cables.

Overhead cooling modules mount directly above the cold aisle, requiring no floor space. Such a module draws in hot air from the hot aisle, then discharges cool air into the cold aisle where the equipment air inlets are located. The rack cooling module mounts vertically on or above the IT rack enclosure, also requiring no floor space. It draws hot air from inside the cabinet or from the hot aisle; it then cools the air and discharges it down to the cold aisle. An in-row cooling module is placed directly in line with the rack enclosures, and requires very little floor space. Air from the hot aisle is drawn in through the rear of the unit, cooled, and then discharged horizontally through the front of the unit into the cold aisle. An in-row system typically is able to cool data centers with racks operating in the range of 30 kW per rack. These systems are designed to allow adaptive and scalable expansion without interruption of cooling operations. They are used as supplemental sensible cooling for high heat density racks or zones.

Typical mid-range systems, such as, but not limited to, the Liebert DS™ Precision Cooling System, operate at capacities in the range of 30 to 100 kW (8 to 30 ton) and allow installation in upflow or downflow configurations, and in air, water, glycol, and Glycool™ models. Also, these CRACs provide flexible cooling capacity via variable-capacity compressors. Typical mid-range systems, such as, but not limited to, the Liebert CW™ Chilled Water Cooled Precision Air Conditioning For Data Centers, are available in downflow and upflow models, in 30- to 180-kW capacities. Upflow and downflow configurations are available to cover both raised floor and non-raised applications.

Design and efficiency issues are revisited in greater detail in Chapter 7.

References

[ADC200601]   ADC KRONE, Designing an Optimized Data Centre, ADC KRONE White paper, November 2006, P.O. Box 335, Wyong NSW 2259, Australia.

[DOE200101]   U.S. Department of Energy, Energy Efficiency and Renewable Energy, Greening Federal Facilities: An Energy, Environmental, and Economic Resource Guide for Federal Facility Managers and Designers, second edition, Pub. DOE/GO-102001-1165, NREL/BK-710-29267, May 2001.

[DOE200901] U.S. Department of Energy, Energy Efficiency and Renewable Energy. Data Center Energy Efficiency Training Materials, 5/19/2009.

[GAR200601]   D. Garday and D. Costello, Air-Cooled High-Performance Data Centers: Case Studies and Best Methods, White paper, Intel Corporation, November 2006.

[GRE201001]   The Green Grid, Glossary and Other Reference Materials, 2010.

[ICA199701]   ICAEN, REHVA and Partex, HVAC Systems for Buildings, THERMIE Programme, EUROPEAN Commission, 1997.

[NRC200901]   National Research Council of Canada — Institute of Research in Construction — Cost-effective Open-Plan Environments (COPE) Project, [http://www.nrc-cnrc.gc.ca/eng/projects/irc/cope.html] Ventilation Principles, [http://irc.nrc-cnrc.gc.ca/ie/cope/05-1-Vent_Principles_e.html].

[PAT200201]   C.D. Patel, R. Sharma, C.E. Bash, and A. Beitelmal, Thermal considerations in cooling large scale high compute density data centers, Itherm2002, The Eighth Intersociety Conference on Thermal and Thermomechanical Phenomena in Electronic Systems, San Diego, CA.

[TRA199901]   Trane, Glossary, ABS-M-11A Operation/Maintenance Manual, 27 April 1999, Ingersoll-Rand Company, Corporate Center, One Centennial Ave., Piscataway, NJ 08854.

Chapter 6

Regulatory and Best Practices Background

This chapter focuses on regulatory activities related to green initiatives that have relevance to green IT in general and green networking in particular. The purpose is to sensitize the reader as to who the key players in this space are and the general activities with which these players are involved. What follows is a partial survey of the space; a large number of organizations are involved in green initiatives from an advocacy and/or standardization perspective, but this chapter addresses only a subset of the stakeholders.

6.1 Alliance for Telecommunications Industry Solutions (ATIS)*

In late 2008, the ATIS (Alliance for Telecommunications Industry Solutions) Board of Directors formed the Exploratory Group on Green (EGG) to address green issues for the Information and Communications Technology (ICT) industry. ATIS is a technical planning and standards development organization committed to the rapid development of global, market-driven standards for the information, entertainment, and communications industries. More than 300 companies actively formulate standards, with 20 committees within ATIS covering issues including IPTV (Internet Protocol Television), Service-Oriented Networks, Home Networking, Energy Efficiency, IP-Based and Wireless Technologies, Quality of Service, Billing, and Operational Support. The organization states that ATIS and its member companies

---

* This section is based directly on materials from ATIS [ATIS200901, ATIS200902].

are committed to advancing environmental sustainability within both its industry sector and the industry as a whole by utilizing its unique position to foster the reduction of greenhouse gas emissions in other sectors through the use of ICT technologies, devices, services, and applications. ATIS is initially focusing on

- Industry-wide operational policies, technical standards, or performance benchmarks to consistently "quantify" conformity to, or levels of, being "green"
- The development of industry-based parameters, criteria, and guidelines, when possible, versus government-driven mandates or programs
- Enacting simple energy-saving practices to reduce the consumption of energy, such as performing energy audits on a per facility basis and outlining actions toward reducing energy consumption, such as retiring and de-powering obsolescent or underutilized equipment
- Proactively seeking ways in which ICT solutions can be readily adopted in reducing the carbon footprint of a number of industries, including transport, power, logistics, and others

The EGG, populated by appointees from ATIS Board Member companies, had the charter to investigate how ATIS and its members could advance environmental sustainability efforts, starting with an investigation and prioritization of items identified as important to ATIS member companies during the ATIS Green Initiative Workshop. The EGG is looking to develop a roadmap for the industry's continued sustainability, and to ensure that those actions are balanced against industry's environmental responsibility, social acceptance, and economic viability. Going forward, applications of certain identified methods can enable industry to improve efficiency and reduce its carbon footprint.

ATIS has been working with equipment vendors and operators concerned with reducing energy consumption, heat generation, and the equipment footprint for a number of years in an effort to increase operational efficiencies and reduce cost. The current, renewed emphasis is to accelerate these initiatives even further. To date, work on these topics has been done primarily in the ATIS Network Interface, Power, and Protection Committee (NIPP), where standards related to power consumption reduction through energy efficiency improvements, reducing power consumption for Digital Subscriber Line (DSL) modems, and the Restriction of the use of certain Hazardous Substances (RoHS) in electronic equipment have been developed. To create a uniform method for measuring telecommunication equipment energy consumption (power), as well as establishing reporting methods, in 2009 ATIS NIPP completed the following series of documents; they have been accepted by the American National Standards Institute (ANSI) and are also being considered by the International Telecommunication Union (ITU):

- ATIS-0600015.2009: Energy Efficiency for Telecommunication Equipment: Methodology For Measurement and Reporting General Requirements (Baseline Document)

- ATIS-0600015.01.2009: Energy Efficiency for Telecommunication Equipment: Methodology for Measurement and Reporting Server Requirements
- ATIS-0600015.02.2009: Energy Efficiency for Telecommunication Equipment: Methodology for Measurement and Reporting Transport Requirements

The Telecommunications Energy Efficiency Ratio (TEER) mentioned in Chapter 1 and Chapter 3 is a measure of network-element efficiency. The standards just listed provide a comprehensive methodology for measuring and reporting energy consumption, and uniformly quantify a network component's ratio of "work performed" to energy consumed. The efficiency standards are specific to equipment type, network location, and classification. Normalizing these ratings by functionality enables "apples-to-apples" equipment comparison. This systemized assessment results in repeatable and comparable energy consumption measurement.*

ATIS-0600015.2009 is the base standard for determining telecommunications energy efficiency. The latter two documents (server requirements and transport system or network configuration requirements) are part of an ongoing series to define the telecommunications energy efficiency of various telecommunications components.

In general, each TEER follows the formula:

$$TEER = Useful\ Work\ /\ Power$$

where Useful Work is defined in the supplemental standard based on the equipment function. Examples could be, but are not limited to, data rate, throughput, processes per second, etc. Power is the power in watts (dependent on the equipment measurement). Typical TEER values range from 1 to 1000. The higher the TEER value, the more energy efficient the equipment is compared to other like equipment.

For the purposes of the standard, equipment is classified based on the application and the location in the network and whether the network is wire-line or wireless. The following equipment classifications are used in TEER:

- *Core Equipment:* Equipment to provide support for the network features and telecommunication services. The support provided includes functionality such as the management of user location information, control of network features and services, and the transfer (switching and transmission) mechanisms for signaling and for user-generated information.
- *Transport Equipment:* Equipment to enable information transfer capabilities between originating and terminating access service facilities.

---

* ICT companies participating in the development of ATIS specifications include 2Wire, ADTRAN, AT&T, Alcatel-Lucent, Cisco Systems, Ericsson, Fujitsu, Huawei, Intel, Intertek, Juniper Networks, Nortel, Qwest Communications, Sun Systems, Tellabs, and Verizon Communications.

- *Access Equipment:* Equipment whose purpose is to connect the core or transport network to the end user. This may be equipment used to transmit data from the core network to a node within the network. This may also be equipment that transmits data from the core or node to the end user's demarcation point. Network Terminating Equipment (NTE) and wireless equipment are included within Access Equipment.
- *Customer Premise Equipment (CPE):* Equipment located on the customer's side of the demarcation point that is used as a part of the premise network. Examples of CPE include, but are not limited to, modems, broadband home routers, power adapters, and set-top boxes. For the purpose of this family of standards, NTE is not considered CPE.
- *Power Equipment:* Power conditioning equipment, including telecommunications type rectifiers, converters, and inverters.

The following items work items were in progress at press time:

- NIPP-TEE-2009-013: Energy Efficiency for Telecommunication Equipment: Methodology for Measurement and Reporting Router and Ethernet Switch Requirements
- NIPP-TEE-2009-015R1: Energy Efficiency for Telecommunication Equipment: Methodology for Measurement and Reporting Direct Current (DC) Power System-Rectifier Requirements

Subsequent documents in the NIPP's series of documents, planned for release over time, are expected to cover other network and consumer equipment and devices including, but not limited to, core network routers and switches, outside plant equipment, gateways, set-top-boxes and other consumer electronic (CE) devices, and power systems. ATIS NIPP also has work underway addressing the potential use of environmentally friendly materials in describing materials used for connectors; airborne contamination (mixed flowing gas and hygroscopic dust) requirements for network equipment in the central office and outside plant environments; and heat dissipation and power consumption requirements for network equipment in central office and outside plant environments, including methods to reduce power consumption for DSL modems at both ends of the line.

6.2 American Society of Heating, Refrigerating and Air Conditioning Engineers (ASHRAE)*

ASHRAE has its mission focused on advancing heating, ventilation, air conditioning, and refrigeration (HVAC&R) to serve humanity and promote a sustainable world through research, standards writing, publishing, and continuing education. ASHRAE was founded in 1894. The organization is committed to the advancement

---

* This section is based directly on ASHRAE materials.

of the arts and sciences of HVAC&R, for the benefit of society through research, technology development and transfer, and education and training.

ASHRAE gained recognition as an organizational authority on energy issues in the 1970s when it was asked to develop a national standard on energy conservation. It met that challenge, and ANSI/ASHRAE/IESNA Standard 90.1 Energy Standard for Buildings Except Low-Rise Residential Buildings continues to influence building designs worldwide. It has become the basis for building codes, and the standard for building design and construction throughout the United States. Buildings built to these standards use significantly less energy than buildings built prior to the development of Standard 90-75. ANSI/ASHRAE/IESNA Standard 90.1 is published as a consensus standard to provide minimum requirements for the energy-efficient design of new and renovated or retrofitted buildings. Standard 90.1-2004 has recently been approved by the DOE as the minimum standard to be met by all states in the United States. It is written in code language intended as minimum requirements, and thus it does not necessarily provide exemplary or state-of-the-art design guidance. ASHRAE Standard 90.1 is on continuous maintenance and is republished on a 3-year cycle. The current version at press time was 90.1-2007.

Among other activities, ASHRAE develops standards for both its members and others professionally concerned with refrigeration processes and the design and maintenance of indoor environments. ASHRAE writes standards for the purpose of establishing consensus for (1) methods of test for use in commerce and (2) performance criteria for use as facilitators with which to guide the industry. ASHRAE publishes the following three types of voluntary consensus standards: Method of Measurement or Test, Standard Design, and Standard Practice. ASHRAE does not write rating standards unless a suitable rating standard will not otherwise be available. Consensus standards are developed and published to define minimum values or acceptable performance, whereas other documents, such as design guides, may be developed and published to encourage enhanced performance.

ASHRAE conducts research in the areas of indoor air quality, new refrigerants, energy efficiency, and air filtration, among others. The organization publishes the *ASHRAE Advanced Energy Design Guides* (*AEDG*). These are a series of publications designed to provide recommendations for achieving energy savings over the minimum code requirements of ANSI/ASHRAE/IESNA Standard 90.1-1999. The *Guides* are intended to provide a simple and easy approach to design and build energy-efficient buildings for use by contractors and designers who design and construct the specific building types represented in the *Guides*.* The *Guides* have

---

* The *Energy Design Guides* are a series of publications designed to provide prescriptive recommendations for achieving 30% energy savings over the minimum requirements in Standard 90.1-1999 in eight U.S. climate zones. They are developed by a committee of experts and undergo peer review but are not developed through a consensus process. They show a way, but not the only way, to achieve 30% savings.

been developed in collaboration with The American Institute of Architects (AIA), the Illuminating Engineering Society of North America (IES), the U.S. Green Building Council (USGBC), and the U.S. Department of Energy (DOE) (the New Building Institute [NBI] participated only in the development of the *Advanced Energy Design Guide for Small Office Buildings*). The initial series of *Guides* has an energy savings target of 30%, which is the first step in the process toward achieving a net zero–energy building—defined as a building that, on an annual basis, draws from outside resources equal or less energy than it provides using on-site renewable energy sources.* Each 30% *Guide* addresses a specific building type. Additional *Guides* for existing buildings and at 50% energy savings toward a net zero energy building are also planned.

ANSI/ASHRAE Standard 127-2001, "Method of Testing for Rating Computer and Data Processing Room Unitary Air-Conditioners," specifies relevant operational parameters for Computer Room Air Conditioner (CRAC) units to be used for data and communications equipment room.

ANSI/ASHRAE/IESNA Standard 90.1-1999, the energy conservation standard, provides the fixed reference point for all of the 30% *Guides* in this series. The primary reason for this choice as a reference point is to maintain a consistent baseline and scale for all of the 30% AEDG series documents. The recommendations in the 30% *Guides* will allow those involved in designing or constructing the various building types to easily achieve advanced levels of energy savings without having to resort to detailed calculations or analyses. All of the energy-saving recommendations for each of the eight U.S. climate zones are contained on a single page, thus facilitating the 30% *Guide's* use. Additional recommendations point out other opportunities to incorporate greater energy savings into the design of the building. The use of the *Guides* provides a prescriptive path to achieving Leadership in Energy and Environmental Design (LEED) v2.2 Energy & Atmosphere credits for New Construction and Major Renovation projects. USGBC's LEED Green Building Rating System™ provides a roadmap for measuring and documenting success in the design, construction, and operation of high-performance green buildings. The LEED for New Construction focuses primarily on office buildings, but the USGBC is also developing LEED Retail for New Construction and LEED for Schools. A number of 30% *Guides* have been completed thus far:

■ The ASHRAE 30% Advanced Energy Design Guide for Small Office Buildings – Office buildings up to 20,000 ft²

---

* No exotic or specialized equipment and/or materials are required to achieve the 30% energy savings. To quantify the expected energy savings, potential envelope, lighting, HVAC, and service, water heating energy saving measures were selected for analysis and recommendation. Only measures and equipment that were deemed both practical and commercially available from at least two manufacturers were considered. Although some of the products may be considered premium, products of similar energy efficiency performance are available from multiple manufacturers.

- The ASHRAE 30% Advanced Energy Design Guide for Small Retail Buildings – Retail spaces up to 20,000 ft^2
- The ASHRAE 30% Advanced Energy Design Guide for K-12 School Buildings – Elementary, Middle, and High School Buildings
- The ASHRAE 30% Advanced Energy Design Guide for Small Warehouses and Self-Storage Buildings – Warehouses up to 50,000 ft^2 and self-storage buildings that use unitary heating and air-conditioning
- The ASHRAE 30% Advanced Energy Design Guide for Highway Lodging — Roadside hotels up to 80 rooms and 4 stories.

Other planned guides include

- The ASHRAE Energy Efficiency Guide for Existing Commercial Buildings: The Business Case for Building Owners and Managers
- The ASHRAE 30% Advanced Energy Design Guide for Small Healthcare Facilities
- Several 50% Advanced Energy Design Guides are planned for the 2009–2011 time period, including Small to Medium Office Buildings, Mid-size Retail Buildings, Highway Lodging, Grocery/Supermarket, and K–12 School Buildings

ASHRAE also publishes the *ASHRAE GreenGuide: The Design, Construction, and Operation of Sustainable Buildings, 2nd ed.* The *ASHRAE GreenGuide* provides guidance to designers of HVAC&R systems on how to participate effectively on design teams charged with producing green buildings. It also covers green design techniques applicable to related technical disciplines such as plumbing and lighting, and it addresses how mechanical and electrical systems may interact with and be influenced by architectural design. It starts with the very earliest stages of a green building design project and carries through to the resulting structure's construction, operation and maintenance, and eventual demolition.

Some of the goals of ongoing ASHRAE research include the following:

- Provide guidance on techniques to achieve 30%, 50%, and 70% reduction in building energy usage based upon the energy codes in place at the turn of the millennium. The goals are to move 30% toward net zero-energy use buildings by 2008, 50% by 2012, and 70% by 2015. (A net zero-energy use building is a building that uses equal or less energy than it produces on an annual basis.)
- Produce by 2015 new residential and light commercial buildings that have 70% less energy use than buildings built at the turn of the millennium according to ASHRAE Standard 90.2.
- Develop economically viable applications of renewable energy that produce 25% reductions in conventional energy use by 2015.

- Develop systems and components that reduce energy use in supermarkets by 30% by 2015.
- Optimize and make consistent ASHRAE Standards 90, 62, and 55 to achieve measured and verified high system energy efficiency with high indoor environmental quality (IEQ) for indoor built environments.
- Develop integrated, best-practice design methods that will allow energy consumption, life-cycle cost, and environmental impact to be minimized, and that will allow system life span and IEQ to be maximized.
- Develop evaluation methods that allow reductions in energy, cost, and emission and improvements in comfort, health, and productivity to be quantitatively measured.
- Establish benchmark data on energy use in industrial refrigeration.
- Establish techniques to improve the energy efficiency and reliability of heating, ventilating, cooling, and refrigeration system components (e.g., heat exchangers, compressors, pumps, fans, distribution systems).
- Continue research into new alternative and natural refrigerants.
- Improve performance and reliability, and minimize the environmental impacts of working fluids and materials.
- Develop reliable, durable, and self-correcting sensor technology for monitoring indoor environmental quality, pollutants, energy conservation, and fault detection and diagnosis.
- Move one or more of the nontraditional technologies that have comparable performance and cost to traditional vapor compression systems to market readiness by 2010.
- Develop techniques that reduce the installed energy use of HVAC&R system auxiliary equipment 50% by 2015.

6.3 Energy Consumption Rating Initiative (ECR)*

The Energy Consumption Rating (ECR) Initiative seeks to develop a framework for first-order approximation of energy efficiency for network and telecom equipment. It is an open initiative that welcomes participants and users from network equipment manufacturers, government agencies, carriers, and enterprises.

In response to the growing interest from national and international standard bodies and businesses to lower the environmental and operational footprint of networking, the ECR developed methodology for reporting, measuring, and regulating energy efficiency of network and telecom components. The ECR metric creates a common energy denominator between different network and telecom systems operating within a single class. A draft specification has been published [ECR200801].

---

* This section is based directly on materials from the Energy Consumption Rating Initiative organization [ECR200901].

The draft specification defines rules for classifying network and telecom equipment into classes and a methodology for measuring energy efficiency within each class. The final "performance-per-energy unit" rating can be reported as a peak (scalar) or synthetic (weighted) metric that takes dynamic power management capabilities into account. This rating can be further utilized to optimize energy consumption for telecom and network equipment.

- *ECR for Businesses:* The ECR defines an energy-efficiency rating applicable to Internet service providers (ISPs) as well as all ICT-dependent enterprises. The ECR ratings can be utilized for enterprise purchase decisions, providing an immediate and direct means of reducing energy relative to expenses in data centers and private network infrastructures. In addition, ISPs can benefit from the ECR specification by reducing the internal test effort required to qualify equipment for internal corporate standards and external government-driven goals and regulations. Telecom products carrying ECR readings can be easily compared against each other during the network design and Request for Proposal (RFP) processing stages at no extra cost or effort. In addition, the ECR methodology provides a turnkey infrastructure for energy and facility planning, allowing operational expenses to be kept under tight control.

- *ECR for Telecommunication Equipment Vendors:* The ECR specification offers a standardized way of estimating the energy efficiency across a vendor's product portfolio in a repeatable and automated way. This can significantly improve the effectiveness of R&D processes. Best-practice energy management engineering efforts, such as dynamic power consumption, slow/fast corner component analysis, and voltage/performance regulation, can be regulated at a high level with the ECR methodology. In addition, the ECR provides a common set of terms to express the operational cost of products to customers.

- *ECR for Government Agencies:* Due to their sheer simplicity and direct physical meaning, ECR metrics can be readily employed to set compliance goals and approval programs. When used for this purpose, the metric syntax may change to reflect the authority name, compliance revision, and product class, for example:
  - ECR-2008, Class C1.1 (green label)—awarded to qualifying platforms within 25% of the reference metric of 15 W/Gbps (or better).
  - ECR-2008, Class C1.1 (yellow label)—awarded to qualifying platforms within 35% of the reference metric of 15 W/Gbps (or better).

- ECR-2008, Class C1.1 (red label)—noncompliant platforms within 45% of the reference metric of 15 W/Gbps (or worse).

  Forward-looking labels can be defined in accordance with the goals of government certification, environment goals (i.e., CO_2 reduction), local business or regulatory needs, that is, ECR-2010, Class C2.2 (green label)—awarded to qualifying platforms within 25% of the reference metric of 8 W/Gbps (or better). The metric constitutes the minimum requirement for equipment to be put in service in 2010 and later.

■ *ECR for Standard Bodies and Organizations:* The ECR can be easily adapted for use in telecom-related energy efficiency standardization efforts. For example, in the goal of standardizing the framework of building a data center, if the upper-level metric is defined as payload-per-energy unit, every technology archetype (power conversion, cooling, computing, telecom) gets its own efficiency vector. The ECR can be used to cover the telecom part of the equation. Likewise, in standardization of the broadband infrastructure, there are consumer and ISP levels of energy consumption; the ECR can be used for the aggregation (telecom) portion of the equation.

The 2008 ECR Draft defines energy efficiency as energy consumption normalized to effective throughput. Such an approach is in line with the high-level methodology suggested in [SAI200801] and [VER200901] documents. The ECR assumes that the more energy-efficient network system is the one that can move more data (in bits) using the same energy budget (in Joules). The ECR definition is best suited for medium- to large-scale network and telecom systems primarily serving data streams. It is less relevant to small office, SoHo (Small office/Home office), and multipurpose devices, where throughput is less relevant and efficiency criteria need to be more complex and involved, such as described in [MET200801].* The measure is applicable to many types of network and telecom equipment, including, but not limited to, routers, L2/L3 switches, optical shelves, security devices, and load balancers.

As we saw in Chapter 3, a primary and a secondary metric are defined. The primary is a peak ECR metric, which is calculated according to the following formula:

$$ECR = E_f / T_f \text{ (expressed in W/10 Gbps)}$$

---

* The ECR methodology is different from METI and EC documents because METI and EC work is primarily targeted at consumer-level equipment where performance is not a differentiator. Instead, METI and EC documents define a fixed set of energy allowances for every product class and functionality option supported. The fact that this option may not operate at line rate typically does not matter in home and SoHo environments. As a result, consumer-level network and telecom equipment can be massively oversubscribed from the bandwidth perspective without noticeable impact on usability. For instance, it does not matter if the home DSL router cannot operate all wireless or wired LAN ports at line rate, as sustained performance is not required for domestic LAN operation. In fact, consumer-grade network devices can be easily compared to a light bulb—it fills a basic need at a fixed energy cost. Carrier-class network and telecom equipment, on the other hand, presents a different case, where functions are delivered across many ports at high speed and revenue generation depends on performance. In the carrier world, an oversubscribed platform is not equal to line-rate device application-wise, and thus it cannot be fairly compared from the energy consumption perspective.

where T_f is the maximum throughput (in Gbps) achieved in the measurement and E_f is the energy consumption (in watts) measured during running test T_f. The ECR is normalized to Watts/10 Gbps and has a physical meaning of energy consumption to move 10 Gbits worth of user data per second. This reflects the best possible platform performance for a fully equipped system within a chosen application and relates to the commonly used interface speed. Sometimes, for convenience purposes, energy efficiency is also reported in Gbps/Watt under the name of EER (energy efficiency ratio); EER = 1/(ECR).

The secondary metric is a weighted (synthetic) metric (i.e., Energy Consumption Rating Weighted; ECRW) that takes idle mode into account. It is used in addition to the primary metric to estimate power management capabilities of the device.

$$ECRW = ((\alpha \times E_f) + (\beta \times E_h) + (\gamma \times E_i)) / T_f \text{ (dimensionless)}$$

where T_f is the maximum throughput (in Gbps) achieved in the measurement; E_f is the energy consumption (in watts) measured during running test T_f; E_h is the energy consumption (in watts) measured during half-load test; E_i is energy consumption (in watts) measured during idle test; and α, β, γ are weight coefficients to reflect the mixed mode of operation (ECR specifies that $\alpha = 0.35$, $\beta = 0.4$, $\gamma = 0.25$).

6.4 Environmental Protection Agency (EPA)*

6.4.1 Overview

The Environmental Protection Agency (EPA) leads the United States' environmental science, research, education, and assessment efforts. The mission of the EPA is to protect human health and the environment. The EPA leads by example by striving to reduce their "environmental footprint" in the following areas (among others):

- EPA facilities
- Energy conservation
- Green power
- Green buildings
- GHG Emissions Inventory
- Water conservation
- Stormwater management
- Greening EPA's fleet
- Green practices

---

* This section is based in its entirety on EPA materials.

- Executive Order (E.O.) 13423
- Waste reduction and recycling
- Green printing and copying
- Sustainability champions

Some of the goals of the EPA in this context include

■ Upgrading existing HVAC systems to make them more energy efficient and environmentally friendly using both appropriated funds and the energy savings performance contracting mechanism to finance these upgrades

■ Incorporating innovative, renewable, and low emission technologies such as photovoltaics, fuel cells, ground-source heat pumps, solar walls, and solar hot water heating systems into EPA facilities

■ Purchasing renewable energy wherever possible. Under Executive Order 13423, purchases of renewable energy may contribute only up to 60% of the annual energy reduction goal for FY 2008 and will gradually be reduced to zero by 2012.

In developing and implementing its long-term energy master planning strategy, the EPA plans to meet its energy reduction goals without netting out its renewable energy purchases. Some of these activities are reviewed briefly here because they (may) have applicability to other firms.

6.4.2 Initiatives

6.4.2.1 Energy Conservation

Energy Independence and Security Act of 2007 (EISA): Signed in December 2007, the EISA aims to increase U.S. energy security, develop renewable fuel production, and improve vehicle fuel economy. The EISA reinforces the energy reduction goals for federal agencies put forth in Executive Order 13423, as well as introduces more aggressive requirements.

■ Energy Efficiency: The EISA requires federal agencies to reduce energy intensity by 3% per year, or 30% by FY 2015 (compared to an FY 2003 baseline):
        9% by FY 2008
        12% by FY 2009
        15% by FY 2010
        18% by FY 2011
        21% by FY 2012
        24% by FY 2013
        27% by FY 2014
        30% by FY 2015

- EISA Section 432—Management of Energy and Water Efficiency in Federal Buildings—establishes a framework for facility project management and benchmarking. Under this requirement, agencies must identify all "covered facilities" that constitute at least 75% of the agency's facility energy use, and an energy manager must be designated at each of these covered facilities.

  Each facility energy manager is responsible for completing comprehensive energy and water evaluations of 25% of covered facilities each year; implementing identified energy and water efficiency measures within years of these evaluations, and following up on these measures.

- Sustainable Buildings: New or renovated agency buildings must be designed to reduce fossil fuel-generated energy consumption by 55% by FY 2010 and 100% by FY 2030 (compared to an FY 2003 baseline).

  New commercial buildings must reach a long-term "zero net–energy" goal by FY 2025; existing commercial buildings must also reach "zero net–energy" by FY 2050. By 2010, federal agencies will be required to lease space that has earned the ENERGY STAR® label in the most recent year.

■ Renewable Energy: At least 30% of hot water demand in new or renovated federal buildings must come from solar hot water heating, if life-cycle cost-effective. Existing buildings with minor renovations must incorporate the most energy-efficient designs, equipment, and controls.

Executive Order 13423: Signed in January 2007, E.O. 13423, "Strengthening Federal Environmental, Energy, and Transportation Management," mandates new sustainability goals for the federal government that match or exceed previous statutory and E.O. requirements. The new order consolidates and strengthens the sustainable practices of five existing E.O.s, including E.O. 13123, which are all now revoked.

■ Energy Efficiency: E.O. 13423 mandates an annual 3% reduction and cumulative 30% reduction in energy intensity by 2015 (compared to an FY 2003 baseline):

> 3% by FY 2006
> 6% by FY 2007
> 9% by FY 2008
> 12% by FY 2009
> 15% by FY 2010
> 18% by FY 2011
> 21% by FY 2012
> 24% by FY 2013
> 27% by FY 2014
> 30% by FY 2015

This requirement seeks to reduce greenhouse gas emissions and achieve in 10 years the same level of energy efficiency improvement that federal agencies achieved in the past 20 years. E.O. 13423 is 50% more stringent than the requirements of the Energy Policy Act of 2005 (EPAct 2005)

■ Renewable Energy: At least 50% of current renewable energy purchases must come from new renewable sources (defined as those that began operation after January 1, 1999). While EPAct 2005 set a renewable energy goal, E.O. 13423 is the first legislation to require a percentage of renewable energy to come from new sources.

EPAct 2005. Signed into law in August 2005, EPAct 2005 requires federal agencies to reduce energy intensity every year in their facilities by 2% per year beginning in FY 2006, up to a cumulative 20% reduction by the end of FY 2015 (compared to an FY 2003 baseline):

> 2% by FY 2006
> 4% by FY 2007
> 6% by FY 2008
> 8% by FY 2009
> 10% by FY 2010
> 12% by FY 2011
> 14% by FY 2012
> 16% by FY 2013
> 18% by FY 2014
> 20% by FY 2015

6.4.2.2 Green Buildings

Designing, constructing, operating, and maintaining buildings involves large amounts of energy, water, and other resources, and creates significant amounts of waste. The building process also impacts the environment and ecosystem surrounding the building site. Even after buildings are constructed, occupants and building managers face a host of challenges as they try to maintain a healthy, efficient, and productive work environment. The EPA addresses these challenges by promoting energy and resource efficiency, waste reduction and pollution prevention practices, indoor air quality standards, and other environmental initiatives for both new construction and existing buildings. Typical green buildings often incorporate the following features:

– Careful site selection to minimize impacts on the surrounding environment and increase alternative transportation options
– Energy conservation to ensure efficient use of natural resources and reduced utility bills
– Water conservation to ensure maximum efficiency and reduced utility bills
– Responsible stormwater management to limit disruption of natural watershed functions and reduce the environmental impacts of stormwater runoff
– Waste reduction, recycling, and use of "green" building materials

- Improved indoor air quality through the use of low volatile organic compound products and careful ventilation practices during construction and renovation
- Reduced urban heat island effect to avoid altering the surrounding air temperatures relative to nearby rural and natural areas

6.4.2.3 Green House Gases

Thanks to a combination of energy efficiency projects and extensive green power purchases, in FY 2008, the EPA reduced the net GHG emissions of its most energy-intensive facilities by nearly 70% compared to an FY 2003 baseline. In FY 2003, the EPA's baseline GHG emissions associated with energy consumption at its reporting facilities totaled 139,007 Metric Tons of Carbon Dioxide equivalent ($MTCO_2e$). As a direct result of agency-wide energy efficiency improvements since FY 2003, these emissions in FY 2008 totaled 122,732 $MTCO_2e$—a reduction of 16,275 $MTCO_2e$ (nearly 12%) from the agency's FY 2003 baseline.

6.4.2.4 Green Information Technology

By now we know that "Green IT" brings environmental thinking to the world of information technology. The EPA encourages the industry to consider how the environment might be affected by IT-related decisions. Green IT encompasses everything from incorporating fewer pollutants in the manufacture of technology, to examining the energy and environmental footprint of the physical buildings that house computer enterprise operations, to recycling older technologies, and much more. IT products and processes may contain or use some substances of potential concern to human health and the environment. Lead, mercury, cadmium, and brominated flame retardants are among the substances of concern in electronics.

By embracing the EPA's environmentally friendly programs for information technology, one can

- Reduce unnecessary energy use and the carbon footprint while saving money
- Select technologies that are environmentally preferable
- Recycle unwanted equipment to conserve natural resources

The overarching areas of Green IT, as considered by the EPA, include Designing Green IT, Buying Green IT, and Using Green IT

Designing Green IT: The Design for the Environment program has engaged with leaders in the electronics industry to explore current practices and potentially safer alternatives in various areas. The EPA works in partnership with a broad range of stakeholders to inform substitution to safer chemicals in such areas as

- Computer displays

- Lead-free solder
- Flame retardants in printed circuit boards
- Printed wiring board

Buying Green IT: ENERGY STAR® is a joint program of the EPA and the Department of Energy (DOE) to help save money, reduce wasteful consumption, and protect the environment through energy-efficient products and practices. The goal of the ENERGY STAR program is to develop performance-based specifications that determine the most efficient products in a particular category. Products that meet the specifications earn the ENERGY STAR label. The EPA has added the trustworthy ENERGY STAR label to more than 50 product categories for the home and office, including information technology:

- Computers
- Monitors
- Printers, scanners, and all-in-ones (imaging products)

For computers and monitors, the penetration of ENERGY STAR–compliant devices in the United States is approximately 97%; the penetration of ENERGY STAR–compliant TV sets is approximately 46% (compliance is expected to increase with the replacement of older CRT-based televisions with newer televisions). ENERGY STAR–qualified power adapters are, on average, 30% more efficient than conventional models, and an ENERGY STAR–compliant set-top box can provide a 30% energy savings compared with a set-top box that is not ENERGY STAR–compliant; hence, it follows that it is desirable for institutions and users to employ STAR-rated products.

Environmentally Preferable Purchasing (EPP) can help green vendors, businesses large and small, and consumers find and evaluate information about green products and services such as electronics and calculate the costs and benefits of their choices.

The Electronic Product Environmental Assessment Tool (EPEAT) is an easy-to-use, on-line tool helping institutional purchasers select and compare computer desktops, laptops, and monitors based on their environmental attributes. EPEAT is managed by the Green Electronics Council and was developed through a grant. The Council's web site provides guidance for purchasers and manufacturers and hosts the database of EPEAT-registered products. EPEAT-registered computer desktops, laptops, and monitors must meet an environmental performance standard for electronic products.

Using Green IT: The EPA has undertaken a number of initiatives in the recent past to enable end users to green their IT use, from desktop to data center. These initiatives provide tools to data center operators as they improve the energy efficiency of their facilities. One can

- Enable a personal computer's or an enterprise computer's power management mode with ENERGY STAR and save $25 to $75 per computer per year
- Work with the EPA and IT industry stakeholders to finalize the specification for ENERGY STAR–qualified enterprise servers
- Collaborate on the development of an energy performance rating system for data center infrastructure through involvement in the National Data Center Energy Efficiency Information Program

The DOE's Save Energy Now program is working with U.S. computer data centers to reduce their energy consumption. It is a national initiative of the Industrial Technologies Program (ITP) to drive a reduction of 25% or more in industrial energy intensity in 10 years.

Electronic Product Environmental Assessment Tool (EPEAT)

Launched in 2006, EPEAT was developed in response to growing demand by institutional purchasers for an easy-to-use evaluation tool enabling them to compare electronic products based on environmental performance, in addition to cost and performance considerations. Creation of EPEAT was guided by electronics manufacturers' expressed need for clear, consistent procurement criteria. Over 600 products manufactured by 20 manufacturers (Apple; CTL Corp.; Dell, Inc.; Fujitsu Computer Systems Corp.; Gateway, Inc.; Hewlett Packard; Lenovo; LG Electronics USA Incorporated; Mind Computer Products; MPC Computers LLC; NEC Display Solutions, Inc.; Northern Micro, Inc.; Panasonic; Phillips Electronics Ltd.; Samsung Electronics America; Sona Computer, Inc.; Sony Electronics Inc.; Toshiba; and ViewSonic Corp.) have EPEAT registered and listed on the EPEAT Product Registry Web page. EPEAT was developed using a grant from the EPA and is managed by the Green Electronics Council (GEC). It is dedicated to informing purchasers of the environmental criteria of electronic products. The GEC's EPEAT web site provides guidance for purchasers and manufacturers and hosts the database of EPEAT-registered products. Purchases of EPEAT-registered laptops, desktops, and monitors over conventional products had the following benefits in 2007 (the last report on this as of press time):

- Reduce the use of primary materials by 75.5 million metric tons, equivalent to the weight of more than 585 million refrigerators.
- Reduce use of toxic materials, including mercury, by 3,220 metric tons, equivalent to the weight of 1.6 million bricks.
- Eliminate use of enough mercury to fill 482,381 household fever thermometers.
- Avoid the disposal of 124,000 metric tons of hazardous waste, equivalent to the weight of 62 million bricks.

In addition, due to EPEAT's requirement that registered products meet ENERGY STAR specifications, these products will consume less energy throughout their useful life, resulting in

- Savings of 42.2 billion kWh of electricity—enough to power 3.7 million U.S. homes for a year
- Elimination of the release of 174 million metric tons of air emissions (including greenhouse gas emissions) and almost 365 thousand metric tons of water pollutant emissions
- Reduction of 3.31 million metric tons of carbon equivalent (MTCE) greenhouse gas emissions—equivalent to removing 2,630,000 U.S. cars from the road for a year.

Green Power Partnership

The EPA's Green Power Partnership is a voluntary program that supports the organizational procurement of green power by offering expert advice, technical support, tools, and resources. Partnering with the EPA can help an organization lower the transaction costs of buying green power, reduce its carbon footprint, and communicate its leadership to key stakeholders. Green power is electricity produced from a subset of renewable resources, such as solar, wind, geothermal, biomass, and low-impact hydro. Buying green power is one of the easiest and most effective ways to improve an organization's environmental performance. The Partnership currently has hundreds of partner organizations voluntarily purchasing billions of kilowatt-hours of green power annually. Partners include a wide variety of leading organizations such as Fortune 500 companies; small and medium-sized businesses; local, state, and federal governments; and colleges and universities.

The organization offers a *Guide to Purchasing Green Power,* which is intended for organizations that are considering the merits of buying green power as well as those that have decided to buy it and want help doing so. The *Guide* was written for a broad audience, including businesses, government agencies, universities, and all organizations wanting to diversify their energy supply and reduce the environmental impact of their electricity use. The *Guide* provides an overview of green power markets and describes the necessary steps to buying green power. There are three types of green power products, as defined by the EPA: renewable electricity, renewable energy certificates, and on-site renewable generation. Renewable electricity is generated using renewable energy resources and delivered through the utility grid; renewable energy certificates (RECs) represent the environmental, social, and other positive attributes of power generated by renewable resources; and on-site renewable generation is electricity generated using renewable energy resources at the end-user's facility.

Standards for Green Products and Services

The EPA works with a variety of nongovernmental standards developers to promote the development of voluntary consensus standards for environmentally preferable goods and services. For example, the Environmentally Preferable Purchasing (EPP) "Database of Environmental Information for Products and Services" contains standards and specifications developed by governmental and nongovernmental organizations on a wide range of products and services. Some, but not all, of these standards were developed through voluntary consensus processes. Standards specific to the design and operation of green buildings are incorporated in the *Federal Green Construction Guide for Specifiers.*

6.5 European Commission Joint Research Centre (JRC)

The European Commission Joint Research Centre (JRC) has released a code of conduct for data centers, "Code of Conduct on Data Centres Energy Efficiency," as a response to the energy challenges that the European Union (EU) faces. With this "Code of Conduct…," the European Union is seeking to address both issues of the reduction of carbon emissions in line with the 1997 Kyoto Protocol and managing energy consumption. The "Code of Conduct…" was created in response to increasing energy consumption in data centers and the need to reduce the related environmental, economic, and energy supply security impacts. Electricity consumed in data centers, including enterprise servers, ICT equipment, cooling equipment, and power equipment, is expected to contribute substantially to the electricity consumed in the EU commercial sector in the near future. Western European electricity consumption of 56 TWh per year was estimated for the year 2007 and is projected to increase to 104 TWh per year by 2020.

Version 1.0 of the "Code of Conduct…" was released in October 2008. The aim of the "Code of Conduct…" is to encourage companies with data centers to cost-effectively reduce energy consumption while ensuring business objectives are met. Adherence to the code is by way of voluntary acceptance and implementation of its principles. The aim is to inform and stimulate data center operators and owners to reduce energy consumption in a cost-effective manner without hampering the mission-critical function of data centers. The "Code of Conduct…" aims to achieve this by improving understanding of energy demand within the data center, raising awareness, and recommending energy-efficient best practice and targets [EUC200801]. To achieve its aims, the "Code …" establishes a basis of recognized best practice and a framework of operation for the design, operation, maintenance, and retiring of data centers. The recommended best practice associated with the "Code of Conduct…" has been grouped into the following distinct areas:

■ Data center utilization, management, and planning

- IT equipment and services
- Cooling
- Data center power equipment
- Other data center equipment
- Data center building
- Monitoring

The "Code of Conduct…" has both an equipment- and system-level scope. At the equipment level, it covers typical equipment used within data centers required to provide data, Internet, and communication services. This includes all energy-using equipment within the data center, such as IT equipment (e.g., rack optimized and non-rack optimized enterprise servers, blade servers, storage, and networking equipment); cooling equipment (e.g., computer room air-conditioning units) and power equipment (e.g., uninterruptible power supplies and power distributions units); and miscellaneous equipment (e.g., lighting). At the system level, the "Code of Conduct…" proposes actions that optimize equipment interaction and the system design (e.g., improved cooling design, correct sizing of cooling, correct air management and temperature settings, correct selection of power distribution) to minimize overall energy consumption.

The document states that participants of the "Code of Conduct…" should endeavor and make all reasonable efforts to ensure

A.1. Data centers are designed so as to minimize energy consumption whilst not impacting business performance;

A.2. Data centre equipment is designed to allow the optimization of energy efficiency while meeting the operational or service targets anticipated;

A.3. Data centers are designed to allow regular and periodic energy monitoring;

A.4. Energy consumption of data centers is monitored; where data centers are part of larger facilities or buildings, the monitoring of the specific data centre consumption may entail the use of additional energy and power metering equipment;

A.5. Data centers and their equipment are designed, specified and procured on the basis of optimizing the TCO within the requirements for reliability, availability and serviceability;

A.6. When the Energy Star program has set specification for servers and other IT equipment, these specifications should be followed by Participants when procuring equipment. For UPS the specifications of the European Code of Conduct on Energy Efficiency and Quality of AC Uninterruptible Power Systems (UPS);

A.7. Data centers should be designed to minimize the energy used, if any, to remove heat from the facility.

As a purely voluntary scheme, there are no penalties incurred if compliance with the code is not achieved, and resignation from the scheme can be taken at any time. A similar approach exists for manufacturers, vendors, consultants, and other interested organizations that have embraced the code. Endorser status applies to organizations that are implementing its ethos into products, or services that are aimed at helping participants to achieve their obligations under the code. Being an endorser is also voluntary, but as participants and "Green-IT"-oriented companies look for vendors with similar aspirations, commercial benefits of being an endorser will become apparent [DIM200901].

By press time, only two telecom companies had signed the "Code of Conduct…" (i.e., Swisscom and TDC Services). Notwithstanding this, several companies are already referring to the energy consumption levels suggested by the code. The ETNO (discussed next) strongly encourages member companies to sign the code and abide by its principles.

6.6 European Telecommunications Network Operators' Association (ETNO)*

The European Telecommunications Network Operators' Association (ETNO) was established in 1992 and has become the principal policy group for European electronic communications network operators. The ETNO's primary purpose is to establish a dialogue between its member companies and decision-makers and other actors involved in the development of the European Information Society, to the benefit of users. The ETNO strongly encourages member companies to sign the EU "Code of Conduct on Data Centres Energy Efficiency" and to abide by its principles.

The ETNO advocates that it is important for European business and industry to seize the opportunity for sustainable innovation, through the use of "clean" technologies and actively supporting climate change actions. Energy consumption is the single largest environmental impact of all telecommunications operators and therefore all ETNO member companies. The bulk of ETNO members' energy use is from electricity consumption, which is used to power, and cool, their communication networks. Therefore it is the responsibility of all operators, through the signing of the ETNO charter, to ensure that energy consumption is kept to a minimum and to seek environmentally friendly alternatives. As noted in passing in Chapter 1, EU legislation is now forcing business to ensure that the design of electrical and electronic equipment minimizes the impact on the environment during its life cycle.

In turn, ETNO member companies seek to ensure that equipment and usage is as energy efficient as possible. This ultimately means removing the least efficient equipment from the networks and encouraging competition between our suppliers to achieve improved products and network equipment and, finally, making it easier for procurement teams to choose the best equipment and suppliers. Ways to do so

---

* This discussion is based in its entirety on ETNO materials, including [ETNO200801].

include minimum standards, voluntary agreements, procurement policy, and better information on product performance.

The energy task team is a sub-group of the ETNO Sustainability Working Group (WG). The task team was established during a meeting of the ETNO WG on Sustainability in June 2004. The objectives are

1. To ensure efficient energy utilization and the reduction of environmental impacts through improved energy management;
2. To contribute to national and global efforts to reduce GHG emissions;
3. To provide opportunities to market environmental practice and demonstrate the viability of voluntary actions;
4. To share knowledge and best practice among all the Association's members;
5. To benchmark among the members and look for best practice;
6. To provide all members with a recommended Energy policy;
7. To put pressure on suppliers with a Code of Conduct;
8. To carry out innovative pilots.

Although telecommunications is perceived as an environmentally friendly technology, in reality carriers and service providers do use large quantities of energy and exert a significant impact on the environment. The ETNO believes that a recommended energy policy will contribute to reducing this impact. They suggest the following (among other initiatives):

- Monitor and measure all types of energy consumption effectively (electricity, gas, and oil) in order to identify areas for improvement and set quantitative improvement targets.
- Identify, monitor, and measure all major GHG emissions from direct and indirect activities related to running a telecommunications business.
- Improve energy and emission efficiency in mainstream processes (networks, buildings, mobility, overhead, etc.) and reduce energy consumption and emissions where practicable relative to business growth.
- Design energy efficiency into all new equipment and services, including terminals, fixed/mobile network elements, buildings, offices, etc.
- Increase, where possible, the use of energy from renewable sources and give preference to energy suppliers with less GHG emissions per energy unit.
- Incorporate energy efficiency criteria in purchasing, supplier selection, and subcontracting processes, and work in partnership with suppliers to minimize equipment energy consumption.
- Educate employees, customers, and partners about energy issues, the impact operators have as major organizations, and what they can do to help.
- Comply with all applicable legal requirements, regulations, and standards.

Upon completion of the work with the European Union's JRC (as discussed above) on the broadband code of conduct, members of the ETNO team started work on two other codes of conduct: the "European Code of Conduct for Digital TV" and the "European Code of Conduct for Data Centres." The "Code of Conduct for Digital TV" is aimed at the design, specification, and procurement of digital TV services equipment. Discussions cover both sophisticated set-top boxes for subscriber services over satellite, cable, Digital Subscriber Line (DSL), and terrestrial, as well as simple digital-to-analog converter boxes for free to air digital transmission. The main aim is to create a broad and shared consensus on the Code of Conduct consumption limits and power management. The "Code of Conduct for Data Centres" is aimed at opportunities to address increasing energy consumption in data centers and the need to reduce the related environmental, economic, and energy supply security impacts. The aim is to coordinate activities by manufacturers, vendors, consultants, utilities, and data center operators/owners to reduce the electricity consumption in a cost-effective manner without hampering the mission and critical function of data centers. The code of conduct aims to achieve this by improving understanding of power demand within the data center, raising awareness, and recommending energy-efficient best practice and targets.

As noted elsewhere in this text, one of the most significant costs for a majority of carriers is the energy bill; energy efficiency can deliver reductions in cost and CO_2 emissions. To address the issues, the ETNO has adopted a benchmark for energy consumption, savings, and cooling for operators. The highlights of the benchmark are discussed next.

Energy consumption and management:

■ Operators should define energy efficiency parameters, relationship with traffic load, and classification for exchanges for the dimensioning of power supply.
■ Operators should define a phase-out plan: phasing out old technology when replaced with new, turning off surplus equipment.
■ Operators should define energy consumption models: developments and variations in the business cases, a model related to network components, and the influence in the choice of suppliers.
■ Operators should develop interest in forming a consortium in order to reduce energy consumption of new equipment.

Environment and energy saving:

■ Operators should define measures to achieve energy savings.
■ Operators should define a corporate environmental policy.
■ Operators should assess the effect of energy-savings.
■ Operators should define what percentage of "brown" versus "green" energy they use.
■ Operators should assess their own generated green energy (if that is the case) and assess the (corporate) policy and type of generated energy.

- Operators should assess the compensatory measures such as CO2 certificates related to the Kyoto Protocol.
- Operators should define low energy consumption and low energy conditioning as a subject of conversation by product and suppliers choice—even to the point of including that in RFPs.

Power and cooling:

- Operators should define availability and reliability at interface A* (in accordance, for example, with European Telecommunications Standards Institute [ETSI] guidelines).
- Operators should define the use of hardware concepts, backup time, redundancy, use of elements for powering and cooling, and policy changes for the future.
- Operators should define differentiation in service and quality to different operators/operator groups.
- Operators should define grounding in accordance with regulatory guidelines (e.g., ETSI guidelines).
- Operators should define design criteria for different types of cooling and heat generation per square meter.
- Operators should assess fresh air values for room temperatures in relation to outside air.
- Operators should assess alarm settings and failure management, managing of system alarms, and remote parameter settings.
- Operators should assess the influence of working conditions on room temperatures and use of mobile cooling/ventilation units.

Additionally, the World Wide Fund for Nature (WWF, formerly the World Wildlife Fund) and ETNO recently launched a joint project to highlight the potential impact of a wider use of ICT on greenhouse gas emissions based on already existing projects. As part of their partnership, the ETNO and WWF published a roadmap that sets concrete targets to ensure that ICT is fully taken into consideration by policy makers in climate change strategies, both at EU and national levels.

6.7 European Telecommunications Standards Institute (ETSI)

The ETSI develops ICT standards for wireline, wireless, broadcast, and Internet networks. It produces globally applicable ICT standards, including fixed, mobile,

---

* Keep in mind that in traditional telephony, the carrier will provide powering across some interfaces, particularly across the analog telephone plant interface to the user.

radio, converged, aeronautical, broadcast, and IP technologies. An independent, not-for-profit association, the ETSI's 766 member companies (at press time) come from 63 countries worldwide. The ETSI is officially recognized by the European Commission as a European Standards Organization.

The ETSI aims at showing how ongoing and planned ETSI standardization activities contribute to reducing the negative impact on climate change and to ensuring a balance between ICT standardization, regulation, and innovative solutions. The ETSI's Green Agenda was a seen as a strategic activity that started in 2008. The short-term goal was to adopt the International Organization for Standardization (ISO) 14001 and 14004 standards, and develop a green checklist for all work on standards.

The Technical Committee on Environmental Engineering (ETSI TC/EE) has been working on issues related to the reduction of energy consumption in telecommunications equipment and related infrastructure. The ETSI recently published TS 102 533 (2008-06): "Measurement Methods and Limits for Energy Consumption in Broadband Telecommunication Networks Equipment," a document that establishes an energy consumption measurement method for broadband telecommunication network equipment including a target energy consumption value for wired broadband equipment including Asymmetric DSL (ADSL) and Very-high-bit-rate Digital Subscriber Line (VDSL). Other recent efforts include

- ICT energy consumption and global energy impact assessment methods (DTR/EE-00002 Work Item TR 102 530, "The Reduction of Energy Consumption in Telecommunications Equipment and Related Infrastructure").
- The use of alternative energy sources in telecommunication installations ("DTR/EE-00004 Work Item TR 102 532, "The Use of Alternative Energy Sources in Telecommunication Installations").
- Reverse powering of small access network node by end-user equipment.
- Power considerations for outdoor equipment (DTS/EE-00006 Work Item, "Environmental Consideration for Equipment Installed in Outdoor Location").
- DSL power optimization (DTR/ATTM-06002 Work Item: "Transmission and Multiplexing (TM): Power Optimization for xDSL Transceivers").
- Energy efficiency of wireless access network equipment (DTS/EE-00007: "Environmental Engineering (EE): Energy Efficiency of Wireless Access Network Equipment"). This work establishes wireless access network energy efficiency metrics, and defines efficiency parameters and measurement methods for wireless access network equipment. In the first phase, GSM/EDGE, WCDMA/HSPA, and WiMAX are addressed.
- Energy efficiency in broadband networks (DTS/EE-00005 Work Item; TS 102 533, "Environmental Engineering (EE): Energy Consumption in BB Telecom Network Equipment").
- Methods to assess the environmental impacts of ICT (DTR/EE-00008: "Environmental Impact Assessment of ICT including the Positive Impact by using ICT Services"). This work defines the methods to assess the

environmental impacts of ICTs, which have two aspects: (1) negative impact caused by the energy consumptions or CO_2 emissions of operators of ICT equipment and sites, including telecom networks, users terminals, data centers for residential and business services; and (2) positive impact caused by energy saving or CO_2 emission saving using ICT services.

6.8 International Organization for Standardization (ISO)*

The ISO (International Organization for Standardization) is the world's largest developer and publisher of international standards. It is a network of the national standards institutes of 162 countries, one member per country, with a Central Secretariat in Geneva, Switzerland, who coordinates the system. The ISO is a nongovernmental organization that forms a bridge between the public and private sectors. On the one hand, many of its member institutes are part of the governmental structure of their countries, or are mandated by their government. On the other hand, other members have their roots uniquely in the private sector, having been set up by national partnerships of industry associations. Therefore, the ISO enables a consensus to be reached on solutions that meet both the requirements of business and the broader needs of society.

The ISO has started to work on green issues in the recent past. It has developed a three-part standard, ISO 14064:2006, to provide a set of globally harmonized, unambiguous, and verifiable requirements or specifications; to provide government and industry with an integrated set of tools for programs aimed at reducing greenhouse gas emissions; as well as for emissions trading. ISO 14064 and ISO 14065:2007 provide an internationally agreed framework for measuring GHG emissions and verifying claims made about them so that "a ton of carbon is always a ton of carbon." The ISO thus supports programs to reduce GHG emissions and also emissions trading programs.

ISO 14064 comprises three standards covering GHG emission quantification, monitoring, and reporting, and for the validation and verification of claims made about GHG emissions. ISO 14064 was developed as a solution to the problems posed by the fact that governments, business corporations, and voluntary initiatives were using a number of approaches to account for organization- (and project-) level GHG emissions and removals with no generally accepted validation or verification protocols.

■ ISO 14064-1:2006: Greenhouse gases—Part 1: Specification with guidance at the organization level for quantification and reporting of greenhouse gas emissions and removals

---

* This section is based in its entirety on ISO materials.

■ ISO 14064-2:2006: Greenhouse gases—Part 2: Specification with guidance at the project level for quantification, monitoring and reporting of greenhouse gas emission reductions or removal enhancements
■ ISO 14064-3:2006: Greenhouse gases—Part 3: Specification with guidance for the validation and verification of greenhouse gas assertions

The complementary standard ISO 14065:2007 provides requirements for bodies that undertake GHG validation or verification using ISO 14064 or other relevant standards or specifications. The aim of GHG validation or verification bodies is to give confidence to parties that rely on a GHG assertion or claim (e.g., regulators or investors) that the bodies providing the declarations are competent to do so, and have systems in place to manage impartially and to provide the required level of assurance on a consistent basis.

■ ISO 14065:2007: Greenhouse gases — Requirements for greenhouse gas validation and verification bodies for use in accreditation or other forms of recognition

6.9 International Telecommunication Union (ITU)*

The International Telecommunication Union – Telecommunications (ITU-T) is a standards-making agency of the United Nations focused on telecommunications. It publishes a gamut of standards to support global interworking, and is the leading agency for information and communication technology issues and the global focal point for governments and the private sector in developing networks and services.

For over 145 years, the ITU has coordinated the shared global use of the radio spectrum, promoted international cooperation in assigning satellite orbits, worked to improve telecommunication infrastructure in the developing world, established the worldwide standards that foster seamless interconnection of a vast range of communications systems, and addressed the global challenges of our times, such as mitigating climate change and strengthening cyber-security. The ITU also organizes worldwide and regional exhibitions and forums, such as *ITU TELECOM WORLD*, bringing together the representatives of government and the telecommunications and ICT industry to exchange ideas, knowledge, and technology for the benefit of the global community, and in particular the developing world. The ITU covers technologies and services such telephony, optical networks, broadband Internet, wireless, aeronautical/maritime navigation, radio astronomy, satellite-based meteorology, convergence in fixed-mobile phone, Internet access, data, voice, and TV broadcasting. The ITU is based in Geneva, Switzerland, and its membership includes 191 member states and more than 700 members.

---

* This section is based in its entirety on ITU materials.

The result of a cooperative effort that sees leading industry players put their competitive rivalries aside in favor of building global consensus on new technologies, ITU-T standards (known as Recommendations) are the underpinning of modern information and communication networks; now serve as the lifeblood of virtually every economic activity. For example, for manufacturers, they facilitate access to global markets and allow for economies of scale in production and distribution, safe in the knowledge that ITU-T-compliant systems will work anywhere in the world: for purchasers from telcos to multinational companies to ordinary consumers, they provide assurances that equipment will integrate effortlessly with other installed systems.

Active in the area of climate change for well over a decade, the ITU is now making climate change a key priority, with strategies to

- Reduce the environmental impact through
  - The creation of a standard methodology for calculating carbon footprint
  - The promotion of Next Generation Networks (NGNs) (reducing power consumption by up to 40%)
  - Online versus print publications
- Harness the power of ICTs through
  - Remote collaboration
  - Intelligent transport systems
  - Sensor-based networks based on RFID and telemetry
- Monitor climate change by
  - Conducting and managing studies on remote sensing
- Providing key climate data via radio-based applications

To solve today's climate problems, the ITU is expected to work closely with its membership to lead efforts in achieving a climate-neutral ICT industry. Dr. Hamadoun I. Touré recently made the following Declaration on Climate Change (partial statement):

> "... There is a strong role for ITU in standards for energy efficiency of the ICT equipment on which our digital economy depends. ITU has always taken the lead in setting high standards for telecommunications and ICTs, and this is another key area in which ITU can make a real difference.
>
> The Resolution passed recently at the World Telecommunication Standardization Assembly (WTSA) in Johannesburg encourages ITU Member States to work towards reductions in Greenhouse Gas (GHG) emissions arising from the use of ICTs, in line with the UN Framework Convention on Climate Change. ITU aims to achieve climate neutrality for its operations within three years, and ITU is at the forefront of this progress compared with many other international organizations.
>
> In the global effort to combat climate change, ITU is continuing to help developing countries to mitigate the effects of climate change,

including the use of emergency telecommunications and alerting systems for disaster relief. ITU, in collaboration with its membership, is identifying the necessary radio-frequency spectrum for climate monitoring and disaster prediction, detection and relief, including a promising cooperation with the World Meteorological Organization (WMO) in the field of remote-sensing applications.

ITU will continue to join efforts in the context of the UN system, in order to "deliver as one" with a principal focus on ICTs and climate change. In 2000, UN Members adopted the Millennium Declaration as a renewed commitment to human development, including the eight Millennium Development Goals (MDGs). However, climate change impacts will tend to offset progress being made to meet the MDGs by 2015, so it is crucial to empower developing countries by facilitating their access to the ICTs needed for climate change adaptation and disaster risk reduction…"

Within the UN system, the ITU has unique competence in the ICT sector, making its work relevant to nearly all aspects of developing a system-wide approach to this issue. The ITU sees itself as being able to contribute to nearly all of the main pillars of work under the Bali framework for negotiations, that is, science and data monitoring, adaptation, mitigation, and technology. At the same time, the ITU must reach out to its membership to assist them in combating climate change and adapting to it, and will engage more fully with other organizations active in these efforts. The draft strategy builds on existing strengths within the ITU relevant to climate change and identifies new opportunities for climate-relevant activities. It proposes four main objectives for an ITU strategy and general orientations to achieve those objectives:

■ *Objective 1: Develop a Knowledge Base and Repository on the Relation between ICTs and Climate Change:* Promote a focused approach to development of product and services in areas where ICTs can readily contribute to reductions in GHG emissions, including more standardized power supplies and batteries, smart devices and buildings, new low-consumption devices, research and development on consumption and power supplies, use of ICTs in travel management, and paperless meetings.
  – Conduct a systematic review of existing ITU treaties, resolutions, and recommendations in the light of climate change, and identify requirements for future work.
  – Conduct and foster further research into the relationship between ICTs and energy efficiency and issue appropriate materials (e.g., a handbook on ICTs and their impact on climate change, a national e-environment toolkit), and organize meetings/symposia on this issue.
  – Participate actively in major meetings on climate change.

- Disseminate information on relevant ICT success stories and best practices, through the ITU web site, handbooks, toolkits, etc.
■ *Objective 2: Position ITU as a strategic leader on ICTs and Climate Change:* Develop, though the membership, a normative framework for addressing the issue of ICTs and climate change. This can include adopting draft resolutions at WTSA-08 on ICT standardization requirements for combating climate change, follow-up to WTDC-06 outcomes, and future resolutions at WTDC-10 to foster the use and disposal of environmentally sound ICTs and PP-10.
  - Implement existing ITU instruments, such as Plenipotentiary Res. 35 (Kyoto, 1994), relevant to climate change, as well as relevant WRC, RRC, WTDC-06, and PP-06 Resolutions.
  - Develop strategic partnerships with Member States, Sector Members, and other organizations (e.g., GeSI, WEF, ETNO, WWF, UNEP, WMO) with an interest in using ICTs to combat climate change.
  - Encourage more Member States to sign and ratify the Tampere Convention on Emergency Telecommunications.
  - Promote the positive effect of introducing new uses for ICTs (e.g., reduction of power consumption, reduction of atmosphere/ionosphere heating by powerful transmitters, videoconferencing, and travel management) in combating climate change at ITU seminars and workshops.
  - Develop and implement technical cooperation projects to assist developing countries to use ICTs to adapt to and mitigate the effects of climate change.
  - In partnership with one or more developing countries, develop and submit projects under the Clean Development Mechanism of the Kyoto Protocol for reducing carbon emissions through the use of ICTs.
■ *Objective 3: Promote a global understanding of the relation between ICTs and climate change through international fora and agreements:* Take an active role in efforts to deliver a ONE UN approach to climate change by contributing to the debate at the Chief Executives Board (CEB) and the High-Level Committee on Programs (HLCP) on climate change through appropriate inputs, and closely monitor their outputs and adopt ITU strategies accordingly.
  - Closely monitor ongoing global negotiations on climate change and participate actively at meetings planned under the Bali Roadmap and organize events to raise the visibility of ITU on this issue.
  - Take an active role in other UN inter-agency mechanisms dealing with climate change, including the Environmental Management Group.
  - Strengthen strategic partnerships with FAO, UNEP, WMO, IPCC, and other UN agencies; the World Bank; the European Commission; international and national agencies and organizations (e.g., meteorological agencies, Group on for Earth Observations (GEO), EUMETSAT, ESA, Space Frequency Coordination Group (SFCG), JAXA, NOAA, NASA, RSA, etc.); NGOs; and the private sector involved in combating climate change.

- Promote the link between ICTs and climate change at other intergovernmental meetings where the issue is raised, (e.g., the WSIS C7 facilitation meetings on e-science and e-environment).
- *Objective 4: Achieve a climate-neutral ITU within three years:* Appoint a focal point in general secretariat for a climate-neutral ITU to manage this campaign and lead a project team, to coordinate all climate change activities at the Union.
  - Engage all staff to generate ideas and initiatives toward a climate-neutral ITU, including encouraging teleworking.
  - Conduct a carbon audit of ITU's premises and activities (both internal and external) and intensify efforts to use ICTs to reduce the carbon footprint of at ITU in its internal operations and technical cooperation activities (e.g., consideration of low carbon technologies in ITU projects.
  - Seek approval from Council for a program of carbon offsets, if needed, to achieve carbon-neutral status, by proposing a strategy on acquisition of carbon offsets through ITU activities in country projects focusing on using ICTs for carbon emissions reduction (under the Clean Development Mechanisms of the Kyoto Protocol).
- Assist the membership, UN entities, and other organizations to use ICTs to become more energy efficient and to implement programs for the disposal of ICT components in an environmentally sound manner.

6.10 Institute of Electrical and Electronics Engineers (IEEE)*

Among other activities IEEE runs conferences related to green technologies and also has formed a Green Initiatives Ad Hoc Committee. The scope of the this Committee may be inward-looking by seeking to develop and recommend strategies for implementing environmentally friendly practices into the IEEE, including facilities, conferences and other meetings, and publications operations; however, the conferences have wider scope.

Through the conferences (e.g., *First International Workshop on Green Communications* held in conjunction with the *IEEE International Conference on Communications* in 2009), the IEEE is looking at energy-efficient network protocols and devices and at energy management with the goal to identify areas for energy saving in the access network, home networks, and end-user devices. Areas of interest include but are not limited to

- Energy-efficient protocols and protocol extensions
- Energy-efficient transmission technologies

---

* This section is based in its entirety on IEEE materials.

- Cross-layer optimizations
- Energy-efficient network device technology
- Energy-efficient switch and base station architectures
- Zero-power sleep mode
- Exploitation of passive network elements
- Energy management in communication networks
- Architectures and frameworks
- Hierarchical and distributed techniques
- Remote power management for terminals
- Harvesting distributed energy generation
- Measurement and profiling of energy consumption
- Instrumentation for energy consumption measurement
- Operator experiences
- Energy efficiency in specific networks

IEEE, together with the university research community, has an opportunity to weigh in on environmentally friendly alternatives and improvements to the traditional energy economy. Wind farms being built and connected to the grid infrastructure, and emerging photovoltaic technologies are approaching cost competitiveness; sustainable design in new construction is providing effective alternatives for energy efficiency; and Green Building certifications are becoming business discriminators.

The *2008 IEEE-USA Annual Meeting*, entitled "Green Engineering: A Push Toward Sustainability," was devoted to raising awareness of and spurring action toward sustainability within the realm of engineering. The concept of sustainable development gained mainstream popularity in 1987 when the World Commission on Environment and Development (WCED) released *Our Common Future*. The WCED defined sustainable development as "development that meets the needs of the present without compromising the ability of future generations to meet their own needs." According to the report, sustainable development is "a process of change in which the exploitation of resources, the direction of investments, the orientation of technological development, and institutional change are all in harmony and enhance both current and future potentials to meet human needs and aspirations" [MEY200801].

At the *2010 IEEE Green Technology Conference*, the IEEE was exploring the following issues:

- Current and emerging technologies in large- and small-scale renewable energy sources—wind, solar, water, geothermal, tidal, piezoelectric, and others—and their integration
- Current and emerging technologies in alternative fuel technologies—biomass fuel production, landfill gas reclamation, nuclear, fuel cells and batteries, and others

■ Current and emerging technologies in alternate vehicular power sources—fuel cells, hybrid electric, plug-in electric and hybrid, and other transportation alternatives

■ Technologies and their integration used for energy usage reduction and conservation—home automation and energy management, commercial building energy management, reduced energy usage lighting, and use and integration of multiple energy alternatives

■ Current and emerging technologies to facilitate the integration and management of renewable and reduced emission energy sources into the electric power grid

■ The social, economic, and political impacts of renewable energies, reduced emissions, and independent certifications

6.11 The Ministry of Economy, Trade and Industry (METI) of Japan and Other Initiatives

We mentioned in Chapter 1 that there is an interest in "savings in IT-related energy consumption" and also "achieving energy conservation through the use of IT." As a backdrop, the Ministry of Economy, Trade and Industry (METI) of Japan estimated that the electricity consumption of IT equipment in 2025 will be five times what it was in 2006, and that it will be twelve times greater in 2050. The "Green IT initiative" proposed in 2007 by METI seeks not only to promote energy saving products and technologies, but also an energy-saving system by introducing more efficient supply chains, including the use of IT. To that end, METI has engaged the New and Renewable Energy Subcommittee of the Advisory Committee for Natural Resources and Energy to deliberate on Japan's future energy policy, and elicited public comments on possible proposals. In 2008 after consideration of the opinions submitted, the subcommittee released the "New and Renewable Energy Subcommittee Interim Report." The document addresses the following issues [ACN200901]:

■ Expanding the introduction of new and renewable energy is significant in relation to not only Japan's energy and environmental policy, but also its industrial policy.

■ To meet the goal of increasing the installed capacity of solar power generation facilities about 20-fold by around 2020, a new system to buy electricity generated by solar power generation facilities should be implemented as soon as possible. For other new and renewable energies, an environment to maximize their introduction must be created by a comprehensive mix of regulations, assistance, and voluntary efforts, according to the characteristics of each energy source.

- Emphasis should also be placed on increasing public awareness (e.g., by publicizing leading-edge activities such as Next-Generation Energy Parks) and developing and promoting innovative technologies.
- Through these activities, Japan should pursue the ambitious target of increasing the renewable energy share of final energy consumption to 20% by around 2020.

Conventionally, the concept of Green IT focuses on how to reduce the energy consumption of IT equipment and systems (including data centers). An interesting aspect of METI's green IT initiatives is that it focuses on both "energy-saving IT equipment and systems" and the engenderment of an "energy-efficient society by IT use." The initial emphasis is on residential solar power generation; infrastructure optimization is a follow-on consideration. Japan has been a leader in developing energy-saving technologies since the 1970s. Recently, METI issued directives relevant to the Law on the Promotion of the Use of Nonfossil Energy Sources and Effective Use of Fossil Energy Materials by Energy Suppliers (Law No. 72 of 2009), including ministerial ordinances and notifications.

Under the auspices of the METI, the Japan Electronics and Information Technology Industries Association (JEITA) declared in 2008 that it would assemble the Green IT Promotion Council. The main activities of this council include (1) advocacy for innovative technologies and (2) statistical analyses of the impact of IT/electronics technologies on energy-saving measures. Pursuant to METI's vision, Green IT is envisioned as a tool to improve mechanisms to encourage environmentally friendly business activities through Supply Chain Management (SCM), all with the goal of reducing energy consumption.

The Green IT Project is designed to achieve a significant reduction in energy consumption for entire network systems, including not only IT devices, but also entire IT systems such as data centers. This program aims at encouraging manufacturers to develop more –energy-efficient equipment. METI's policies and research strategies are carried out by the New Energy and Industrial Technology Development Organization (NEDO). NEDO has three new research programs focusing on green IT projects (see Table 6.1): (1) nanobit technology, (2) large organic light-emitting diode (OLED) technology, and (3) green network system technology. In 2007, the Council for the Science and Technology Policy (CSTP) discussed the need for entire networks (including associated data centers, servers, and network equipment) to employ a mid- to long-term research approach for effective energy saving. NEDO's "Strategy for Energy-Saving Technology in 2007" was published to improve efficiency of energy consumption up to 30% by 2030 through taking priority over energy-saving technologies and high-speed and large-capacity network technology. Green network system technology reflects this concept [MYO200801].

Major Japanese ICT companies, including Hitachi, Fujitsu, NEC, and others, have all made recent strides in the green IT/green networks arena. See Table 6.2 for some relatively recent activities [MYO200801].

Table 6.1   METI's Green IT Projects and Allocated Budgets (FY 2008–FY 2012)

Project/Budget	Goal	Research Topics
1) Nanobit Technology for Ultra-High Density Magnetic Recording Budget: 909 million yen	Reduce power consumption per unit storage volume to 1/50 through HDD recording density innovation	Ultra-high density nanobit magnetic technology Ultra-high performance magnetic head (MH) Ultra-high precision nano Addressing technology HDD system technology
2) Large OLED Technology Budget: 668 million yen	Reduce power consumption of a 40-inch display panel to 40 W in the large display manufacturing process	Low-damage, large-area electrode formation technology Large-area transparent-sealing technology Large-area organic film technology Verification for large-display production
3) Green Network System Technology Budget: 1,283 million yen	Reduce network power consumption by 30% using traffic control technology Reduce data center power consumption by 30% using cooling technology, etc.	Basic technology of data center for optimization of energy use Technology for innovative energy-saving network and router

Note: 1 million yen ~U.S. $11,000.

Source: NEDO.

6.12 Network Equipment Building System (NEBS)*

Through NEBS™ (formerly Network Equipment Building Systems), Verizon has published a Technical Purchasing Requirement (TPR) document to provide a Telecommunications Equipment Energy Efficiency Rating (TEEER) methodology for procuring equipment assemblies used in Verizon's telecommunications networks

---

* This section is based in its entirety on Verizon materials.

Table 6.2   Green IT/Network Activities by Some Japanese ICT Companies

Company	Activity
Hitachi	As early as 2007, Hitachi assembled a special group, the Hitachi Group CEnO (Chief Environmental Strategy Officer), to assume all responsibility for environmental strategies in the company. The Environment Strategy Office was also established in 2008. The three main activities in the strategy office are (1) working on environmental business management and policy-making suggestions, (2) networking activities to tackle global environment issues, and (3) clarifying the direction of mid- and long-term technologies and business development. Hitachi reportedly intends to reduce CO_2 emissions resulting from the manufacture of its products to 100 million tons per year worldwide before 2025. By the year 2010, Hitachi proposed an agenda to boost the sales of environmentally friendly products to 6.6 trillion yen (which is twice the current level). Hitachi reportedly promotes a "Cool Centre 50" initiative, which has a target of reducing power consumption in its data centers by 50% over 5 years. Hitachi's key environmentally friendly technologies for power reduction are the so-called virtualization technologies for server operation and "Massive Array of Idle Disks (MAID)" for data storage, as well as small-size and large-capacity HDDs (Hard Disk Drives) for other devices. Hitachi estimates that MAID will be able to reduce power consumption by 40%.
NEC	As early as 2007, NEC proposed the "Real IT Cool Project" with which NEC reportedly intends to reduce power consumption through its energy-saving platform, energy-saving control software, and energy-saving facility service. In its "Real IT Cool Project," NEC decided to commit itself to a plan of developing technologies capable of realizing energy savings in its IT platforms, manufacturing and services. The project spans the following areas:

1. Energy-Saving Platform: To realize weight savings, space savings, and energy savings as a result of high-density packaging and cooling technologies.

2. Energy-Saving Control Software: To realize energy savings through the addition of special functions to IT equipment in data centers, machine rooms, and by use of management software.

3. Energy-Saving Facility Service: To provide total service of facility design and construction throughout survey on energy, design, set-up, and administer in data center and machine room. |

Table 6.2   Green IT/Network Activities by Some Japanese ICT Companies (Continued)

Fujitsu	In 2007, Fujitsu announced an ambitious project called "Green Policy Innovation," the objective of which reportedly was a 700-million-ton (or greater) reduction in CO2 emissions during the 4 years from 2007 to 2010. Specifically, Fujitsu aims to achieve a 0.76-million-ton reduction in the output of CO2 by its IT infrastructure and a 6.3-million-ton reduction in CO2 emissions resulting from IT utilization. Fujitsu has already made significant achievements with respect to energy-saving products. For example, it has developed technology for the visualization of environment-related activities within a company. The XML (eXtensible Mark-up Language)-based data of database, so-called the integrated DB of environment, centralizes environmental information in order to make it easy to analyze decision making and environmental management efficiently. The integrated environment database is also capable of visualizing, for instance, green procurement, design support, and life-cycle assessment.

[VER200901]. The TPR is the operative Verizon's document defining the minimum requirements for the purchase of telecommunications equipment for NEBS compliance. The intent is to foster the creation of more energy efficient telecommunications equipment by Verizon's supplier community, thereby reducing the energy requirements in Verizon networks. This methodology is applicable, but not limited to shelf, frame, and cabinet mounted DC powered network equipment to be installed in environmentally controlled environments and equipment located at the customer premises. AC-powered customer premises equipment (CPE) and AC-powered data center type equipment will also be covered in the TPR. TEEER and the types of equipment that the TPR document covers was discussed in Chapter 3.

Verizon has defined "Minimum TEEER Pass/Fail Requirements." The TEEER value is expected to meet the minimum TEEER value allowable as defined in Table 6.3. The pass/fail criteria are based on averages of typical equipment located in Verizon equipment rooms with an additional 20% improvement value. As of 1 January 2009, all equipment provided to Verizon had to meet the minimum TEEER values.

6.13 U.S. Department of Energy (DOE)*

The DOE supports a multitude of initiatives related to energy (also noted above when discussing the EPA). In the context of energy efficiency, one can identify the following groups and offices:

---

* This section is based in its entirety on DOE materials.

Table 6.3   Pass/Fail Criteria

Telecom Equipment Type	Minimum TEEER Allowable
Transport	7.54
Optical and video	7.54
Point-to-point microwave	5.75
Switch/router	7.67
Media gateway	6.54
Access	2.50
Power	9.20
Rectifier	9.20
Converter	9.10
Inverter	9.00
Power amplifier (wireless)	1.05
Data Center Equipment Type	Minimum TEEER Allowable
Server	6.53
Equipment Type	Minimum TEEER Allowable
Optical line terminal (OLT) Power Supply	7.20

- *Office of Energy Efficiency and Renewable Energy (EERE):* The office aims at strengthening the United States' energy security, environmental quality, and economic vitality in public-private partnerships.
- *Federal Energy Management Program (FEMP):* The program aims at helping government agencies save energy and demonstrate leadership with responsible, cleaner energy choices.
- *Industrial Technologies Programs (ITP) Best Practices:* These practices aim at improving the energy efficiency of the U.S. industrial sector through a coordinated program of research and development, validation, and dissemination of energy-efficient technologies and practices. The department implements and disseminates best practices in energy management and helps identify opportunities to save substantial amounts of energy in industrial manufacturing plants.
- *Save Energy Now:* A national initiative of the ITP to drive a 25% reduction in industrial energy intensity within 10 years.

The DOE has also established specific energy savings goals for data centers and is partnering with government and industry organizations to reach those targets.

Specifically, the ITP is working with U.S. computer data centers to reduce data center energy consumption 10% by 2011. Building on the successful Save Energy Now energy assessment model, ITP provides tools and resources to help data center owners and operators benchmark data center energy use, identify opportunities to reduce energy, and adopt energy-efficient practices. Some of the concepts and recommendations for data centers/IT are discussed in Chapter 7.

The U.S. Government is leading by example to alleviate the burden of growing data center energy consumption. FEMP plays a critical role in the execution of that leadership; FEMP participates in the DOE data center initiative, which also includes the DOE Industrial Technologies Program's Save Energy Now in Data Centers Program and the Lawrence Berkeley National Laboratory. This group develops tools and resources to make data centers more efficient throughout the United States. FEMP supports data center efficiency initiatives by encouraging federal agencies to adopt best practices, construct energy-efficient data centers, and educate energy managers and information technology professionals.

The Federal Partnership for Green Data Centers was created in 2009 to advance the energy efficiency and renewable energy use goals of the Energy Independence and Security Act of 2007 (EISA 2007) as they relate to data centers. The group also works to reduce the federal government's cost burden by promoting data center energy efficiency. The group meets these goals by facilitating communication between member agencies. The group works to spread best practices for data center energy efficiency across the federal government.

The DOE is also partnering with industry to develop a Data Center Certified Energy Practitioner (DC-CEP) Program to accelerate energy savings in the dynamic and energy-intensive marketplace of data centers. The DOE has set a goal to have at least 200 practitioners certified by 2011. Significant knowledge, training, and skills are required to perform accurate energy assessments in data centers. DC-CEPs will

- Be qualified to identify and evaluate energy efficiency opportunities in data centers
- Demonstrate proficiency in the use of the DOE's DC Pro software tool suite
- Address energy opportunities in electrical systems, air management, HVAC, IT equipment, and on-site generation
- Receive training on conducting data center assessments
- Be required to pass a certification exam and be recertified every 2 to 3 years

Some of the DOE resources that the reader should consult include, but are not limited to

- DOE Energy Efficiency and Renewable Energy: http://www.eere.energy.gov
- DOE: Building Energy Software Tools directory: http://www.energytools-directory.gov
- DOE: Energy Plus Simulation Software: http://www.energyplus.gov

■ DOE: High Performance Buildings: http://www.eere.energy.gov/buildings/
highperformance

And for FEMP, with the mission of facilitating the federal government's implementation of sound, cost-effective energy management and investment practices to
enhance the nation's energy security and environmental stewardship:

■ Renewable energy: www.eere.energy.gov/femp/technologies/renewable_energy.
cfm
■ Renewable purchasing: www.eere.energy.gov/femp/technologies/renewable_
purchasepower.cfm
■ Design assistance: www.eere.energy.gov/femp/services/projectassistance.cfm
■ Training: www.eere.energy.gov/femp/technologies/renewable_training.cfm
■ Financing: www.eere.energy.gov/femp/services/project_facilitation.cfm

6.14 Others (Short List)

The Consumer Electronics Association (CEA). The Association is addressing the issue
of sustainability through two initiatives related to (1) electronic product recycling
and (2) energy efficiency of consumer products [ATI200901, ATI200902].

The program on electronic recycling seeks to promote a national solution and
thus avoid an impractical web of state rules. In 2008, the CEA launched the web
site "myGreenElectronics.org" to inform consumers about electronics recycling
and provide information on local electronics recycling centers. The web site also
allows product manufacturers to register "green" products in a database. The CEA
also supports market-driven environmental design initiatives, such as federal and
state government programs on purchasing of environmentally preferable devices.

The CEA Energy Efficiency Working Group (EEWG) leads the organization's policy on energy efficiencies. The CEA supports voluntary, market-oriented
programs and initiatives, including industry-led standards, which highlight and
sustain energy efficiency in the consumer electronics industry and opposes government mandates. CEA standards on energy efficiency include the following:

■ CEA-2013-A: Digital Set-Top-Box (STB) Background Power Consumption
■ CEA-2022: Digital STB Active Power Consumption Measurement

Global e-Sustainability Initiative (GeSI). This Initiative was launched in 2001
to advance sustainable development in the ICT sector. GeSI advocates open cooperation and members' voluntary actions to improve sustainability performance. At
the same time, the GeSI set out to identify the key areas where the ICT sector can
make the biggest contribution to sustainability. GeSI's goals are to [GES200801]

- Create an open and global forum for the improvement and promotion of products, services, and access to ICT for the benefit of human development and sustainable development
- Stimulate international and multi-stakeholder cooperation for the ICT sector
- Encourage continual improvement in sustainability management and share best practice
- Encourage companies in developing countries to join and share benefits of GeSI
- Promote and support partner regional initiatives and liaise with other international activities
- Promote and support greater awareness, accountability, and transparency

GeSI's Climate Change Group is working to identify the carbon impact of ICT. At press time they were also developing tools to quantify carbon credits for using video teleconferencing as an alternative to business travel. In 2008, GeSI published the report "SMART 2020: Enabling the Low Carbon Economy in the Information Age," describing how, through enabling other sectors to reduce their emissions, the ICT industry could reduce global emissions by as much as 15% by 2020. This report quantifies the direct emissions from ICT products and services based on expected growth in the sector. It also looks at where ICT could enable significant reductions in emissions in other sectors of the economy such as smart motor systems, smart logistics, smart buildings, and smart grids, and quantified these in terms of CO_2e emission savings and cost savings.

The Green Grid. The Green Grid endeavors to advance energy efficiency in data center and business computing environments. It is a global consortium of IT companies and professionals seeking to improve energy efficiency in data centers. The organization seeks to unite global industry efforts to standardize on a common set of metrics, processes, methods, and new technologies to further its common goals. Recently it has focused on defining

- Models and metrics
- Measurement methods
- Processes and evolving technologies to improve and optimize data center performance

We discussed the metrics they have advanced in both Chapter 1 and Chapter 3.

World Research Institute/Greenhouse Gas Protocol. The Greenhouse Gas Protocol (GHG Protocol) is the most widely used international accounting tool for government and business leaders to understand, quantify, and manage greenhouse gas emissions. A decade-long partnership between the World Resources Institute (WRI) and the World Business Council for Sustainable Development (WBCSD), the GHG Protocol is working with businesses, governments, and environmental groups around the world to build a new generation of credible and effective green programs. This Protocol seeks to provide standards and guidance for companies and organizations preparing a

GHG emissions inventory. It serves as the foundation for nearly every GHG standard and program in the world—from the International Organization for Standardization (ISO) to The Climate Registry—as well as hundreds of GHG inventories prepared by individual companies. The GHG Protocol also offers developing countries an internationally accepted management tool to help their businesses compete in the global marketplace and their governments to make informed decisions about climate change. Many industries have used the GHG Protocol Initiative to develop industry-specific calculation tools, which account for both direct and indirect emissions:

- *Direct emissions:* Emissions from sources that are owned or controlled by the organization in question.
- *Indirect emissions:* Emissions that are a consequence of the activities of the organization in question, but occur at sources owned or controlled by another organization.

In 2006, the ISO adopted the Corporate Standard as the basis for its ISO 14064-I: Specification with Guidance at the Organization Level for Quantification and Reporting of Greenhouse Gas Emissions and Removals. This milestone highlighted the role of the GHG Protocol's Corporate Standard as the international standard for corporate and organizational GHG accounting and reporting.

Climate Savers Computing Initiative. The Climate Savers Computing Initiative promotes efforts to increase the use and effectiveness of power-management features by educating computer users on the benefits of these tools and by working with software vendors and IT departments to implement best practices. Member companies commit to purchasing energy-efficient PCs and servers for new IT purchases, and to broadly deploying power management. By publicly declaring their support for this important effort, companies demonstrate their commitment to the "greening" of IT and join other industry-leading companies and organizations blazing new trails in corporate social responsibility and sustainable IT.

System manufacturers participating in the Climate Savers Computing Initiative have committed to working to develop products that meet or exceed the Initiative's Program Criteria. The following system types are currently included in the Climate Savers Computing specifications [CSC200901]:

- Clients (typically featuring multi-output Power Supply Units [PSUs])
- Workstations (typically featuring multi-output PSUs)
- Single, dual, and four-socket-servers in pedestal and rack form factors (single and multi-output PSUs)
- Blade servers capable of having up to four processors

Note that Climate Savers Computing has not defined a separate client specification for laptop computers. Laptops are covered by the ENERGY STAR external PSU requirements.

Table 6.4 outlines the required efficiency levels for PSUs (both multi-output and single-output) at various loading conditions and power factor levels.

Table 6.4 Climate Savers Computing Initiative Goals

Multi-Output Power Supply Unit (PSU)[a]

Loading Condition (%)	Base Target Efficiency Level Starting July 2007		Bronze Target Efficiency Level Starting July 2008		Silver Target Efficiency Level Starting July 2009		Gold Target Efficiency Level Starting July 2010	
	Efficiency (%)	Power Factor	Efficiency (%)	Power Factor	Efficiency (%)	Power Factor	Efficiency (%)	Power Factor
20	80		82		85		87	
50	80		85	0.9	88	0.9	90	0.9
100	80	0.9	82		85		87	

Single-Output Power Supply Unit (PSU)[b]

Loading Condition (%)	Bronze Target Efficiency Level Starting June 2007		Silver Target Efficiency Level Starting June 2008		Gold Target Efficiency Level Starting June 2010		Platinum Recommended Target; Not a Member Purchase Requirement	
	Efficiency (%)	Power Factor	Efficiency (%)	Power Factor	Efficiency (%)	Power Factor	Efficiency (%)	Power Factor
20	81		85		88		90	
50	85		89	0.9	92	0.9	94	0.9
100	81	0.9	85		88		91	

[a] Multi-output PSU refers to desktop and server application power supplies in nonredundant applications.
[b] Single-output PSU typically refers to volume servers power supplies in redundant configurations (1U/2U single, dual, four-socket, and blade servers).

Telcordia Technologies, Inc., Piscataway, NJ, has made a contribution to the field by publishing the document GR-3028-CORE: Thermal Management in Telecommunications Central Offices (Telcordia, 2001). The purpose of the document was to establish requirements for thermal management in central offices to ensure network integrity. The basic goal was to document a common well-defined language for equipment manufacturers and service providers by utilizing equipment-cooling (EC)/rack cooling (RC) classes, environmental criteria, heat-release targets, and equipment specifications. The EC-class describes where the entry and exit points for the cooling air are located on the equipment envelope; an optimal class moves air from the cold front aisle to the rear hot aisle, conserving the hot and cold aisles. GR-3028-CORE was developed in collaboration with 20 major equipment manufacturers and service providers as well as HVAC and chip manufacturers, including 3Com, AFC, Alcatel, AT&T, C-Mac, Calix, Ciena, Cisco, Compaq, Fujitsu, Intel, Liebert, Lucent, Marconi, Nokia, Qwest, SBC, Siemens, Turin, and Verizon.

References

[ACN200901]   New And Renewable Energy Subcommittee Interim Report, Released by Advisory Committee for Natural Resources And Energy, States News Service, Monday, 31 August 2009.

[ATIS200901] Alliance for Telecommunications Industry Solutions (ATIS) Promotional information. ATIS, 1200 G Street, NW, Suite 500, Washington, DC 20005.

[ATIS200902]   ATIS, ATIS Report on Environmental Sustainability, March 2009, A Report by the ATIS Exploratory Group on Green, ATIS, 1200 G Street, NW, Suite 500, Washington, DC 20005.

[CSC200901]   Climate Savers Computing Initiative. Beaverton, OR, http://www.climatesaverscomputing.org/.

[DIM200901]   EU Code of Conduct on Data Centre Energy Efficiency, Position Materials, Dimension 85 Ltd., 2nd Floor, 145-157 St John Street, London.

[ECR200801]   Network and Telecom Equipment - Energy and Performance Assessment, Test Procedure and Measurement Methodology, Draft 1.0.2, 14 October 2008.

[ECR200901]   The Energy Consumption Rating (ECR) Initiative, IXIA, Juniper Networks, http://www.ecrinitiative.org.

[ETNO200801]   Energy Task Team, European Telecommunications Network Operators' Association, The Hague, London, April 2008.

[EUC200801]   EU Code of Conduct on Data Centers Energy Efficiency — Version 1.0, 30 October 2008.

[GES200801]   SMART 2020: Enabling the Low Carbon Economy in the Information Age, Report by the Global e-Sustainability Initiative (GeSI), 2008.

[MET200801]   METI Final Report (Small Routers, L2 Switches) by Router, etc. Evaluation Standard Subcommittee, Energy Efficiency Standards Subcommittee of the Advisory Committee for Natural Resources and Energy. http://www.eccj.or.jp/top_runner/pdf/tr_small_routers-apr_2008.pdf.

[MEY200801]   P. Meyer, The 2008 IEEE-USA Annual Meeting: Fostering Sustainability Initiatives among Engineers & Technical Professionals, *IEEE-USA Today's Engineer Online,* August 2008.

[MYO200801]   Y. Myoken, Japan's Green IT Initiative", Science and Innovation Section, British Embassy, June 2008.

[SAI200801]   L. Ceuppens, D. Kharitonov, and A. Sardella, Power saving strategies and technologies in network equipment opportunities and challenges, risk and rewards, *Proceedings of the 2008 IEEE Symposium on Applications and the* Internet.

[VER200901]   Verizon, Verizon NEBSTM Compliance: Energy Efficiency Requirements for Telecommunications Equipment, Verizon Technical Purchasing Requirements, VZ.TPR.9205, Issue 4, August 2009.

Chapter 7

Approaches for Green Networks and Data Centers: Environmentals

As we have seen, "green computing" refers to environmentally sustainable Information and Communications Technology (ICT) technologies and operations. Recognizing that networks and data centers support mission-critical computing and connectivity functions essential to U.S. economic, scientific, and technological organizations, one still must note that these assets consume relatively large amounts of energy. The energy is consumed not only to operate computer systems, servers, and networking equipment such as switches, routers, firewalls, repeaters, and transmission gear, but also to operate air-conditioning systems, fire-suppression systems, redundant and backup power supplies, and perimeter security systems. To that end, planners need to apply Best Practices, which allow organizations to improve overall efficiency. Analytical methods (and tools) to model energy management and heat disposition in data centers and telecom nodes are still complex and costly. Therefore, common-sense and heuristic techniques are often employed to achieve efficiency improvements.

This chapter examines Best Practices that have been advanced in the industry in the recent past to address greening goals. The content of this chapter is based on industry documentation, as noted throughout. The chapter is segmented into two main sections: one covering data center issues and the other covering networking issues. Several techniques discussed in the data center section also (can) apply to networks, and vice versa.

As noted in Chapter 1, in 2007, ICT represented approximately 2% of all Greenhouse Gas (GHG) emissions worldwide and, by 2020, the sector's GHG output may increase to 1.4 $GtCO_2e$. Over half of the emissions (58%) are associated with devices in the home such as personal computers and peripherals; the telecom industry is generally responsible for 26% of ICT emissions, and data centers are generally responsible for 16%. This data shows that the telecom/networking contribution to emissions is growing from 0.14 $GtCO_2e$ in 2002 to 0.35 $GtCO_2e$ in 2007—this is slightly more than a twofold increase and it represents a 5% CAGR (Compound Annual Growth Rate) over the period in question. While the absolute value of the emissions is relatively small, the trend has a nontrivial positive slope. It follows that a set of prudent plans by enterprise and carrier administrators is advisable in order to hold or improve these trends.

Recall that System Load is consumption efficiency of the equipment in the data center or telecom node while the Facilities Load is the mechanical and electrical systems that support the IT/networking electrical load such as cooling systems (e.g., chiller plant, fans, pumps), air-conditioning units, Uninterruptible Power Supplies, Power Distribution Units, and so on. Many (but not all) of the suggestions offered in this chapter deal with the Facilities Load; the suggestions of Chapter 8 (e.g., virtualization) effectively impact the System Load. As much as 50% of a data center's energy bill is from infrastructure (power and cooling equipment). There are two kinds of energy consumption reductions that may be pursued [ATI200902]: (1) those that avoid energy consumption but do not reduce power capacity requirements are referred to as "temporary consumption avoidances," and (2) those that also allow the reduction of installed power capacity are referred to as "structural consumption avoidance." In general, structural consumption avoidance is more valuable than temporary consumption avoidance.

Some general recommendations follow [WOG200801, TUC200901, CSC200901]:

◼ Telecommunications operators of all types should look into network-power-saving techniques (e.g., optimize cooling requirements, "power down" when not in use), and also consider the refresh of the central office and of the outside plant infrastructure (e.g., powering for repeaters, cell towers, etc.), and of the systems (Network Element) at a point in the life cycle that makes logical and business sense. Carriers need to urge Network Element and component manufacturers to develop products that are more energy efficient (such as Verizon did, as discussed in Chapter 6). Figure 7.1 provides a graphical view of possible greening actions for carriers for various portions of the network.

◼ Institutions with large data centers should apply correct cooling techniques to optimize data center power use, and refresh both the infrastructure and the systems at a point in the life cycle that makes logical and business sense. Some initiatives may include the following:

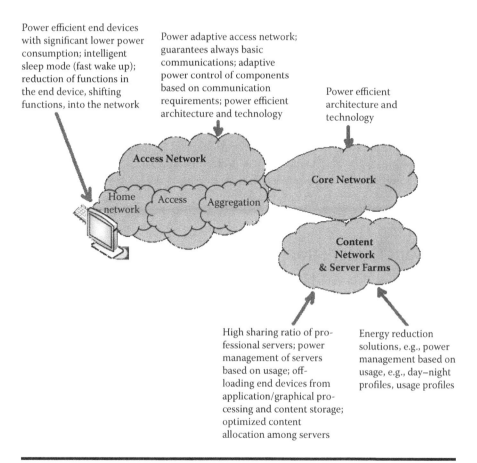

Power efficient end devices with significant lower power consumption; intelligent sleep mode (fast wake up); reduction of functions in the end device, shifting functions, into the network

Power adaptive access network; guarantees always basic communications; adaptive power control of components based on communication requirements; power efficient architecture and technology

Power efficient architecture and technology

Access Network

Core Network

Home network

Access

Aggregation

Content Network & Server Farms

High sharing ratio of professional servers; power management of servers based on usage; off-loading end devices from application/graphical processing and content storage; optimized content allocation among servers

Energy reduction solutions, e.g., power management based on usage, e.g., day–night profiles, usage profiles

Figure 7.1 Possible greening actions for carriers for various portions of the network.

- By changing the arrangement of windows and ventilation openings, firms can reduce the temperature of rooms and data centers.
- By changing the locations of data centers and power plants (if possible), firms can reduce wasted energy spent on cooling and power transmission.

These institutions should also consider virtualization and Cloud Computing. Virtualization of a data center provides an abstraction layer between physical servers and physical storage elements and virtual machines and virtual storage; it enables a single server to appear to be many machines to different applications and thus allows for pooling of shared resources (this topic is discussed in Chapter 8).

■ System buyers, including corporations, government institutions, universities, and energy companies, are encouraged to select systems that meet or exceed the latest ENERGY STAR® specification for a majority of their PC and volume server computer purchases, and to employ power-management tools on PCs.

■ Computer and component manufacturers are encouraged to develop products that meet or exceed the latest ENERGY STAR specification. Such products include PCs, thin clients, workstations, and servers.

■ Consumers are encouraged to use power-management capabilities on their PCs and by selecting systems that meet or exceed the latest ENERGY STAR specification for new PC purchases.

7.1 Data Center Strategies

In recent years, data centers have increasingly deployed more compact but more energy-intensive servers to support the growing demand for processing power. Hence, new techniques are needed to properly design and manage such data centers. Energy-saving techniques must be updated continuously because computer technology evolves rapidly; new ICT product requirements change faster than manufacturing processes. Also, even in the best case, ICT equipment energy use is difficult to forecast because as we have seen in previous chapters one must take into account loading, configuration, and operating environment (heat loads in a data center can also vary with time due to the addition or removal of equipment or racks and changing computing workloads). According to some reports, for every $1.00 spent on new data center hardware, an additional $0.50 is spent on power and cooling; the goal is to reduce those costs both for basic financial reasons as well as to reduce the carbon footprint.

A high-density data center may have thousands of racks, each with multiple 1 rack unit or blade computing units. The heat dissipation from a rack containing such computing units can range from 10 to 30 kW. A data center with 1000 racks, over 30,000 square feet (ft^2), requires 10 to 30 MW of power for the computing infrastructure. A 100,000 ft^2 data center of the future may require 50 to 125 MW of power for the computing infrastructure. The energy required to dissipate this heat will be an additional 20 to 50 MW. A 100,000 ft^2 data center, with 5,000 10 kW racks, would cost about $44 million per year (at $100/MWh) to power the servers and $18 million per year to power the cooling infrastructure for the data center; if the racks require 25 kW, it would cost about $100 million per year to power the servers and $45 million to cool. At a macro level, the benefits of an efficiency optimization program include [EER200901]

■ Manage electricity demand and associated carbon emissions.
■ Protect data and computing functions vital to the economy.

- Reduce risk of power outages and increase regional electricity reliability.
- Postpone need to build new electricity generation capacity.
- Support replication of energy-efficient practices across the sector.
- Free up power and cooling.

Data center infrastructure strategies focus on improved cooling and better power distribution, which we discuss below. In terms of system load, studies have shown that there is a cumulative savings effect when saving downstream power: For every watt of reduced data center equipment energy consumption, there is a reduction of 0.3 to 2 W of facility support equipment energy consumption; thus, there is an overall reduction of anywhere from 1.3 to 3 W of overall energy consumption (the specific figure depends on the energy efficiency of the facility, power distribution system, etc.). See Figure 7.2. Some ways to achieve these savings include (1) a

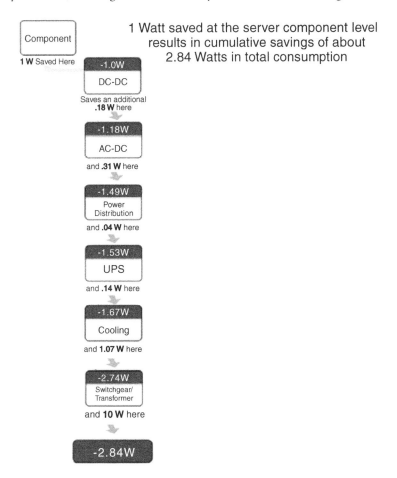

Figure 7.2   Cumulative savings effect. (Courtesy of Emerson Network Power.)

technology refresh, (2) a technology leap-thru (e.g., replace pedestal servers with blade servers), (3) virtualization, or (4) conservation/power management (turning off a device when not in use).

7.1.1 Data Center Cooling Strategies

Data center cooling entails air delivery, air movement, and heat rejection. In this text we have described traditional cooling system for computer rooms, which can be summarized as follows [42U201001]:

- Computer Room Air Conditioner (CRAC):
  - Refrigerant-based (DX), installed within the data center floor and connected to outside condensing units
  - Moves air throughout the data center via fan system: delivers cool air to the servers, returns exhaust air from the room
- Computer Room Air Handler (CRAH):
  - Chilled water based, installed on data center floor and connected to outside chiller plant
  - Moves air throughout the data center via fan system: delivers cool air to the servers, returns exhaust air from the room
- Humidifier:
  - Usually installed within CRAC/CRAH and replaces water loss before the air exits the AC units; also available in stand-alone units
  - Ensures that humidity levels fall within the American Society of Heating, Refrigerating and Air-Conditioning Engineers' (ASHRAE) recommended range
- Chiller:
  - Produces chilled water via refrigeration process
  - Delivers chilled water via pumps to CRAH

An under-provisioned CRAC unit is one where the cooling load is higher than the capacity of the unit, while an over-provisioned CRAC unit is one where the load is well within its capacity. Studies have shown that mal-provisioning will lead to waste of energy, and a properly provisioned environment will lead to efficient use of energy resources [PAT200201].

What follows are some Best Practices to optimize the use of CRACs, CRAHs, humidifiers, chillers, and hot/cold aisle isolation. Hot aisle/cold aisle aims at achieving air separation within the server room; this design principle, which seeks to align data center cabinets into alternating rows, is the first basic step in improving airflow management.

At this point in time, most new data centers have been designed to support an average density of 100–200 watts per square foot; the average cabinet in these data centers uses about 4–5 kW, but those cabinets in the data center supporting high-density blade servers require 25–30 kW. We have noted elsewhere in the text that the average power of the typical cabinet has been increasing in the recent past. Some large enterprise establishments now average 8–10 kW per rack, and some high-end establishments have an average as high as 12–15 kW per rack. Using well-engineered airflow arrangements a CRAC can support some significant portion of racks in a data center that operates up to 25–30 kW. It is difficult to go above that rack power without using liquid-based cooling.

A CRAC consumes about 1–1.5 kW/ton. A typical (non-blade) rack requires 1–1.5 tons of AC. A 20 ton unit (supporting 14–20 racks) operating 24 × 365 consumes about 22,000 kWh/month in electrical energy; this typically equates to $2,200/month, or $25,000/year. A one degree increase in the data center temperature can save 4% in energy costs. The potential savings yearly for a 10 degree temperature adjustment are around 9,000 kWh/month or about 108,000 KWh/year, which equate to a ballpark figure of $11,000/year. This is a CO_2 reduction of 57 tons/year based on 1.1 lbs of CO_2/kWh.

7.1.1.1 Data Center Best Practices

Possible data center strategies include

- IT efficiencies:
    - Virtualization
    - Dynamic power management
    - Storage consolidation
- Data center cooling optimization:
    - Improved co-location operations
    - Air containment
    - Efficiencies through increased temperatures
    - Dynamic CRAH unit control
    - Variable frequency drives (VFDs) on CRAC
    - Hot/cold aisle isolation
    - System comparisons
    - Cold aisle pressure control
    - Improved power distribution

Concepts that are applicable to cooling include the following [VAR201001, 42U201001]:

- VFDs reduce the capacity of centrifugal chillers and thus save energy. They include electronic control of the fan motor's speed and torque to continually match fan speed with changing building-load conditions. Electronic control of the fan speed and airflow can replace inefficient mechanical controls, such as inlet vanes or outlet dampers. A VFD is an electronic controller that adjusts the speed of an electric motor by regulating the power being delivered. VFDs provide continuous control, matching motor speed to the specific demands of the work being performed. VFDs allow operators to fine-tune processes while reducing costs for energy and equipment maintenance. For example, by lowering fan or pump speed by 15% to 20%, shaft power can be reduced by as much as 30%.

▪ Cold Aisle Containment (CAC) is a system that directs cooled air from air-conditioning equipment to the inlet side of racks in an efficient manner.
  – Cold Aisle is an aisle where rack fronts face into the aisle. Chilled airflow is directed into this aisle so that it can then enter the fronts of the racks in an efficient manner.
  – Cold Spot is an area where ambient air temperature is below acceptable levels. Typically caused by cooling equipment capacity exceeding heat generation.

▪ Hot Aisle Containment (HAC) is a system that directs heated air from the outlet side of racks to air conditioning equipment return ducts in a highly efficient manner.
  – Hot Aisle is an aisle where rack backs face into the aisle. Heated exhaust air from the equipment in the racks enters this aisle and is then directed to the CRAC return vents.
  – Hotspot is an area, typically related to a rack or set of racks, where ambient air temperature is above acceptable levels. Typically caused by heat generation in excess of cooling equipment capacity.

▪ In-Row Cooling is a cooling technology installed between server racks in a row that delivers cooled air.

We noted in Chapter 5 that there may be localized hotspots within the data center and some servers may be ineffectively drawing in hot air from the hot aisles. To address these issues, one has to reorient the equipment (if needed) such that cabinets are always oriented within rows in a uniform way so that two adjacent rows of cabinets draw their cooling air from a common cold aisle and expel their hot air into a common hot aisle. Cold aisles should be cold, hot aisles can be hot, and one should seek to increase the return air temperature. Typically, data centers are located in buildings with a nominal ceiling height of 14 to 17 ft, below which is a dropped acoustic ceiling located 9 to 10 ft above the raised metal floor (RMF). With a 2 ft RMF, this means that there is a 2 to 6 ft zone above the dropped ceiling that one may not be using effectively; one efficiency approach is to convert this dead air zone into a return air plenum. For example, one can [GAR200601]

▪ Place return air grills in the ceiling above the hot aisles to provide a path for the rising hot air to escape into the return plenum.
▪ Fit the CRAC units with a duct on top of the air inlet so that they draw hot return air out of the plenum instead of drawing it from the room. This

changes the hot air return path: Instead of flowing horizontally across the room, hot air moves vertically up into the return plenum. This reduces hot air re-entrainment.

The following results were shown by [GAR200601]:

- *Barriers:* A symmetrical airflow system that creates even pressure differentials everywhere in the room is needed. Barriers are required between hot and cold aisles. Hot return air needs to flow through grills in the hot aisles into a return air plenum above the drop ceiling. The best approach to preventing cold and hot air from mixing is to place barriers above and below the cabinets, isolating the hot aisles from the cold aisles. This technique significantly reduces both the amount of hot air that infiltrates back into the cold aisle (re-entrainment) and the cold air that goes directly from the floor diffusers to the return air grills (bypass).
- *Cabling:* The telecommunications cable trays and power wireways sometimes obstruct airflow up through floor diffusers in the cold aisles. One should locate them under the solid tiles of the hot aisles. Under-floor cables traditionally enter cabinets through RMF cutouts in the hot aisles; these cutouts can be yet another source of leaks because they allow air to short-circuit the equipment. One should replace them with brush-lined openings that allow the cabling to pass through, yet block excess leakage.
- *Cabinet rows parallel with airflows:* In data centers built on an RMF, the diffusers closest to the CRAC units tend to emit the least air or, in some cases, actually draw air from the room. This means that if we place rows of cabinets perpendicular to the airflow from the CRAC units, then the cold aisles closest to the CRACs receive the least airflow, while cold aisles in the center of the room receive the most. To reduce this effect, one should place rows parallel with the airflow; this also reduces airflow obstructions by orienting the cables under the hot aisles parallel with the airflow.
- *Makeup air handler (MAH):* The MAH filters and conditions the makeup air, either adding or removing water vapor. If there is no coordination between the CRACs, capacity will be wasted. Sometimes one CRAC can be cooling while its neighbor can be heating. One CRAC may be humidifying while another is dehumidifying. One needs to adopt a coordinated, centralized approach for using MAHs to manage humidity control for the entire data center.

Designing a data center for blade servers without a raised metal floor is challenging. Here one places cabinets directly on the slab and provides all cooling air from ductwork and diffusers above the cold aisles (telecom and power cable typically make use of dual overhead busways). The design should ascertain that one is able to share airflow from multiple RAHs in a common supply air plenum and is capable of continuing to operate in the event of a single RAH failure. One should place high-capacity RAHs on a mezzanine, allowing them to draw the hot return air from above and supply air at the level of the dropped ceiling. Here one can use a mixture of chimney cabinets and hot aisle enclosures, which enables one to reuse existing open-backed cabinets that may lack chimneys. This design works well in a building 25 (or more) ft tall, which can easily accommodate the mezzanine with the RAHs (note that office buildings generally have a ceiling height of only 17 ft).

Future high-performance data center design has the goal of building a green greenfield data center for air-cooled data center with large WPSF to fit into an office building with a ceiling height of only 17 ft (30 kW per cabinet and 1,250 WPSF over the RMF area) and supporting, say, 220 cabinets (the 17 ft height can accommodate a 3 ft RMF, 10 ft from the top of the RMF to the drop ceiling, and 4 ft for the return air plenum). Here, concerns about total airflow and noise become a tangible consideration. Each cabinet needs 1,600 standard cubic feet of air per minute (SCFM)* to pass through the blades at a 60°F ΔT. The result is a total flow for the 220 cabinets per room of around 500,000 SCFM. Such a design is based on a 60 × 88 ft RMF area using an air handling unit (AHU) containing several large RAH units with VFDs. The use of passive chimney cabinets is indicated. Warming the 65°F supply air by 50°F as it passes through the cabinets will result in return air at 115°F. One may decide to mix this with an additional 20% bypass air (67,000 SCFM), thereby tempering the return air to 108°F to match the capabilities of the RAH coils. The designer may opt to use a makeup air handler, a symmetrical layout, and high-flow floor diffusers with approximately 50% free area. The walls separating the RMF (a 36 inch-high RMF works best) server area from the AHU rooms should extend from the RMF to the structure and build the RAHs in a vertical downflow style, sized to return air through this wall. The vertical height available allows space for airflow silencers on both sides of the fan, if required. No single factor is responsible for allowing one to double the power density. Instead, one should include the following:

■ Improved airflow management allows one to supply air at 65°F to 70°F instead of the traditional 50°F to 55°F, yet improve temperature distribution and cooling to the servers. This creates energy savings.

■ Higher supply air temperatures result in the air handlers performing only sensible cooling—rather than also performing latent cooling that dehumidifies the data center, creating a relative humidity that is too low for some types of IT equipment.

■ Raising the supply air temperature also raises the return air temperature, allowing one to get more cooling capacity from each coil at the same chilled water temperature.

---

* SCFM is a measure of free airflow into the intake filter of an air compressor. It is the volumetric flow rate of a gas representing a mass flow rate corrected to "standardized" conditions of pressure, temperature, and relative humidity.

■ Increasing the airflow per square foot also increases cooling. Larger-capacity RAHs designed in a vertical arrangement provide more airflow and better utilize space.

■ The higher ΔT of blade servers means that more heat transfer per CFM of air, which reduces the fan horsepower required.

■ Eliminating wasted cold air bypass allows one to use that portion of the total airflow to cool equipment.

It is critical to achieve the highest (mechanical) cooling efficiency, because the power system that powers the AC may not be in itself efficient, thus compounding the problem. A 2008 study found that a typical AC system in today's data center has an Uninterruptible Power Supply (UPS) that is about 85% efficient, and power supplies around 73% efficient [42U201001] (the topic of power distribution is discussed next).

Variable-capacity cooling is a cost-effective greening strategy. Data center systems are designed to handle peak loads, which rarely exist; consequently, operating efficiency at full load is rarely a good indication of actual operating efficiency. Newer technologies, such as Digital Scroll compressors and variable frequency drives in CRACs, allow high efficiencies to be maintained at partial loads. Digital Scroll compressors allow the capacity of room air conditioners to be matched exactly to room conditions without turning compressors on and off. Typically, CRAC fans run at a constant speed and deliver a constant volume of airflow. Converting these fans to variable frequency drive fans allows for a reduction in fan speed and power draw as the load decreases. A 20% reduction in fan speed provides a 50% reduction in fan power consumption. These drives are available in retrofit kits that make it easy to upgrade existing CRACs with a payback typically realized in less than a year [EME201001].

Table 7.1 provides a summary of suggested practices offered by DOE and others [DOE200901, MIT201001].

7.1.2 Improved Power Distribution

There is interest in improved AC power distribution within the data center. The existing, legacy-components 480 V to 208 V power distribution architecture typical of a data center is relatively inefficient. There are multiple points, including the UPS and server power supplies, where incoming AC power is converted to DC. Losses occur during each conversion, the severity of which depends on the equipment. The components (UPS, PDU [power distribution unit], and server power supply) can lose upward of 40% of the incoming electricity during transformation and distribution. For this reason, administrators are investigating new power technologies and distribution designs: modern architectures can achieve overall efficiencies between 80 and 90%; in addition, performance curves for newer equipment are steeper and better sustained across the loading curve, making them significantly superior

Table 7.1   Data Center Energy Efficiency Recommendations (DOE and others)

Optimal Equipment-Cooling (EC) Classes	EC classes have been defined to describe where the entry and exit points for the cooling air are located on the equipment envelope. For example, a F2-R2 class has two inlets in the cold front aisle and two outlets in the hot rear aisle. An optimal class moves air from the cold front aisle to the rear hot aisle, conserving the hot and cold aisles.
Perforated Floor Tiles	Perforated floor tiles (or overhead diffusers) should be placed only in the cold aisles to match the consumption of air by the electronic equipment. Too little or too much supplied air results in poor overall conditions. The hot aisles are supposed to be hot and perforated tiles should not be placed in those areas. A rigorous program should be in place to maintain the hot and cold aisle configuration of perforated floor tiles. The number and position of perforated tiles in the cold aisles need to be adjusted to optimize airflow around carefully sealed racks that house blade servers/blade storage racks.
Cable Congestion	Cable congestion in raised-floor plenums can sharply reduce the total airflow as well as degrade the airflow distribution through the perforated floor tiles.
Maintain Tight Raised Floors	Cable congestion in raised-floor plenums can sharply reduce the total airflow as well as degrade the airflow distribution through the perforated floor tiles.
Raised-Floor Plenum Pressure	A high raised-floor plenum static pressure means high floor leakage and by-pass air. A mid-range static pressure may be considered because it allows relatively high tile airflow rates but caps the floor leakage.
Managing Blanking Panels	Managing blanking panels and unbroken equipment lineups is especially important in hot- and cold-aisle environments. Any opening between the aisles will degrade the separation of hot and cold air. A rigorous program should be in place to maintain the panels.
Vary Supply Airflow Not Supply Temperature	Traditionally, few CRAC units had the capability of varying the airflow in real time, and adjusting the temperature was the only option. With variable speed drives, capacity control should be modified to improve the cooling effectiveness of the electronic equipment as well as save fan and cooling energy. The supply airflow should closely match the equipment airflow; too little or too much supply air degrades performance.

Table 7.1   Data Center Energy Efficiency Recommendations (DOE and others) (Continued)

Air Balancing	When changes are made to the electronic equipment inventory, the air distribution system eventually needs to be rebalanced. An out-of-balance system results in a degraded thermal equipment environment, and often higher airflow rates and energy costs to combat hot spots. Relatively high pressure drops at the diffuser level improve the chances for successful balancing.
Use Containment	The traditional simple hot/cold aisle approach is not always optimal. Containment is the strategy of creating high-density zones on the floor where doors seal off the ends of either the hot or cold aisles (barriers may also be used on the top of each row of cabinets to prevent hot and cold air from mixing). This strategy creates a segregation of cold and hot, eliminating the mixing of air while optimizing the airflow (these techniques are indicated for environments with a large number of racks needing 25 kW.
Increase the Data Center Ambient Temperature	It is possible to raise the data center temperature. Current equipment is designed to operate at a maximum temperature of 81°F; this is higher than the operating temperature for equipment built prior to 2004 when the official specification set by the ASHRAE Technical Committee 9.9 was 72°F. At the higher temperatures outside air can be used, but cannot support dense floor layouts. However, current vinatage equipment has limitations. Furthermore, raising the operating temperature of servers and other data center gear does not always save on cooling costs. To keep the processor and other component temperatures constant most IT manufacturers increase fan speeds for servers and other equipment as temperatures exceed about 77°F. At temperatures above 77°F, the speed of fans in most servers sold today increases significantly and processors suffer high leakage currents. Power consumption increases as the cube of the fan speed—so if speed increases by 10% that means a 33% increase in power. At temperatures above 81°F servers are increasing power usage at a faster rate than what is saved in the rest of the data center infrastructure [MIT202001]. Some studies show a 4% energy saving for every 1°F in temperature increase (moving from 68°F to 77°F would save 36% in energy costs).

Continued

Table 7.1 Data Center Energy Efficiency Recommendations (DOE and others) (Continued)

Liquid Cooling	Some data centers have already begun to move to liquid cooling to address high-density hot spots in data centers. The most common technique, closely couple cooling, involves piping chilled liquid, usual water or glycol, into the middle of the raised floor space to supply air-to-water heat exchangers within a row or rack. High-end enterprise users employ this type of liquid cooling for at least some high-density racks. These closely coupled cooling devices may be installed in a cabinet in the middle of a row of server racks, as data center vendor APC does with it InRow chilled water units, or they can attach directly onto each cabinet, as IBM does with its rear door heat eXchanger [MIT201001]. Some see a bifurcation where high-performance computing will make use of water-based cooling while the rest of the enterprise's data center will stay with air but (when possible) relocate into locations where space and power costs are cheaper.

at lighter loads [42U201001]. Table 7.2 depicts some power distribution schemes, based on reference [42U201001], also see Figure 7.3. In recent years there has also been renewed interest in distributing Direct Current (DC) power throughout the data center. Proponents see DC power distribution as a way to minimize electrical losses, thus achieving a more efficient design, but the dearth of trained personnel in high-voltage DC for data center applications could present an obstacle to deployment. DC is typically the method used by telecom carriers to support their central offices/POPs (Points Of Presence).

New, enhanced technologies are available that improve efficiencies in data center power distribution. Contemporary UPS systems and PDUs achieve efficiencies in the high nineties, while some server power supplies have reached or exceeded the ninetieth percentile. Likely, these best-of-breed technologies are not the equipment currently deployed in a typical data center. Upgrading the raw infrastructure is often more expensive and operationally complex than upgrading the systems. Hence, a server-level hardware refresh may provide some easy-to-implement approaches to power savings (also see Chapter 8). DC distribution is the most efficient design, but

Figure 7.3 Multiple power conversions in a data center.

Table 7.2   Data Center Power Distribution Schemes

480 V–208 V Power Distribution (traditional/legacy)	This traditional 480 V–208 V distribution design accepts 480 VAC (voltage alternating current) from the grid and passes that through a UPS (Uninterruptible Power Supply) where a double conversion takes place. The output is still 480 VAC. This stage is followed by an isolated PDU (Power Distribution Unit) that outputs 208 VAC. This is followed by an isolated PSU (Power Supply Unit) that converts power to 12 (voltage direct current) VDC. Figure 7.3 illustrates the process.
480 V–208 V Power Distribution (present-day)	The present day 480 V–208 V distribution design is the same as its predecessor, except that it uses the highest efficiency components on the market today. The overall efficiency, depending on the load, ranges between 80 and 85%, a significant improvement over the legacy design. One can say that this system uses an enhanced UPS, enhanced PDUs, and enhanced PSUs.
600 V–208 V Power Distribution	600 V–208 V is most commonly used in Canada and closely resembles the present-day 480 V–208 V design. The only difference is the higher input voltage into the UPS and then into the PDUs (600 V). The overall efficiency of the design also ranges between 80 and 85%.
480 V/277 V Power Distribution	This power scheme distributes electricity in a three-phase Wye configuration at 277 V—the phase to neutral voltage in 480 V three-phase power. Instead of stepping down to 208 V, the voltage remains constant through the PDU, improving the efficiency of design. The input to the UPS is 480 V/277 V; the input to the PDUs is the same; the input to the PSU is 277 V. This power distribution scheme achieves efficiencies between 85 and 90%, depending on the load and is considered the U.S. equivalent of 400/230 V and 415/240 V, although currently server power supplies are not yet readily available to accept 277 V.
415 V/240 V Power Distribution	This design is typically employed outside the United States. Because most U.S. facilities have 480V/277 V available, they would have to convert to 415 V at the UPS level. The input to the UPS is 480 V/277 V; the input to the PDUs is 415 V/240 V; the input to the PSU is 240 V. This scheme is another high-efficiency approach, achieving 85% to 90% efficiency, based on the load.

Continued

Table 7.2   Data Center Power Distribution Schemes (Continued)

480 VAC to 48 VDC Power Distribution	The 48 VDC power distribution design is most common in telecom deployments. It achieves efficiencies of 85% to 90% based on the load. The input to the UPS is 480 V/277 V; the input to the PDUs is 48 VDC; the input to the PSU (which is the output of the PDU) is 48 VDC.
480 V AC–575 VDC–48 VDC Power Distribution	This design distributes 575 VDC power from the UPS to minimize distribution cabling losses and cost. The efficiency levels, like most other present-day designs, ranges from 80% to 85%, depending on the load. The input to the UPS is 480 V/277 V; the input to the PDUs is 575 VDC; the input to the PSU (which is the output of the PDU) is 48 VDC.
480 VAC–380 VDC Power Distribution	This design distributes 380 VDC to the server power supply, achieving greater efficiency and minimizing losses in the distribution cabling. It achieves 90% at 30% to 50% of capacity, with the overall efficiency decreasing slightly as the load nears full capacity. The input to the UPS is 480 V/277 V; the input to the PDUs is 380 VDC; the input to the PSU (which is the output of the PDU) is 48 VDC.

an infrastructure change to support DC distribution is impractical for an existing data center; consequently, DC distribution currently appears to be most applicable for specialty data center applications and for the telecom segment. Perhaps at some point in the future, server power supplies will support 277 VAC, which would eliminate the intermediate transformer loss between the UPS and the load (as we saw in Chapter 4, 277 V is the phase-to-neutral voltage of 480 V three-phase power).

7.1.3 Overall Best Practices

As noted in Chapter 6, the U.S. Department of Energy's Office of Energy Efficiency and Renewable Energy (EERE) is working to evaluate energy efficiency opportunities in data centers. The EERE has developed assessments designed to help data center professionals identify energy-saving measures that are most likely to yield the greatest energy savings. The assessments are not intended to be a complete energy audit but rather the process is meant to educate data center staff and managers on an approach that can be used to identify potential energy-saving opportunities that can be further investigated. Figure 7.4 depicts the recommended steps [EER200902].

The listing that follows, developed by the EERE, can be considered a list of data center greening Best Practices. The interested reader should consult [EER200903] for additional information.

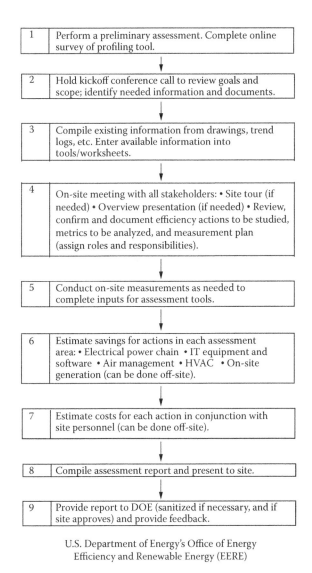

1	Perform a preliminary assessment. Complete online survey of profiling tool.

2	Hold kickoff conference call to review goals and scope; identify needed information and documents.

3	Compile existing information from drawings, trend logs, etc. Enter available information into tools/worksheets.

4	On-site meeting with all stakeholders: • Site tour (if needed) • Overview presentation (if needed) • Review, confirm and document efficiency actions to be studied, metrics to be analyzed, and measurement plan (assign roles and responsibilities).

5	Conduct on-site measurements as needed to complete inputs for assessment tools.

6	Estimate savings for actions in each assessment area: • Electrical power chain • IT equipment and software • Air management • HVAC • On-site generation (can be done off-site).

7	Estimate costs for each action in conjunction with site personnel (can be done off-site).

8	Compile assessment report and present to site.

9	Provide report to DOE (sanitized if necessary, and if site approves) and provide feedback.

U.S. Department of Energy's Office of Energy
Efficiency and Renewable Energy (EERE)

Figure 7.4 DOE EERE assessments for data centers.

- Global Practices:
  - GL-001: Upgrade All Cooling Supply Fan, Pump, and Cooling Tower Fan Motors to Premium Efficiency
- Energy Management Practices:
  - EM-001: Perform an Energy Audit
  - EM-002: Create an Energy Management Plan
  - EM-003: Assign an Energy Manager

- EM-004: Engage Upper Management with a Compelling Life-Cycle Cost Case
- EM-005: Implement an Energy Measurement and Calibration Program
- EM-006: Conduct Regular Preventive Maintenance
- EM-007: Sub-Meter End-Use Loads and Track Over Time
- EM-008: Review Full System Operation and Efficiency on a Regular Basis
- EM-009: Install Monitoring Equipment to Measure System Efficiency and Performance
- EM-010: Raise Awareness and Develop Understanding among Data Center Staff about the Financial and Environment Impact of Energy Savings
- EM-011: Train/Raise Awareness of Data Center Designers in the Latest Energy Management Best Practices and Tools
- EM-012: Use Life Cycle Cost to Make Decisions
- EM-013: Install Peak Shaving Devices on Lighting Systems
- EM-014: Implement a Continuous Commissioning Plan

■ IT Equipment Practices:
- IT-001: Monitor Utilization of Servers, Storage, and Networks
- IT-002: Perform an Audit to Ensure All Operational Servers Are Still in Active Use
- IT-003: Evaluate the Potential Savings from Upgrading to Newer IT Equipment
- IT-004: Implement Server Virtualization
- IT-005: Consolidate to Network-Attached Storage and Diskless Servers
- IT-006: Assess Data Storage Usage
- IT-007: Reduce the Capacity Requirements of Data Storage Systems
- IT-008: Evaluate Alternative Financing Methods to Enable Faster Technology Refresh
- IT-009: Enable Power Management Features on Servers
- IT-010: Consolidate User Data
- IT-011: Automate Data Retention and Deletion Policies
- IT-012: Obtain Realistic Estimates of IT Equipment Actual Power Use
- IT-013: Use Vendor Programs to Dispose of Old Servers
- IT-014: Specify More Efficient Power Supplies in IT Equipment.
- IT-015: Specify Computing Performance Metrics for New IT Equipment

■ Environmental Conditions Practices:
- EC-001: Measure the Return Temperature Index (RTI) and Rack Cooling Index (RCI)
- EC-002: Increase the Supply Air Temperature
- EC-003: Provide Warmer Temperatures at the IT Equipment Intakes
- EC-004: Place Temperature and Humidity Sensors to Mimic the IT Equipment Intake Conditions
- EC-005: Recalibrate the Temperature and Humidity Sensors

- EC-006: Network the CRAC/CRAH Controls
- EC-007: Add Personnel and Cable Grounding to Allow Lower IT Equipment Intake Humidity
- EC-008: Disable or Eliminate Humidification Controls or Decrease the Humidification Setpoint
- EC-009: Disable or Eliminate Dehumidification Controls or Increase the Dehumidification Setpoint
- EC-010: Change the Type of Humidifier
- EC-011: Change Cooling Unit Air Temperature Setpoints Based on IT Equipment Thermal Demand
- EC-012: Use an Enthalpy Sensor to Control the Airside Economizer

■ Air Management Practices:
  - AM-001: Ensure Adequate Match between Heat Load and Raised-Floor Plenum Height
  - AM-002: Provide Adequate Ceiling Supply/Return Plenum Height
  - AM-003: Provide Adequate Clear Ceiling
  - AM-004: Use Existing Dropped Ceiling as Return Plenum
  - AM-005: Remove Abandoned Cable and Other Obstructions
  - AM-006: Implement a Cable Mining Program
  - AM-007: Implement Alternating Hot and Cold Aisles
  - AM-008: Provide Physical Separation of Hot and Cold Air
  - AM-009: Convert to Variable Speed Fans
  - AM-010: Configure Equipment in Straight Rows
  - AM-011: Place Supply Air Devices in Cold Aisles Only
  - AM-012: Implement a Tile/Diffuser Location Program
  - AM-013: Use Appropriate Overhead Diffusers
  - AM-014: Place Air Returns at High Elevation
  - AM-015: Take Return Air from Hot Aisles
  - AM-016: Provide Adequate Floor Plenum Pressure
  - AM-017: Seal Floor Leaks
  - AM-018: Implement a Floor Tightness Program
  - AM-019: Use Supplemental Cooling
  - AM-020: Line up CRAC/CRAH Units with Hot Aisles
  - AM-021: Ensure an Adequate Ratio of System Flow to Rack Flow
  - AM-022: Balance the Air Distribution System
  - AM-023: Use IT Equipment with High Design Temperature Rise
  - AM-024: Use IT Equipment with Front to Rear or Front to Top Cooling Airflow
  - AM-025: Remove Cosmetic Doors from IT Equipment Racks
  - AM-026: Provide Adequate Free Area on Rack Doors for Air Movement
  - AM-027: Maintain Tight Racks and Rows
  - AM-028: Implement a Rack and Row Tightness Program
  - AM-029: Maintain Unbroken Rows

- AM-030: Shut off CRAC/CRAH Units
- AM-031: Implement an Air Balancing Program
- AM-032: Control All Supply Fans in Parallel
- AM-033: Get Rid of Pre-Filters
- AM-034: Change Filters to Appropriate MERV Rating
- AM-035: Seal Ducts or Casings to Reduce Leakage
- AM-036: Fix System Effects in Air Distribution System
- AM-037: Change CRAC/CRAH/AHU Fan Motors to Premium Efficiency
- AM-038: Add an Airside Economizer to the AHU
- AM-039: Retrocommission the Airside Economizers
- AM-040: Replace the Existing CRACs/CRAHs/AHUs with More Efficient Equipment
- AM-041: Replace Dirty CRAC/CRAH/AHU Filters
◼ Cooling Plant Practices:
- CP-001: Add Variable Speed Drives (VSDs) to Cooling Tower Fans
- CP-002: Convert Cooling Towers from Series Staging to Parallel Staging
- CP-003: Evaluate Chillers for Replacement.
- CP-004: Optimize Cooling Plant Controls
- CP-005: Convert all 3-Way Chilled Water (CHW) Valves to 2-Way
- CP-007: Select Chiller for High Full and Part-Load Efficiency
- CP-008: Select High Efficiency Cooling Towers
- CP-009: Convert Primary/Secondary CHW Pumping System to Primary/Series Secondary
- CP-010: Monitor System Efficiency
- CP-011: Right-Size the Cooling Plant
- CP-012: Recover Waste Heat for Heating Uses in Other Spaces.
- CP-013: Add Integrated Waterside Economizer to CHW Plant
- CP-014: Improve Cooling Tower Water Treatment to Reduce Energy Use
- CP-015: Recalibrate Chilled Water Supply Temperature Sensors.
- CP-016: Recalibrate the Condenser Water Supply Temperature Sensors
- CP-017: Convert Air-Cooled DX CRACs to Water-Cooled DX CRACs
- CP-018: Retrofit Constant-Speed Chiller with VSD
- CP-019: Trim Pump Impeller and Open Triple Duty Valve
- CP-020: Remove Suction Diffusers Where Possible.
- CP-021: Convert Primary/Secondary Chilled Water Pumping System to Primary-Only.
- CP-022: Install High Efficiency Pumps
- CP-023: Specify an Untrimmed Impeller, Use a VFD to Limit Pump Speed, Match the Pump Motor Size to the Design Flow Rate
- CP-024: Reduce the Chilled Water Supply Pressure Setpoint
- CP-025: Implement a Chilled Water Pumping Pressure Setpoint Reset
- CP-026: Optimize the Number of Pumps Running in a Bank of Variable-Speed Pumps

- CP-027: Use VFD to Adjust Condenser Water Flow Rate
- CP-028: Increase Cooling Tower Capacity
- CP-029: Specify High Efficiency DX Cooling Units
- CP-030: Specify a High Efficiency Air-Cooled Chiller
- CP-032: Specify a Low Approach Temperature Cooling Tower
- CP-033: Specify an Evaporatively Cooled-Chiller
- CP-034: Replace the DX Cooling System with a Chilled Water Cooling System
- CP-035: Implement a Chilled Water Storage System
- CP-036: Decrease the Condenser Water Temperature Setpoint
- CP-037: Eliminate Low Chilled Water Delta-T Syndrome if Present
- CP-038: Implement Variable Condenser Water Flow
- CP-039: Increase the Chilled Water Supply Temperature Setpoint
- CP-040: Reduce the Condenser Water Flow Rate

■ IT Power Distribution Chain Practices:
- ED-001: Reconfigure the UPS Topology for More Efficient Operation
- ED-002: Install a Modular UPS
- ED-003: Shut Down UPS Modules and PDUs when Redundancy Level Is High Enough
- ED-004: Use UPS without Input Filters
- ED-005: Right-Size the Datacenter Power Equipment
- ED-006: Use High Efficiency MV and LV Transformers
- ED-007: Reduce the Number of Transformers Upstream and Downstream of the UPS
- ED-008: Locate Transformers Outside the Datacenter
- ED-009: Maintain Total Harmonic Distortion at Main Feeder Panel at Less than 8%
- ED-010: Maintain Power Factor at Main Feeder Panel at 0.90 or Higher
- ED-011: Retrofit IT Equipment to Maintain High Power Factor and Low Total Harmonic Distortion
- ED-012: Use 480V instead of 208V Static Switches
- ED-013: Use Alternate Power Source to Warm Generator Blocks
- ED-014: Use Chilled Water Return to Warm Generator Blocks
- ED-015: Apply Thermostat Control to Generator Block Heaters
- ED-016: Install Power Analyzer Meters at Critical Components
- ED-017: Specify High Efficiency Power Supplies
- ED-018: Eliminate Redundant Power Supplies
- ED-019: Eliminate UPS Systems
- ED-020: Bypass the UPS
- ED-021: Supply DC Voltage to IT Rack
- ED-022: Perform an Infra-Red Test
- ED-023: Perform Routine Maintenance and Testing of the Electric Distribution System

- ED-024: Maintain Balanced PDU Loads
- ED-025: Change UPS DC Capacitors
- ED-026: Look for the Simple Energy Saving Actions First
- ED-027: Account for Operating Cost As Well As First Cost
■ Lighting Practices:
- LT-001: Install Energy-Efficient Lamps and Ballasts
- LT-002: Install Occupancy Sensors to Control Lights
■ On-Site Power Generation

7.1.4 Case Study Examples

Three case studies documented by the EERE are included here for illustrative purposes [EER200901, EER200902, EER200903].

7.1.4.1 Case Study 1: Lucasfilm

Lucasfilm identified annual savings of 3,109,200 kWh, which resulted in savings of $343,500. In 2007, the company participated in a U.S. DOE Industrial Technologies Program energy assessment to examine the energy performance of its data center. The assessment was conducted through Save Energy Now, a national initiative to drive a 25% reduction in industrial energy intensity in 10 years, which we alluded to in Chapter 6. Lucasfilm's data center has the following statistics:

■ 13,500 ft^2
■ Houses a render farm (cluster of computers that work around the clock to process digital images), file servers, and storage systems
■ More than 4,300 AMD processors
■ Utilizes standard, constant-speed computer room air conditioners
■ Cooled by a central chilled-water plant serving all the Lucasfilm buildings
■ Receives backup power through UPS systems
■ Minimal use of outside air for cooling

Recommended solutions included the following:

■ Remove redundant rack-mounted UPS systems that impose an additional energy burden, take up rack space, and are useful only if the main UPS system fails to provide backup during a power outage.
■ Turn off servers in between major movie projects when computation loads are significantly reduced.
■ Stage chillers to maintain a high load factor by programming the control system to delay staging additional chillers (during periods of increasing load) until the chillers running at the time reach a higher load factor. This helps ensure that the chillers operate at the highest efficiency possible.

- Operate UPS in switched bypass mode to avoid the conversion losses of going from AC to DC and back to AC. Reconfigure the UPS with an automatically switched bypass to save energy.
- Improve airflow by isolating hot and cold airflow within the center, enabling the cooling system to work more efficiently.
- Implement water-side economizer to capture water produced by cooling towers during periods of low wet-bulb temperature (often at night), eliminating a portion of the load from the chillers and potential points of failure inherent in chilled-water systems.
- Install lighting controls to save both the electricity consumed by lights, which are left on even when the data center is unoccupied, and the HVAC required to offset the heat produced by the lights.

Table 7.3 depicts what the savings were.

7.1.4.2 Case Study 2: Sybase

In 2005, Sybase conducted an energy audit that revealed that its data center N+1 cooling capacity was at risk due to the center's rapid growth. The company's energy managers took action to optimize existing power and cooling resources and free up capacity.

Table 7.3 Case Study 1: Lucasfilm

Measure	Energy Savings kWh/Year	Cost Savings/ Year ($)	Capital Cost ($)	Simple Payback (Years)
Remove redundant rack-mounted UPS	109,500	12,000	0	Immediate
Turn off servers during downtime/power management	273,800	30,000	10,000	0.3
Stage chillers to maintain high load factor	92,800	10,000	4,000	0.4
Operate UPS in switched bypass mode	887,300	98,000	100,000	1.0
Improve airflow	806,700	89,000	113,000	1.3
Implement water-side economizer	928,600	103,000	200,000	1.9
Install lighting controls	10,500	1,000	2,500	2.5
Total for all measures	3,109,200	343,000	429,500	1.2

With the help of local utility company incentives, Sybase is saving nearly 2.3 million kWh of energy per year and has regained cooling capacity to meet growth demands. Implementation of cooling plant and air management measures at Sybase allowed raising the chilled water temperature from 43°F to 52°F, which saved at least 15% of the chiller energy. The room temperature was also raised from 69°F to 74°F. These initiatives saved the firm $262,000. The Sybase data center has the following stats:

- 16,000 ft², 440 racks, and 100 cabinets
- Utility power via a dedicated 2,500 kVA transformer 13 kV/480 VAC 3-phase to two redundant 500 kVA UPSs
- Two 1,000 kVA generators plus one 800 kVA generator
- Three dedicated chillers: one centrifugal with tower and two air-cooled screw chillers

Sybase determined that the following nine energy-saving measures were practical based on estimated implementation costs and payback periods:

- Cooling Plant:
  - Install a high-efficiency base-load chiller and cooling tower with VFD fans and temperature/humidity (enthalpy) controls.
  - Implemented controls to optimize chiller loading for various chiller run scenarios as each chiller has different efficiency curves.
- Air Management:
  - Relocate perforated tiles to where needed, or replace with solid tiles because "closed" adjustable tiles leak about 35 ft³/min. Adjusted tile dampers to match rack airflow requirements.
  - Seal raised-floor penetrations, cable ways, conduits, equipment stands, and ramp skirts; seal unused cut-outs inside the CRAH units to stop most by-pass air.
  - Install VFDs on all 20 CRAH units and create four zones for pressure feedback control. Reduce fan power by 83% to remove the same heat.
  - Install flow diverters on the discharge side of racks with the highest airflow or temperature, reducing the impact on the next lineup's intake temperatures (back-to-front lineup arrangement).
  - Install enthalpy air-side economizer with air ducted into the space. Economizer also serves as emergency cooling system. Add heat recovery to the economizer to heat other spaces.
  - Pending key measure: install blanking panels inside racks.
- Lighting:
  - Eliminated unnecessary lighting using the building management system to control lights from local switches with a 30 minute countdown.

Table 7.4 depicts what the savings were.

Table 7.4   Case Study 2: Sybase

Measure	Energy Savings (kWh/ year)	Cost Savings ($/year)	Capital Cost ($)	Simple Payback (Years)
Cooling Plant:				
Install high-efficiency base-load chiller	476,000	54,000	510,000 (*rebate 54,000*)	8.4
Implement custom control program	75,000	9,000	6,000	0.7
Air Management:				
Relocate perforated tiles	112,000	13,000	0	0
Seal raised floor	150,000	17,000	0	0
Install variable frequency drives (VFD) on 20 CRAHs	866,000	99,000	123,000 (*rebate 52,000*)	0.8
Install partial air-side economizer	313,000	36,000	53,000 (*rebate 24,000*)	0.8
Add heat recovery to air-side economizer	65,000	7,000	1,000	0.1
Lighting:				
Control lights with 30-minute enabled zones	238,000	27,000	17,000	0.6
Total for all measures	**2,295,000**	**262,000**	**710,000 (rebate 130,000)**	**2.2 (with rebates)**

7.1.4.3 Case Study 3: Verizon

In 2008, Verizon participated in a U.S. DOE Industrial Technologies Program energy assessment to examine the energy performance of one of its data centers. The assessment was conducted through Save Energy Now, a national initiative to drive a 25% reduction in industrial energy intensity in 10 years. Verizon serves more than 114 million U.S. customers through its various operations, including Verizon Wireless and Verizon Business, which operates more than 200 data centers in 23 countries. Verizon achieved annual energy savings of 1,540,700 kWh, equating to $181,500 in savings. The Verizon data center under study has the following stats:

- 24,804 ft²
- Receives commercial power from building distribution via a 1,500 kVA transformer 13.2 kV/480 VAC 3-phase, to two redundant 500 kVA UPSs
- Four building-level chiller plants serve more than 750,000 ft² for cooling
- Cooling demand impacted by seasonal temperatures

The following approaches were identified:

- Cooling Plant:
  - Raise chilled-water setpoint from 42°F to 48°F to save chiller energy.
  - Repair water-side economizer to reduce the chilled-water plant power consumption and increase reliability by reducing the wear and tear on the chillers.
  - Install VFDs on the condenser water and chilled-water pumps to reduce pumping energy and maintenance, and decrease potential points of failure.
- Air Management:
  - Raise data center temperature setpoint on the return to the CRAH units until temperatures entering the IT equipment reach the mid- to upper-end of ASHRAE and manufacturer recommended temperature range of 68°F to 78°F.
  - Broaden humidity setpoint to take advantage of the ASHRAE allowable minimum of 20% humidity, resulting in more efficient cooling coil and chilled water system operation. Additional fan energy savings are also possible but would require converting the computer room air conditioners to include variable speed fans.
  - Improve CRAH unit efficiency by adding variable-speed fan controls or consider replacing old units with newer, more energy-efficient equipment.
  - Shut down three CRAH units as sufficient cooling can still be provided with these units turned off.
- Electrical System:
  - Install lighting controls to reduce power use when rooms are not occupied (most of the time). Although small in terms of overall data center power use, the controls are relatively easy to implement and savings are compounded through reduction in cooling.
  - Reduce engine generator heater temperature setpoint to 80°F (from 140°F). Block heaters for the standby generators should be controlled to operate only when the temperature conditions warrant it.

Table 7.5 depicts what the savings were for Verizon.

Table 7.5 Case Study 3: Verizon

Measure	Energy Savings (kWh/year)	Cost Savings ($/Year)	Capital Cost ($)	Simple Payback (Years)
Cooling Plant	**1,273,300**	**150,000**	**150,000**	**1**
Raise chilled-water setpoint				
Repair water-side economizer				
Install VFDs on the condenser water and chilled water pumps				
Air Management	**254,700**	**30,000**	**80,000**	**2.7**
Raise data center temperature				
Broaden humidity setpoint				
Improve CRAH unit efficiency				
Shut down three CRAH units				
Electrical System	**12,700**	**1,500**	**5,000**	**3.3**
Install lighting controls				
Reduce engine generator heater temperature setpoint				
Total for all measures	**1,540,700**	**181,500**	**235,000**	**1.3**

7.1.5 Blade Technology Issues

A blade server is a modular server—containing one, two, or more microprocessors and memory—that is intended for a single, dedicated application (such as serving Web pages) and that can be easily inserted into a space-saving rack with many similar servers. One product offering, for example, makes it possible to install up to 280 blade server modules vertically in a single, floor-standing cabinet. We have discussed some of the issues associated with blade technology in various other places in the text, but extend those observations herewith.

We have already noted that designing a data center for blade servers without a raised metal floor is challenging. Fortunately, most data centers do have raised floors—however, traditional telco rooms often do not. Blade servers can generate a heat load of 14 to 25 kW per cabinet. Servers in high-density arrangements typically exacerbate the power/cooling problem because stacking servers into a small footprint requires more power per rack and, consequently, more cooling. It follows that special considerations must be taken into account when designing a data center to accommodate a large number of racks housing blade servers.

Blades are currently the fastest-growing segment in the server market, but still represent a small portion of the overall market. Market research firms project a compound annual growth rate of 19% for blade shipments from 2007 through 2012; however, this growth does not translate to market domination—in 2007, blades represented 10% of shipments, and we are forecasting that this will rise to only 20% in 2012. Blade technology has undergone major changes during the past 5 years; we expect more-significant changes during the next 5 years. Blades and chassis will not be standardized across vendors in the short term, but the following advancements are anticipated [BUT200801]:

- By 2010, logically joining blades to create a single server from multiple physical blades will become a standard blade server capability.
- By 2011, increased virtualization adoption will make input/output (I/O) quality-of-service controls a standard feature of blade servers.
- By 2012, memory and chassis aggregation will enable blade technology to address large, vertically scaling workloads.

The concept of blade servers emerged in 2000 when a number of small, specialized vendors (e.g., RLX Technologies and FiberCycle Networks) developed first-generation products for large Internet data centers. When the service-provider market collapsed, blade manufacturers started to take their products to the broader enterprise data center market. First-generation blades were rudimentary devices that typically leveraged low-power CPU technology with limited performance; these products lacked key enterprise-class features, such as error-corrective memory and hardware management functions. In recent years, blade performance has risen dramatically because the same multicore CPU technology is now deployed in blades and rack-optimized servers, and new workload deployment has increasingly favored a more node-based approach. As a result, the long-term promise of blade computing has increasingly attracted the attention of the main ×86 server vendors. Dell, HP, Compaq (which was acquired by HP during this period), IBM, Intel, and others invested in designs for blade servers for introduction around the 2002 time frame that were more suited to enterprise needs, with investment in higher-performance CPUs and an emerging focus on management tools; this can be seen as a second generation of blade computing. By 2005, HP (now merged with Compaq) and IBM blade designs evolved further, and a third generation of blade computing emerged. This third generation of blade technology is marked by more focus on performance, better manageability technology, and improved interoperability with network and storage infrastructure. In late 2006 and early 2007, HP, IBM, Dell, and Sun Microsystems introduced new blade designs that further addressed the needs of the data center, including better I/O management and better thermal management. These products represented the beginning of a fourth generation of blade products. Although all the vendors' blade products share similar goals to provide better density, power/cooling, and agility, each of the vendors' products is unique, and there is virtually no interoperability among them. Chassis size varies from vendor to vendor, and the blades, storage modules, switches, power supplies, and other key components are unique to each vendor. Sun, HP, and Rackable Systems are reportedly investing in container-based mobile data center solutions that are typically leveraged on blade technology compared with the same number of pedestal servers [BUT200801].

Blade servers, which share a common high-speed bus, are designed to create less heat and thus save energy costs as well as space; however, heat is more physically concentrated than in the case of pedestal servers [GRE201001]. We can define Delta-T (ΔT) as the difference in temperature between air entering and air exiting the servers. Studies have shown that blade servers generate a 60°F ΔT under peak load—more than double the 26.5°F of 1U servers. Blade servers can generate a heat load of 15 to 30 kW per cabinet, representing a density of more than 300 WPSF over the total data center area. However, many existing and planned data centers support only a fraction of this density; one analysis of 19 data centers estimated the average density at the end of 2005 to be only 32 WPSF [GAR200601]. Climate Savers Computing goals for systems with single-output power supplies (these systems are typically volume servers, including 1S, 2S, 4S, and blade servers) range from/to [CSC200901]

- Climate Savers Computing Bronze: Volume servers must have 85% minimum efficiency rating for the PSU at 50% of rated output (and 81% minimum efficiency at 20% and 100% of rated output), and a power factor of at least 0.9 at 100% of rated output.
- Climate Savers Computing Gold: Volume servers must have 92% minimum efficiency rating for the PSU at 50% of rated output (and 88% minimum efficiency at 20% and 100% of rated output).

7.1.6 Other IT Greening Techniques

This section provides a short list of other possible near-term or longer-term greening techniques.

7.1.6.1 Data Center Power Management Systems

Data centers are often designed for peak usage. Multi-time-zone operations and globalization may extend the workday considerably (even to 24 hours) and thus extend the peak usage period. Nonetheless, in many instances, data center resources are underutilized as application load fluctuates during the course of a day or a week. To better manage power during potential low utilization periods, product architecture and design of data center infrastructure equipment must evolve to support the control of power consumption of individual components. Management systems must be put in place that monitor utilization and dynamically bring on- and off-line components of the data center Infrastructure based on workload [ATI200901].

7.1.6.2 Collocated Facilities for Data Centers

In recent years, IT management has consolidated its IT infrastructure as much as possible, including a trend toward consolidating computing assets into raised-floor, secured, centralized data center facilities. Collocation is a type of data center where multiple customers locate network, server, and storage equipment and interconnect to a variety of telecommunications and other network service provider(s). Collocation typically results in cost savings (and carbon footprint reduction) because of economy of scale related to large power and mechanical systems (collocation facilities may be 50,000 to 100,000 ft² in size). Traditionally, facilities power costs are shared across customers, depending on their infrastructure and space requirements but individual rack-metering has become available of late [ATI200901].

7.1.6.3 Outside Air Cooling

Outside air cooling can be used in some cases. Intel has documented a proof of concept on the outside air cooling idea, indicating that the company was able to cool the data center almost exclusively with outside air that was 90°F or cooler. Intel reportedly concludes that a correctly placed 10 MW data center could save $2.87 million annually in power costs. In Intel's proof of concept, a data center with 900 blade servers was set up in a temperate climate, half with a traditional air conditioner-chiller setup, half with the air economizer. Intel did not attempt to control the humidity in the air economizer setup and did minimal air filtering. The test ran for about 1 year, and the air economizer setup was used to cool the 450 blade servers on the one side of the data center 91% of the time, resulting in a 67% savings in electricity costs compared to the other half of the data center, which used normal AC [MOR200801].

7.1.6.4 Geometric Symmetry/Asymmetry

Minor geometric asymmetry in the cabinet layout in a data center can impact cooling efficiency. Perturbation of the rack layout disturbs the airflow pattern by changing pressure drop in the aisles and creating positive pressure gradients at the end of aisles. Positive pressure gradients force the air to recirculate in the room, interfering with the overall airflow pattern and creating regions of high temperature in the data center. The presence of such recirculation zones can aggravate the problem by forcing hot air back into the servers. As a result, the cooling of data centers can be very inefficient if proper measures are not taken to ensure an optimal layout. Aided by an optimized airflow pattern, data center performance depends on how heat is extracted, maintaining suitable room-level air temperatures at all times. Obviously it should be noted that from practical construction considerations, a symmetric layout with reference to air-conditioning units is not achievable in most cases. Therefore, exact flow optimization with uniformly balanced cooling load based on layout can be difficult to implement in practice [PAT200201]. Efforts should be made, however, to take symmetry under consideration.

7.1.6.5 Cold Computing

Cold computing refers to the refrigeration of electronics and opto-electronics (e.g., with thermoelectric refrigeration); these refrigeration technologies typically require only small-scale or localized spot cooling of small components that do not impose a large heat load. The advantages of "cold computing" have been known for a while; for example [SLO199601] states that "speed gains of 30% to 200% are achievable in some CMOS computer processors" and that "cooling is the fundamental limit to electronic system performance." Cooling of laser diodes and infrared detectors to temperatures 10 K < T < 200K would greatly improve performance and sensitivity [TRI200201]. This technology is not yet commercially available.

7.2 Network Strategies

In the networking/carrier environment, greening strategies include improving System Load and Facilities Load. System Load improvements include the removal of obsolescent equipment and adoption of Operational Best Practices. Improving the energy efficiency of individual Network Elements (NEs) by replacing them with newer/better products offers the key opportunity to reduce energy consumption and energy cost; improved equipment efficiency includes high-efficiency power supplies and high-efficiency fans. Other approaches include improved facility utilization (e.g., better traffic concentration). Facilities Load improvements include deploying optimal power distribution architecture, achieving cooling efficiencies (e.g., with variable capacity cooling), and introducing facility operational best practices (e.g., proactive monitoring may reduce the number of truck rolls). *Clean energy,* such as photocells (photovoltaics), wind, fuel cells, and biodiesel, holds promise, especially for wireless networks, but is not yet economically attractive in most instances.

The customer premises equipment (CPE), the outside plant (OSP), and the network equipment provide opportunities for greening and for the introduction of eco-sustainable solutions. Equipment-level measures of efficiency include the ATIS Telecommunications Energy Efficiency Rating (TEER), which is a ratio of work performed divided by power dissipated or energy consumed over time, as well as others discussed in Chapter 3. Facility-level measures of efficiency include the Green Grid Power Usage Effectiveness (PUE), which is a ratio of the amount of power entering a data center divided by the power used to run the computer infrastructure within it:

PUE = Total Facility Power or Energy/ICT Equipment Power or Energy

as well as others discussed in Chapter 3. In the networking context, the Total Facility Power includes the NE load and (sub)systems that support the NE equipment load, including building electrical and mechanical systems, DC power plant, the primary and secondary distribution, lights, elevators, and so on.

ICT has the potential to help to reduce energy consumption in businesses and homes (by 15% by 2020) by enabling teleworking, telemedicine, smart grids, and smart building services.

7.2.1 Retire and Replace Obsolescent Equipment

As is the case for data center Best Practices, service-provider planners should schedule retirement and removal of obsolescent equipment. For example, 130-V power plants have been used in Central Offices to support T1 systems, but these copper-based systems have mostly been replaced by fiber-based transport technologies; some copper-based T1 systems are still used in the Outside Plant but these loops are short and do not require 130-V power—therefore, by upgrading the T1 repeaters, the 130-V power plant can be retired. Coin phones, Asynchronous Fiber-Optic Terminals, and older Digital Loop Carrier systems are other examples in this category. Telephony services based on wireline access line are eroding, due to wireless and cable substitution and adoption of Voice over IP (VoIP) services. At some point in the (near) future, traditional Class 5 Time Division Multiplexing (TDM) switches will become underutilized, if they are not already so. Consequently, consolidating switch modules will lower energy demand and, in turn, lower the cooling and DC power plant capacity requirements.

In general, the rectifiers utilized in telecommunications DC power plants are reasonably energy efficient. However, poor utilization of these rectifiers can result in energy waste. As a minimum, carriers should seek to improve the utilization of older ferro-resonant rectifier (ferro) power plants to at least 50%. Hybrid power plants are defined as DC power plants that contain both ferro and Switch Mode Rectifiers (SMRs). Hybrid plants will also generally include both legacy and modern primary distribution bays. By implementing a hybrid power plant, both reliability and energy efficiency can be improved. Strategies to improve utilization of ferro rectifier power plants include the following [ATI200901]:

- Turn off superfluous capacity rectifiers until the utilization is ≥50%
- Turn on manufacturer ferro rectifier energy management algorithms for plants with <50% utilization
- Replace ferro rectifiers with SMRs.

- Implement a hybrid power plant (mixture of ferros and SMRs).
- Older outside plant Controlled Environmental Vaults (CEVs) and cabinets make use of ferro-resonant rectifiers; therefore, planners should consider upgrading those ferro-resonant power supplies that are poorly utilized or at their end of service life with newer technology.

7.2.2 Wireless Networks

A number of techniques and technologies are available to reduce the energy consumption of wireless access networks, especially the Radio Base Station (RBS) sites. Reductions up to 30% are achievable. ATIS suggests the following [ATIS200901]:

- Shut down redundant units or resources not needed during low traffic hours using sleep mode functions.
- Avoid unnecessary DC/DC conversion typically saves about 15% in energy consumption and cost.
- An outdoor combined RBS and Battery Backup Unit (BBU) forms a streamlined site with a common cabinet/shelter that minimizes hardware and energy consumption.
- Use remote power schemes, which enable several sites to be powered from one central point, as a result of advances in power transmission technology. The user sites could be RBS sites, handset charging stations, or village power. The central site produces power using generator sets and distributes it to other RBS sites over specially –designed, low-cost, copper-free cable in star, ring, or chain configurations, over several tens of kilometers.
- On the infrastructure side, cooling costs can be reduced by optimizing air-conditioner use or, preferably, by migrating to a more passive approach.

Some other possibilities include

- Use solar-powered devices for wireless sensor networks.
- Use solar-powered devices for wireless sensor networks supporting Intelligent Transportation Systems (ITS); ITS systems support effective traffic management through traffic flow data gathering and wireless vehicle identification [TAV200901].
- Use solar-powered devices for microwave relays and repeaters for remote or challenging regions.

7.2.3 Redundancy

Another opportunity to reduce power relates to the system redundancy architecture. Typically, reliability and redundancy architectures give rise to low equipment utilization with ensuing poor energy efficiency. A "2N" reliability architecture, by definition, results in utilization that is less than 50%; an "N + 1" architecture may provide sufficient reliability, result in better utilization, improved energy efficiency, and also entail decreased capital expenditures.

7.2.4 Access Networks*

We alluded to "eco-services" in Chapter 1. Many of these services depend on broadband connectivity, typically through fiber-deep access networks such as fiber-to-the-node (FTTN), fiber-to-the-building (FTTB), and fiber-to-the-home (FTTH).

Product life-cycle analyses of GPON (gigabit passive optical networking) and VDSL (very high bit-rate digital subscriber line) technologies confirm that the access network's greatest environmental footprint occurs in the home. An Alcatel-Lucent environment study found that 82% of the GPON-related life-cycle eco-impact is attributable to the operational phase of the optical network terminal (ONT) equipment. For VDSL, CPE accounts for 67% of the life-cycle eco-impact. This data implies that efforts should be made to reduce CPE energy use. Even though CPE energy consumption does not typically have a direct cost for operators, it is reportedly on the minds of customers, and solutions that reduce the power requirements of DSL modems, ONTs, remotes, and other in-home technologies may, therefore, have consumer appeal.† The power consumption of ONT and CPE has relatively little direct effect on a service provider's operating expenditures (OPEX), while the power consumption of network equipment has a direct impact on OPEX and is therefore important from a service provider's point of view.

When fiber or other cable is run to a home, some environmental disturbance is unavoidable. Yet the eco-services that are implemented as a direct result of that deployment eventually benefit the environment. According to studies by the FTTH Councils in Europe and Americas, the positive impact of FTTH on ICT emissions should be evident in the United States in 6 years and in the European Union in 15 years (the difference in the two time scales stems from the fact that buried cable is typically used in the EU, while aerial fiber is most often chosen in the United States).

---

* Portions of this section are based on Alcatel-Lucent materials [ALC200902].

† That is (more) typically the case where there are demonstrated cost savings to the consumer. If the payback time to the consumer for some green choice is 10 to 15 years, the practical interest may be limited. To furnish a concrete example in a somewhat related area, this author was responsible for building the business case for either a new building air conditioning or the relocation/re-use of a 3-year-old unit from another building. The new AC was more efficient and had a monthly operational cost of $51,000; the old AC had a higher operational cost of $68,000. But when all was said and done, the payback of installing the new unit versus installing the old unit was 15 years. Naturally, if the cost of energy was higher, then the payback would be shorter.

Power reduction is an ongoing effort
All access technologies have made significant progress

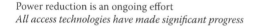

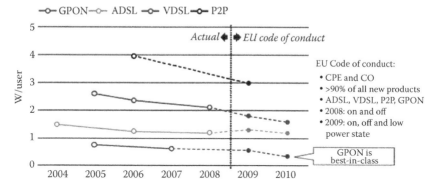

FTTx is greener...
VDSL, GPON and P2P are power efficient per Mbls

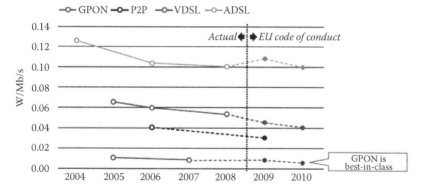

Figure 7.5   Access technologies and power consumption.

The power efficiency of all access technologies has improved in recent years and will continue to do so. GPON, xDSL, and Point-to-Point (P2P) fiber* have all evolved to comply with the targets of the EU Code of Conduct on Energy Consumption of Broadband Equipment (see Figure 7.5). These improvements have been achieved through greater integration of components, and the use of system-on-a-chip technologies and more efficient transceivers. GPON is the most

---

* More specifically, P2P Ethernet over fiber. P2P fiber provides symmetrical bandwidth and dedicates a single fiber for each end user. This provides service transparency: the ability to provide any service irrespective of other network users. Also known as active Ethernet or EFM (Ethernet in the First Mile), it employs an Active Optical Network (AON), in which each CPE is connected to an active port in the switching equipment. P2P fiber is a mature technology that enables dedicated symmetric broadband speeds of 100 Mbps per end user moving toward 1 Gbps (e.g., see [ERI200801]).

eco-efficient access technology, delivering the greatest capacity at the lowest Total Cost of Ownership (TCO). Alcatel-Lucent analysis of lifetime energy OPEX for FTTH in a typical European country reveals that GPON eco-efficiency savings are close to €30 per customer over 5 years compared to P2P (see Figure 7.6).

For Digital Subscriber Line (DSL) technology, two technologies exist and have been productized: (1) the L2 low power mode for Very High bit-rate Digital Subscriber Line (ADSL2/ADSL2+), (2) and MaxSNRM (maximum signal-to-noise ratio margin) for ADSLx and VDSL2. The L2 power mode switches a line to a lower bitrate and power level when user activity drops below a certain threshold, while MaxSNRM caps the noise margin (by reducing power) when it exceeds a certain threshold. These two technologies have existed for some time but have not been widely deployed until now, as power fluctuations can generate varying cross-talk, which in turn can make the network unstable. Alcatel-Lucent asserts that it offers the first stabilized and deployable L2/MaxSNRM end-to-end solution on the market and realizes power savings of up to 25%.

There is also interest in minimizing power consumption in FTTN deployments. One approach is to use the concept of distributed DSLAM (Distributed Digital Subscriber Line Access Multiplexer). With no dedicated controller per node and one central network terminal for 24 Remote Expansion Modules (REMs)/Sealed Expansion Modules (SEMs), the distributed DSLAM concept reduces power consumption. Comparing distributed DSLAM to a traditional stand-alone remote

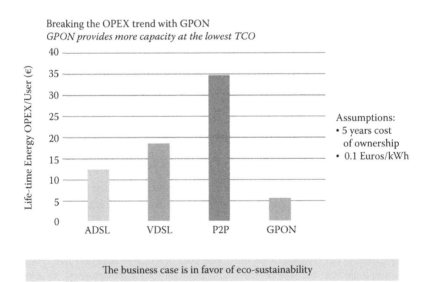

Figure 7.6   OPEX and GPON.

DSLAM scenario, assuming 48 port nodes and a 6-year period, the distributed alternative saves as much as 60% in aggregation CAPEX (CAPital EXpenditure), 40% in OPEX, and up to 20% in power consumption.

According to a life-cycle analysis of FTTH conducted by the FTTH Council Europe, 83% of network-related CO_2 emissions result from the deployment of passive fiber, particularly because of the requirement for digging trenches and laying fiber. Duct sharing and microtrenching can reduce these impacts.

- ■ *Duct sharing:* Duct sharing minimizes the civil work required for fiber deployment and lowers the CAPEX barrier to new rollouts because multiple operators share the cost of deployment. In an Alcatel-Lucent case study of a dense urban area with multi-dwelling units (MDUs) of 20 apartments, duct sharing reduced the FTTH costs of fiber deployment by 15% and the FTTB costs of copper deployment by the same amount. By reusing the existing infrastructure, environmental impact can be reduced by up to 40%, depending on the number of ducts or sewers that are shared among different service providers.
- ■ *Microtrenching:* Deploying outside the plant is time-consuming, expensive, and environmentally disruptive because it typically involves digging, trenching, and fiber deployment; hence, any alternative approaches that minimize the effort or the impact of this process would appeal to operators. Microtrenching is the process of making small, "surgical" incisions in the ground to lay fiber. As well as being eco-friendly, it is also aesthetically preferable for the urban or suburban landscape. Because of its precision and minimized scale, microtrenching cuts deployment costs by about 30% and deployment time by 70%.

Maintaining proper temperature control of network equipment is essential for optimal performance, yet traditional air conditioning is expensive and environmentally taxing. Heat exchange and direct air cooling are more eco-efficient than air conditioning, and both are used in outdoor cabinet solutions. Direct air cooling is the most eco-efficient option, but requires more maintenance (dust filters must be changed on a regular basis, requiring a site visit by a vehicle—which does add to the carbon footprint). Heat exchange technology provides an efficient alternative, especially when using variable-speed fans; however, it typically requires that the enclosed equipment be capable of a wider operating temperature range than when cooled by air conditioning. Overall, using a heat exchanger instead of a conventional air conditioner can save roughly 20% in power. This technology further strengthens NE products, which in many cases does not typically require air conditioning due to the hardening of the equipment by design (many network products can withstand temperatures of up to 65°C).

7.2.5 Satellite Operators

Satellite Operators may wish to upgrade their plants to use solid-state power amplifiers (SSPAs) that are smaller, weigh less, have higher linear power levels, cost less, and draw much less power than the traditional HPAs (high-power amplifiers) such as Traveling Wave Tube Amplifiers (TWTAs) and klystrons (often used today).

There are 22,638 uplink chains in approximately 1,781 commercial and broadcast teleports worldwide [WTA200701]. Electricity demand in the commercial sector will increase 38% between 2007 and 2030, and electricity prices are expected to rise steadily after 2015 because fuel prices will increase, and the need for new capacity will grow [EIA200901]. It then follows that switching to next-generation SSPAs has the potential for reducing annual energy costs.

Table 7.6 provides an example from a recent trade show. This example depicts the savings in power for a teleport with 100 HPAs (50 in 1:1 redundant mode) [BOY200901].

While at first brush the savings appear impressive, a quick analysis shows that the payback period may be fairly long. Consider an existing teleport with 50 1:1 HPAs that are already deployed. Assume indeed that the yearly operational cost is as shown in Table 7.5 for SSPA "A". The equipment cost for a 500 W 1:1 HPA is generally in the $75,000 range, plus installation cost, say, for a total of $85,000. The cost to deploy 50 1:1 units would be 50 × $85,000 = $4,250,000. Table 7.7

Table 7.6   Power Use: Large Teleport – 50 Uplinks 1:1 Configuration

Power Draw	SSPA "A"	SSPA "B"	TWTA	Klystron "A"	Klystron "B"	Next-Gen SSPA
Primary (kW)	185	245	120	440	275	120
Stand By (kW)	10	10	80	335	45	10
Total Draw (kW)	195	255	200	775	320	130
Yearly Electric Costs (@ $0.08/kWh)	$136,656	$178,704	$140,160	$543,120	$224,256	$91,104
Yearly Electric Costs (@ $0.16/kWh)	$273,312	$357,408	$280,320	$1,086,240	$448,512	$182,208
Tons of Coal Burned (Yearly)	854	1,117	876	3,395	1,402	569

Table 7.7 Comparison of Solutions, Including Energy Consumption and Maintenance Costs

Year	New	Existing	New	Existing
	(0.08/kWh)	*(0.08/kWh)*	*(0.16/kWh)*	*(0.16/kWh)*
0	$4,250,000	$0	$4,250,000	$0
1	$4,553,604	$561,656	$4,644,708	$698,312
2	$4,857,208	$1,123,312	$5,039,416	$1,396,624
3	$5,160,812	$1,684,968	$5,434,124	$2,094,936
4	$5,464,416	$2,246,624	$5,828,832	$2,793,248
5	$5,768,020	$2,808,280	$6,223,540	$3,491,560
6	$6,071,624	$3,369,936	$6,618,248	$4,189,872
7	$6,375,228	$3,931,592	$7,012,956	$4,888,184
8	$6,678,832	$4,493,248	$7,407,664	$5,586,496
9	$6,982,436	$5,054,904	$7,802,372	$6,284,808
10	$7,286,040	$5,616,560	$8,197,080	$6,983,120
11	$7,589,644	$6,178,216	$8,591,788	$7,681,432
12	$7,893,248	$6,739,872	$8,986,496	$8,379,744
13	$8,196,852	$7,301,528	$9,381,204	$9,078,056
14	$8,500,456	$7,863,184	$9,775,912	$9,776,368
15	$8,804,060	$8,424,840	$10,170,620	$10,474,680
16	$9,107,664	$8,986,496	$10,565,328	$11,172,992
17	$9,411,268	$9,548,152	$10,960,036	$11,871,304
18	$9,714,872	$10,109,808	$11,354,744	$12,569,616
19	$10,018,476	$10,671,464	$11,749,452	$13,267,928
20	$10,322,080	$11,233,120	$12,144,160	$13,966,240

depicts the TCO for both the existing environment and for the new environment as a function of the year (for the new equipment, it is assumed that the maintenance fee is 5% of the initial cost; for the existing system, it is assumed to be 10% of $4,250,000 or $425,000). As observed, the payback at $0.08/kWh is 16 years and the payback at $0.16/kWh is 13 years). It may be difficult for businesses to justify these paybacks, but if a government program subsidized the substitution, it might make more sense. Also, of course, newer technology would also make sense for new construction and greenfield applications.

It should be noted that what drives the payback figure is more the higher maintenance cost on an older piece of equipment than the energy component. Table 7.8 shows a pure energy consumption comparison (without any maintenance costs included). As can be seen, the new equipment would never pay for itself.

7.2.6 Short List of Other Possible Initiatives

This section could actually be long because many ideas have been surfaced. However, we list just a small set because some of the ideas in this category are not yet totally practical or cost-effective.

The conventional cooling techniques discussed in Section 7.1 are generally applicable to the carrier environment, with the exception of the Outside Plant, where specialized techniques for small shelters, outdoor equipment racks, and so on may be needed. Table 7.9 depicts some newer/nonconventional techniques being investigated by some carriers, particularly in Europe [ETN200801].

In a different direction, some suggest the development of more efficient protocols, particularly in the network management area [ZUK200901], for example, optimize interactions with managed systems and simplify interactions. The deployment of transmission protocols with power management is also advocated or useful; for example, protocols used in sensor networks, including ZigBee, have power

Table 7.8 Comparison of Solutions, Including Energy Consumption but Excluding Maintenance Costs

Year	New	Existing	New	Existing
	(0.08/kWh)	(0.08/kWh)	(0.16/kWh)	(0.16/kWh)
0	$4,250,000	$0	$4,250,000	$0
1	$4,341,104	$136,656	$4,432,208	$273,312
2	$4,432,208	$273,312	$4,614,416	$546,624
3	$4,523,312	$409,968	$4,796,624	$819,936
4	$4,614,416	$546,624	$4,978,832	$1,093,248
5	$4,705,520	$683,280	$5,161,040	$1,366,560
6	$4,796,624	$819,936	$5,343,248	$1,639,872

Table 7.8   Comparison of Solutions, Including Energy Consumption but Excluding Maintenance Costs (Continued)

Year	New	Existing	New	Existing
	(0.08/kWh)	*(0.08/kWh)*	*(0.16/kWh)*	*(0.16/kWh)*
7	$4,887,728	$956,592	$5,525,456	$1,913,184
8	$4,978,832	$1,093,248	$5,707,664	$2,186,496
9	$5,069,936	$1,229,904	$5,889,872	$2,459,808
10	$5,161,040	$1,366,560	$6,072,080	$2,733,120
11	$5,252,144	$1,503,216	$6,254,288	$3,006,432
12	$5,343,248	$1,639,872	$6,436,496	$3,279,744
13	$5,434,352	$1,776,528	$6,618,704	$3,553,056
14	$5,525,456	$1,913,184	$6,800,912	$3,826,368
15	$5,616,560	$2,049,840	$6,983,120	$4,099,680
16	$5,707,664	$2,186,496	$7,165,328	$4,372,992
17	$5,798,768	$2,323,152	$7,347,536	$4,646,304
18	$5,889,872	$2,459,808	$7,529,744	$4,919,616
19	$5,980,976	$2,596,464	$7,711,952	$5,192,928
20	$6,072,080	$2,733,120	$7,894,160	$5,466,240
21	$6,163,184	$2,869,776	$8,076,368	$5,739,552
22	$6,254,288	$3,006,432	$8,258,576	$6,012,864
23	$6,345,392	$3,143,088	$8,440,784	$6,286,176
24	$6,436,496	$3,279,744	$8,622,992	$6,559,488
25	$6,527,600	$3,416,400	$8,805,200	$6,832,800
26	$6,618,704	$3,553,056	$8,987,408	$7,106,112
27	$6,709,808	$3,689,712	$9,169,616	$7,379,424
28	$6,800,912	$3,826,368	$9,351,824	$7,652,736
29	$6,892,016	$3,963,024	$9,534,032	$7,926,048
30	$6,983,120	$4,099,680	$9,716,240	$8,199,360

Table 7.9 Newer/Nonconventional Cooling Techniques Being Investigated by Some Carriers

Fresh air cooling	A new method for cooling telecom equipment based on the use of fresh air year-round (no chillers) was developed in 2005 at Swisscom. Simulations and trial results have shown that using solely fresh air to cool the network equipment is feasible, particularly for geographic cooler regions. More than 120 sites are already equipped with this solution. This cooling system basically consists of two single-stage exhaust fans located at the top of the room. Each fan provides 50% of the rated airflow volume. Should one fan fail, more than half of the cooling power would still be available (over 50% redundancy). If lower system availability is required (e.g., at sites with low thermal load), only one fan providing 100% of the rated airflow volume can be used. The first fan is switched on at 24°C and switched off at 22°C; the second one at 26°C and 24°C, respectively. Fresh air cooling is being used by operators such as Swisscom, BT, France Telecom Orange, among others.
Liquid pressure amplification (LPA)	Modern refrigeration systems typically work at half their theoretical performance efficiency; therefore, add-on benefits can be considered. Liquid Pressure Amplification (LPA) is a new technology designed to increase the efficiency of refrigeration systems by reducing flash gas. By installing LPA booster pumps into the liquid line of suitable chillers at main offices, significant energy reductions are possible. The benefits depend on the local ambient temperatures, runtimes, and the size and inherent efficiency of the machine. In the majority of cases, the benefit is 30% of the running cost. In addition to the energy savings, the compressor load decreases, which increases reliability and reduces noise.
District cooling	District cooling is implemented at one of KPN's largest switches (in Amsterdam). This has resulted in a structural savings in the energy usage needed for the cooling of the site. This sustainable cooling concept has won international acclaim and was nominated for the Green Award for Data Centres 2007 (organized by Data Centres Dynamics). The technique reached the shortlist of four European nominations. The system comprises a pumping station that pumps up water from a lake on the outskirts of Amsterdam (Nieuwe Meer) from a depth of approximately 40 meters. The water is transferred to a production building via a cold/heat exchanger. In addition to cold/heat exchangers, the building accommodates a number of cooling machines. The machines ensure that, if necessary, the temperature of the incoming water is always between 5°C and 6°C.

Table 7.9   Newer/Nonconventional Cooling Techniques Being Investigated by Some Carriers (Continued)

Borehole cooling (Telia-Sonera)	Mälarhöjden LX is one of TeliaSonera's 30 sites that are chilled by closed ground collector systems for peak cooling. The cooling system is designed on the basis of a principle minimizing the number of moving parts with 100% free cooling during all hours of the year. Mälarhöjden LX has no chillers and no Freon installed for balancing the process load. The system at Mälarhöjden has a nominal capacity of 150 kW (systems with nominal outputs of 100, 150, and 250 kW are also possible).
Soil cooling system for small sites	KPN is developing a low-energy cooling system for sites in its mobile network where equipment is installed next to the transmitter masts for mobile networks. To ensure correct operation, the equipment must be cooled. However, using traditional cooling (by means of air-conditioning systems) consumes a relatively large amount of energy, and it uses a fan or compressor that can cause noise nuisance. KPN has launched a trial using an alternative cooling system to remove the heat generated by the equipment. The system indirectly uses the cooling capacity of groundwater in the soil below the site. At the site, one or more pipes are driven vertically into the ground to a depth of up to 75 meters and water is made to flow through them. This system is called a Vertical Soil Heat Exchanger (VSHE). The water cooled by the soil is routed back to the cooling unit in the compartment where the technical equipment is installed. The cooling unit has a heat exchanger that is connected to the VSHE. This makes it possible to cool air heated by the equipment (up to 35°C). A fan in the cooling unit recirculates the air from the room and routes it across the heat exchanger. Utilizing the closed system of loops in the soil, the heated water is sent back to the groundwater in the soil to be recooled.
Heat pipe cooling concept in junction boxes	KPN plans install between 15,000 and 25,000 junction boxes all over the country in the coming years. The hardware in the boxes produces heat. The heat density in the junction box will be somewhere between 1200 and 2000 W, depending on the IT configuration in the cabinets. It is necessary to cool the junction boxes to keep the equipment within the ETSI-climate chart. KPN seeks to use a cooling method in the boxes that uses three to four times less energy than conventional cooling methods. A solution is to install a heat pipe to bring the ambient temperature in the junction box within the ETSI-climate conditions. A heat pipe is a closed pipe that takes away heat by

Continued

Table 7.9 Newer/Nonconventional Cooling Techniques Being Investigated by Some Carriers (Continued)

	means of condensation. The principle underlying the Heat Pipe has been used in the past. What is new is that KPN will install the pipe partly in the soil beneath the junction box. This allows use of the cold stored in the earth. That is, in theory, what makes this cooling method exceptionally efficient. Only a small fan is needed to blow the hot air into the duct with the heat pipe to a depth of approximately 1.5 meters. The cooled air flows back into the box.

control.* Also, self-healing networks may reduce truck rolls and save power. Some other areas of research include the following [ARM200901] (the reader should also refer to Table 1.3):

■ Dynamic, all-optical networks with solar- or wind-powered optical repeaters
■ Wireless mesh ad-hoc networks with mini-solar panels at nodes
■ New shortest energy path Internet architectures with servers, computers, and storage collocated at remote renewable energy sites such as hydro dams, windmill farms, etc.

---

* Just to sensitize the reader to the topic of Wireless Sensor Networks (WSNs), note that WSNs have unique characteristics, such as power constraints, battery life, redundant data, low duty cycle, and many-to-one flows. Consequently, new design methodologies are needed across a set of disciplines, including, but not limited to, information transport, network and operational management, and in-network/local processing. In particular, a lightweight protocol stack is desired; also, a very large number of client units (say, 64k or more) must be supported by the system and by the addressing apparatus. Low power consumption is a key factor in ensuring long operating horizons for non-power-fed systems. Power efficiency in WSNs is generally accomplished in three ways: (1) low duty cycle operation; (2) local/in-network processing to reduce data volume (and, hence, transmission time); and (3) multihop networking (this reduces the requirement for long-range transmission because signal path loss is an inverse power with range/distance)—each node in the sensor network can act as a repeater, thereby reducing the link range coverage required, and, in turn, the transmission power). Sensor networks can be as small as one room or the human body. For example, on-body medical sensors, using what has been called Body Area Networks (BANs), require new ultra-low-power RF technologies. RF integrated circuit (RFIC) technology can now offer low power, reduced external component count, and higher levels of integration. MEMS (micro electro-mechanical system) technology has been used for the development of energy scavengers to power autonomous medical systems. An example is the energy scavengers that generate micropower from body heat, based on the conversion of thermal energy into electrical energy. Because this energy source is continuous, the systems can be always-on and have an almost infinite lifetime. The challenge will be to prove that such devices can extract enough power from the human body to supply the systems in future. See [MIN200701] for a discussion of WSNs.

- Topology and architecture issues to stretch the network and move routers and switches from major intersections
- New routing and resiliency architectures for wired and wireless networks for massively disruptive topology changes due to setting sun or waning winds that power routers and servers
- New stats and measurement analysis of bits per carbon (bpc) utilization, optimized "carbon" routing tables, etc.
- There is a trend toward wanting to "manage flows instead of individual packets, which dramatically reduces energy consumption" [ZUK200902].
- There is research on "Green Radio" that seeks to develop power-efficient radio systems* "derived from greener manufacturing and environmentally-friendly system architectures" [SWE200901].
- There is research within the field of nanotechnology to develop more efficient batteries. Clearly, batteries have wide applicability for wireless systems (including phones, Personal Digital Assistants (PDAs), and sensor nodes) and for data center/telco rooms for UPSs.

Observers state that Internet traffic (now) doubles every 2 years; this causes the Internet energy consumption to increase at a much higher rate than the consumption of other industry sectors—these other sectors normally increase energy consumption in line with population or Gross National Product (GNP) rates. Reducing the growth of energy consumption has been identified as being very important not only due to the GHG effect, the increasing cost of energy, and the expected introduction of a carbon tax (in various countries in the world), but "also because it can limit the growth of the Internet itself because of the unavailability of power and the growing difficulty in cooling off the Internet routers" [ZUK200902]. In today's Internet, packets are processed (including repeated buffering and routing table lookups) individually at each router. Even if the packet processing is done in hardware, it is still highly energy intensive. In circuit switching, there is no need for individual treatment of packets; the switch is configured such that the bits are switched from their input port to the output port(s) with little or no buffering, which is more energy efficient and thus more scalable, according to some. These proponents state that significant reduction of more than an order of magnitude in energy consumption can be achieved by (fractional-)lambda/time-driven switching compared to all-IP networks [OFE200901].

---

* There is now considerable research in the area of cognitive radios. A cognitive radio is a software-based radio whose control processes make use of situational knowledge and intelligent processing to support the needs of the user, application, or network; for example, a radio could autonomously detect and exploit empty spectrum (opportunistic spectrum utilization). It appears at face value that a cognitive radio is a green radio.

7.3 Enterprise Equipment

Enterprises use ICT equipment to run their business. Telecommunications equipment, such as desk and mobile telephones, fax machines, PBXs, routers, switches, teleconference audio, and video facilities, is in common use. IT equipment typically includes servers, computers, PCs and laptops, monitors, printers, data storage, and chargers. The "Retire and Replace Obsolescent/Inefficient Equipment" principle should apply here. According to ATIS, in order to progressively lead to a reduction in GHG, enterprises must take the steps necessary to reduce their own individual carbon footprints through the use of such advanced technology and processes as, but not limited to [ATI200902]:

- Reduction of paper reports and memos
- Use of reusable storage devices
- Use of power management tools
- Encouragement for PC power-down and sleep modes
- Use of more energy efficient display devices rather than Cathode Ray Tubes (CRTs)
- Use of smart motors and variable-speed drive control systems
- Use of solar and other renewable energy resources for running their ICT equipment
- Facilitating employee teleworking and telecommuting
- Use of flexible fleet management, delivery services software, and associated hardware
- Encouraging IT partners to employ energy-efficient techniques for data centers and networking equipment such as routers, switches, servers, and storage devices

The Climate Savers Computing Initiative (CSCI) recommends activating power management after 20 to 30 minutes of inactivity. In addition to efficiency specifications, companies should utilize advanced power management features such as the "sleep" or "hibernate" settings on client PCs. Typical power management policies require PCs to turn off the display and hard drive after 15 minutes of inactivity, and put the system into "sleep" mode after 30 minutes of inactivity. Furthermore, desktop and laptop computers must enable power management such that the power management features place monitors and computers into a low-power S3 or "sleep" state after a defined period of inactivity. The computer is reactivated or "wakes up" when the mouse or keyboard is touched [CSC200901]. CSCI goals for desktop PCs, thin clients, and workstations range from/to

- Climate Savers Computing Base: ENERGY STAR® 4.0 qualified. This specification includes an 80% minimum efficiency for the power supply unit (PSU) at 20%, 50%, and 100% of rated output; a power factor of at least 0.9

at 100% of rated output; and meeting the maximum power requirements in standby, sleep, and idle modes.

■ Climate Savers Computing Gold: ENERGY STAR 4.0 qualified PLUS a 90% minimum efficiency rating for the PSU at 50% of rated output (and 87% minimum efficiency at 20% and 100% of rated output)

In Chapter 1 we noted that a 120 W device (such as a PC or stationary laptop not being used, but left on in sleep mode) for (say) 16 hours a day for 365 days consumes about 700 kWh a year, which is about $100. Table 7.10 is a synthetically developed table (i.e., with no input at all from the companies shown) that illustrates how much money companies could be saving just by handling power management for PCs.

7.4 Home Networking and Communications Equipment

Home networking and communications equipment, whether owned by the user or provided by the carrier in the form of Customer Premises Equipment (CPE), can contribute adversely to the carbon footprint, as we have noted in several instances in this text and also above. There are numerous entities that are addressing the issue, including two key ones: ENERGY STAR and the Home Gateway Initiative (HGI) Energy Task Force.

■ *ENERGY STAR.* As we noted elsewhere in this text (Chapter 6), ENERGY STAR is a joint program of the U.S. Environmental Protection Agency and the U.S. Department of Energy to promote energy-efficient computers and 50 product groupings; however, devices such as routers, modems, printers/scanners, and mobile devices are not included in the scope of the current ENERGY STAR program.

■ *Home Gateway Initiative (HGI) Energy Task Force.* The Home Gateway Initiative Energy Task aims at identifying targets for power consumption for components of the home gateway. The group was founded by nine telecom service providers in 2004. Their charter also includes establishing home gateway-related technical and interoperability specifications and provide input to ITU-T, the ETSI (European Telecommunications Standards Institute), and others.

We noted earlier that 58% of emissions generated by the ICT industry are associated with devices in the home, such as personal computers and peripherals. It follows that a set of prudent plans by enterprise and carrier administrators is advisable in order to address the "end-to-end" carbon footprint. Indeed, Multiple System Operators (MSOs) may be in a position to assist the broader telecommunications industry address carbon footprint reductions because it has been documented that

Table 7.10 Synthetically Developed Table Illustrating How Much Money a Company Could Save with Power Management for PCs

Company	Employees (2006)	PCs per Employee (Estimate)	Number of PCs (Estimate)	Yearly Savings (Estimate) ($)	Industry
Citigroup	303,000	0.7	212,100	21,210,000	Banks: commercial and savings
Fortis	54,245	0.7	37,972	3,797,150	Banks: commercial and savings
Crédit Agricole	136,848	0.7	95,794	9,579,360	Banks: commercial and savings
HSBC Holdings	284,000	0.7	198,800	19,880,000	Banks: commercial and savings
General Electric	316,000	0.7	221,200	22,120,000	Diversified financials
Siemens	461,000	0.7	322,700	32,270,000	Electronics, electrical equipment
Carrefour	440,479	0.3	132,144	13,214,370	Food & drug stores
Wal-Mart Stores	1,800,000	0.1	180,000	18,000,000	General merchandisers
ING Group	115,300	0.8	92,240	9,224,000	Insurance: life, health
AXA	78,800	0.8	63,040	6,304,000	Insurance: life, health
American Intl. Group	97,000	0.8	77,600	7,760,000	Insurance: life, health
Assicurazioni Generali	61,561	0.8	49,249	4,924,880	Insurance: life, health

Aviva	54,791	0.8	43,833	4,383,280	Insurance: life, health
Allianz	177,625	0.8	142,100	14,210,000	Insurance: P & C
General Motors	335,000	0.3	100,500	10,050,000	Motor vehicles & parts
DaimlerChrysler	382,724	0.3	114,817	11,481,720	Motor vehicles & parts
Toyota Motor	285,977	0.3	85,793	8,579,310	Motor vehicles & parts
Ford Motor	300,000	0.3	90,000	9,000,000	Motor vehicles & parts
Volkswagen	344,902	0.3	103,471	10,347,060	Motor vehicles & parts
Royal Dutch Shell	109,000	0.3	32,700	3,270,000	Petroleum refining
BP	96,200	0.3	28,860	2,886,000	Petroleum refining
Chevron	59,000	0.3	17,700	1,770,000	Petroleum refining
ConocoPhillips	35,600	0.3	10,680	1,068,000	Petroleum refining
Sinopec	730,800	0.3	219,240	21,924,000	Petroleum refining
ENI	72,258	0.3	21,677	2,167,740	Petroleum refining
NTT	199,113	0.8	159,290	15,929,040	Telecommunications

an ENERGY STAR–compliant Set-Top Box (STB) can provide a 30% energy savings compared with an STB that is not ENERGY STAR® compliant; hence, it follows that it is desirable for MSOs to deploy green/ENERGY STAR–compliant products. Consumer devices (television sets and STBs) have a substantial overall impact on the carbon footprint due to the large volume of units involved; the shorter product life (compared to that of a carrier's network infrastructure) is also a perennial source of electronic equipment refuse [OBE200901]. Related to the former point, a recent study on Digital Terrestrial Transmission (DTT) of television signals concluded that the transmission portion represents (in the United Kingdom) 0.01% of total U.K. CO_2 emissions, compared to about 3.54% of emissions originating from TV equipment. The bottom line is that consumer equipment power consumption is the main energy and carbon impact for TV broadcasting.*

To that end, the Society of Cable Telecommunications Engineers (SCTE) featured a SCTE Green Pavilion at SCTE Cable-Tec Expo™ 2009. Under the auspices of an MSO Steering Council, the SCTE Green Pavilion assembled a set of exhibitors, each focused on at least one of three specific areas: (1) powering improvements for facilities and plant, (2) fleet enhancements involving smart routing and other fleet management options, and (3) operational upgrades and improvements to enable additional efficiencies. Cisco Systems discussed reduced power consumption of STBs and higher head-end densities; other exhibitors extolled the value of network-based DVR (Digital Video Recording), which can be viewed as a form of Cloud Computing for (a specific) video application (namely, consumer content recording.)

Some related concepts and targets were discussed in Chapter 3.

At press time California (U.S.) regulators gave final approval to the nation's first mandatory energy curbs on television sets, a growing but often overlooked power drain that accounts for 10% of home electric bills in that state. Supporters say the measure will save California consumers more than $8 billion over 10 years in electricity costs and enough energy to power 864,000 homes. The rules require all new TVs sold in California to consume 33% less energy than current sets starting with the 2011 model year, and 50% less starting with 2013 models [GOR200901].

References

[42U201001]   42U, Data center power management, Denver, CO. 42U provides data center efficiency solutions for data center and facilities managers.

[ALC200902]   Alcatel-Lucent, Eco-efficiency in action — Alcatel-Lucent sustainability solutions for access networks, June 2009.

[ALL200901]   All about circuits, three-phase y and delta configurations, On-line resource, www.allaboutcircuits.com.

---

* This implies that a substitution of DTT with either satellite transmission or IPTV will not impact the overall emissions due to television services.

[ARM200901]   B. St. Arnaud, Clouds and optical networks, *OFC/NFOEC 2009* presentation, 21–25 March 2009, San Diego, CA.

[ATI200901]   Alliance for Telecommunications Industry Solutions (ATIS) promotional information. ATIS, 1200 G Street, NW, Suite 500, Washington, DC 2005.

[ATI200902]   ATIS, ATIS Report on Environmental Sustainability, A report by the ATIS Exploratory Group on Green, March 2009, 1200 G Street, NW, Suite 500, Washington, DC 20005.

[BOY200901]   C. Boyd, To get more, use less: Next generation solid state amplifiers, Satcon, New York, 14–16 October 2009.

[BUT200801]   A. Butler and J. Enck, Blade Servers: The Five-Year Outlook, Gartner, 16 June 2008, ID Number G00157910.

[CSC200901]   Climate Savers Computing Initiative. Beaverton, Oregon, http://www.climatesaverscomputing.org/.

[DOE200901]   U.S. Department of Energy, Energy Efficiency and Renewable Energy. Data Center Energy Efficiency Training Materials, 5/19/2009.

[EER200901]   U.S. Department of Energy, Energy Efficiency and Renewable Energy: Industrial Technologies Program — Saving Energy in Data Centers. http://www1.eere.energy.gov/industry/datacenters/.

[EER200902]   U.S. Department of Energy Save Energy Now, Data Center Assessment Process.

[EER200903]   U.S. Department of Energy Save Energy Now, Data Center Assessment Tool Suite "DC Pro," Master List of Energy Efficiency Actions, 6 June 2008.

[EIA200901]   Energy Information Administration, International Energy Outlook 2009, 27 May 2009.

[EME201001]   Emerson Network Power, Variable-Capacity Cooling, White paper, Columbus, OH. Emerson Network Power is a business of Emerson.

[ERI200801]   Ericsson, Ericsson Deep Fiber Access: A Focused Offering within Full Service Broadband, White paper, LZT 108 9807, Ericsson AB, 2008.

[ETNO200801]   Energy Task Team, European Telecommunications Network Operators' Association, The Hague, London, April 2008.

[GAR200601]   D. Garday and D. Costello, Air-Cooled High-Performance Data Centers: Case Studies and Best Methods, White paper, Intel Corporation November 2006.

[GOR200901]   S. Gorman, California sets limits on energy-gulping TVs, Reuters, November 18, 2009.

[GRE201001]   The Green Grid. Glossary and Other Reference Materials. 2010.

[MIN200701]   D. Minoli, K. Sohraby, and T. Znati, *Wireless Sensor Networks* (Wiley 2007).

[MIT202001]   R. L. Mitchell, Data Center Density Hits the Wall. Inside the New Data Center: Power and Cooling for the Modern Era, Computerworld White paper, 2/2/2010.

[MOR200801]   T.P. Morgan, Data centers embrace The Great Outdoors, *Enterprise,* 22 September 2008.

[OBE200901]   G. Oberst, The environmental impact of satellite communications, *Via Satellite Magazine,* 1 September 2009, pp. 16 ff, Satellite Today.com.

[OFE200901]   Y. Ofek and M. Baldi, Time for a "greener" Internet, *Proc. IEEE ICC Workshop,* 2009, Dresden, Germany, June 2009.

[PAT200201] C.D. Patel, R. Sharma, C.E. Bash, and A. Beitelmal, Thermal considerations in cooling large scale high compute density data centers, *Itherm2002, The Eighth Intersociety Conference on Thermal and Thermomechanical Phenomena in Electronic Systems,* San Diego, CA, 2002.

[SLO199601] J. Sloan, *Superconductor Industry,* Fall 1996, p. 32.

[SWE200901] A. Sweeney and L. Hanzo, IEEE WNCN 2009 explores newest advances in wireless communications and cooperative systems, *IEEE Communications Magazine,* November 2009, Global Communications Newsletter Insert.

[TAV200901] J. Taverna and A. Tisot, Power up with solar, *Mission Critical Communications Magazine*, April 2009, pp. 28 ff.

[TRI200201] T. Tritt, Overview of various strategies and promising new bulk materials for potential thermoelectric applications, *Materials Research Society Symposium Proceedings*, Vol. 691, 2002, Materials Research Society.

[TUC200901] R. Tucker, Energy / cost benefits of cloud computing, *OFC/NFOEC 2009* presentation, 21–25 March 2009, San Diego, CA.

[VAR201001] Variable Frequency Drive Co., Variable Frequency Drive Basics, Fremont, CA 94538.

[WOG200801] 2050 Working Group, Proposing a New Societal System for Cutting CO_2 Emissions by Half, 10 October 2008, Ministry of Economy, Trade and Industry, Commerce and Information Policy Bureau, Information Economy Division.

[WTA200701] World Teleport Association, Sizing the World Teleport Market, 2007.

[ZUK200901] D.N. Zuckerman, Workshop on Green Communications, *IEEE ICC 2009*, Dresden, Germany, June 2009.

[ZUK200902] M. Zukerman, Back to the future, *IEEE Communications Magazine,* November 2009, pp. 36ff.

Chapter 8

Approaches for Green Networks and Data Centers: Virtualization, Web Services, Cloud/Network Computing, Hosting, and Packetization

At the enterprise level, an approach to greening data centers and networks is to virtualize ICT (Information and Communications Technology) functions. The advantages of such an approach include the following:

- Increased utilization (and thus increased efficiency) of servers, storage, and other computing devices due to traffic aggregation
- Increased energy savings due to consolidations and higher intrinsic efficiencies related to economies of scale
- Improved application of Best Practices because large Network ("Cloud")-based Computing centers are better able to research and apply best-in-class design principles

The specific approaches that are applicable to these initiatives are as follows (in somewhat increasing level of virtualization):

1. A transition from dedicated traditional time division-multiplexing (TDM) network connectivity to an all-IP packet-based connectivity for carrier networks, including the use of voice and video compression technologies (e.g., IP-based 3/4G networks, VoIP carrier services, IPTV services), as well as for enterprise networks (e.g., use of VPNs [Virtual Private Networks]; use of packet services—such as MPLS [MultiProtocol Label Switching]—for the intranet, the wide area network, and for the extranet; use of corporate VoIP; etc.).
2. Technology refreshment with more efficient/powerful devices.
3. Use of IT virtualization mechanisms (e.g., VMware) within a data center to virtualize servers and/or storage. Server virtualization improves resource utilization on each server on which it is deployed.
4. Hosting of IT function in the network such as Web Hosting.
5. Generic Network-based Computing, also known as Grid Computing or (more recently) Cloud Computing.
6. Web-based Computing, known as Web Services (WS)—some of this may be supported by an Application Service Provider (ASP).

Note: Some may consider items #5 and #6 the same.

We believe that the benefits achievable by the first item in this list are self-evident and, as such, only receive limited discussion herein. The entire reason for the emergence of packet technologies in the late 1960s and 1970s (with IP being the premiere example), to be followed eventually by the Internet/Web infrastructure, was for the purpose of sharing transmission and switching resources, thereby achieving greater utilization, greater efficiency, cost reductions, and intrinsically (what we now call) lower carbon footprint. Packetization per se increases resource utilization; when coupled with compression (whether for digital cellular telephone service, for voice-over-IP (VoIP) transport, or digital television with MPEG-2 or MPEG-4), it increases the effective traffic carrying capacity by x orders of magnitude, (x > 1), where x depends on the situation and/or the perspective.* Table 8.1 depicts some of the key compression algorithms now in common use.

As implied in item #2 above, one approach to power reduction is technology refreshment. Some organizations faced with budget challenges put off capital expenditures (CAPEX) and seek alternatives, such as extending server life cycles and extending software licenses. However, studies apparently reveal that refreshing

---

* If one considers that prior to compression, an HDTV signal requires about 1 Gbps and after compression it requires about 10 Mbps, the improvement is 100 to 1. If one considers that before DTV one couple places one analog signal on a 6-MHz channel and with DTV, one can place 6 SDTV channels on the 19.2 Mbps container, the improvement is 6 to 1.

Table 8.1   Compression Technologies

Video Compression	ISO/IEC	MJPEG • Motion JPEG 2000 • MPEG-1 • MPEG-2 (Part 2) • MPEG-4 (Part 2/ASP • Part 10/AVC) • HVC
	ITU-T	H.120 • H.261 • H.262 • H.263 • H.264 • H.265
	Others	AMV • AVS • Bink • CineForm • Cinepak • Dirac • DV • Indeo • OMS Video • Pixlet • RealVideo • RTVideo • SheerVideo • Smacker • Sorenson Video • Theora • VC-1 • VC-3 • VP6 • VP7 • VP8 • WMV • XVD
Audio Compression	ISO/IEC	MPEG-1 Layer III (MP3) • MPEG-1 Layer II • MPEG-1 Layer I • AAC • HE-AAC • MPEG-4 ALS • MPEG-4 SLS • MPEG-4 DST
	ITU-T	G.711 • G.718 • G.719 • G.722 • G.722.1 • G.722.2 • G.723 • G.723.1 • G.726 • G.728 • G.729 • G.729.1
	Others	AC3 • AMR • AMR-WB • AMR-WB+ • Apple Lossless • ATRAC • DRA • FLAC • GSM-FR • GSM-EFR • iLBC • Monkey's Audio • μ-law • Musepack • Nellymoser • OptimFROG • RealAudio • RTAudio • SHN • SILK • Siren • Speex • TwinVQ • Vorbis • WavPack • WMA • TAK • True Audio
Image Compression	ISO/IEC/ ITU-T	JPEG • JPEG 2000 • JPEG XR • lossless JPEG • JBIG • JBIG2 • PNG • WBMP
	Others	APNG • BMP • DjVu • EXR • GIF • ICER • ILBM • MNG • PCX • PGF • TGA • TIFF • JPEG XR / HD Photo
Media Containers	General	3GP • ASF • AVI • Bink • BXF • DMF • DPX • EVO • FLV • GXF • M2TS • Matroska • MPEG-PS • MPEG-TS • MP4 • MXF • Ogg • QuickTime • RealMedia • RIFF • Smacker • VOB • VVF
	Audio only	AIFF • AU • WAV

server infrastructure on pace with newer technology (e.g., every 2 years) can reduce 6-year server costs by 33% compared with buying and holding servers for those 6 years. This occurs not only because today's servers can do 16 times the work at almost half the power requirements, but also because of the reduced maintenance overhead and IT labor costs associated with advanced technologies [EAS200901].

The remainder of this chapter focuses on the other items listed above. The reader will make the mental connection with the greening advantages of these solutions, as already highlighted above in general terms. Virtualization is an example of greening the enterprise data center (more than being applicable to networks themselves), while Grid/Cloud Computing is an example of how networking can help other areas (IT in this specific case) become greener. A basic approach to achieve green IT is improving the utilization of servers, storage systems, appliances, blade servers, and networking devices (e.g., switches and routers). The traditional data center architecture of "one application per machine" approach to server provisioning leads to an over-provisioned underutilized data center. In this type of architecture, servers typically operate at only 20% to 30% of their capability [ATI200901]. Clustering, virtualization, and Cloud Computing are techniques that invariably improve server utilization.

8.1 Press-Time Snapshot of Cloud/ Grid/Network Computing

The term "Cloud Computing" is currently in vogue. While there are some technical nuances and marketing focus* on various terms, such as Distributed Computing, Network Computing, Grid Computing, Utility Computing, and Cloud Computing, fundamentally there are two approaches to virtualization [MIN200501, MIN200601, MIN200701]:

1. Using a middleware layer deployed on the organization's own set of assets (computing, storage, etc.) to decouple the machine-cycle requestor (application) from the machine-cycle suppliers (the raw hardware) in the data center. All of the assets are still managed, deployed, powered, cooled, and housed by the organization.
2. Using a gateway middleware layer that accesses a "peer middleware layer" deployed by a service provider's on its set of assets (computing, storage, etc.) to decouple the machine-cycle requestor (specifically, applications of the organization in question) from the machine-cycle supplier(s) (the raw hardware) in the remote network/cloud/data center. The key IT assets are now managed, deployed, powered, cooled, and housed by the service provider.

Given the somewhat nuanced (tenuous?) distinction between the terms listed above, we proceed in this chapter to use the term "Cloud/Grid Computing" when discussing the gateway middleware approach, except where specifically noted. The pure term "Grid Computing" was used in the earlier part of the decade of the 2000s. Cloud/Grid Computing† (or, more precisely, a "Cloud/Grid Computing system")

---

* The word "hype" may be applicable here.
† Some portions of this material are updated from the book *Enterprise Architecture A to Z*, by this author, published by CRC Press/Taylor & Francis Group, 2008.

is a virtualized Distributed Computing environment. Such an environment aims at enabling the dynamic "runtime" selection, sharing, and aggregation of (geographically) distributed autonomous resources based on the availability, capability, performance, and cost of these computing resources, and, simultaneously, also based on an organization's specific baseline and/or burst processing requirements [MIN200701]. When people think of a grid, the idea of an interconnected system for the distribution of electricity, especially a network of high-tension cables and power stations, comes to mind. In the mid-1990s, the grid metaphor was (re)applied to computing, by extending and advancing the 1960s concept of "computer time sharing." The grid metaphor strongly illustrates the relation to, and the dependency on, a highly interconnected networking infrastructure. Off-the-shelf products supporting virtualization and server consolidation are now available and are deployed in many Fortune 500 companies. An evolutionary step is to place the computing/ storage assets in the network and use these to deliver general (e.g., e-mail) and/or specific (e.g., an app) service to the user organization. As noted, the term "Cloud Computing" has come into vogue in the recent past. The following is a reasonably crisp definition of Cloud Computing cited from [HAR200901]:

> Cloud Computing overlaps some of the concepts of distributed, grid, and Utility Computing; however, it does have its own meaning if contextually used correctly. The conceptual overlap is partly due to technology changes, usages, and implementations over the years. Trends in usage of the terms from Google searches show that Cloud Computing is a relatively new term introduced in 2007. There has also been a decline in general interest of Grid, Utility, and Distributed Computing. Likely, these terms will be around in usage for a while. But Cloud Computing has become the new buzzword, driven largely by marketing and service offerings from big corporate players such as Google, IBM, and Amazon.
>
> The term "Cloud Computing" probably comes from (at least partly) the use of a cloud image to represent the Internet or some large networked environment. We do not care much what is in the cloud or what goes on there except that we depend on reliably sending data to and receiving data from it. Cloud Computing is now associated with a higher-level abstraction of the cloud. Instead of there being data pipes, routers, and servers, there are now services. The underlying hardware and software of networking is of course still there but there are now higher-level service capabilities available to build applications. Behind the services are data and compute resources. A user of the service does not necessarily care about how it is implemented, what technologies are used, or how it is managed; that user only cares that there is access to it and it has a level of reliability necessary to meet the application requirements.

In essence, this is Distributed Computing. An application is built using the resource from multiple services, potentially from multiple locations. At this point, typically one still needs to know the endpoint to access the services rather than having the cloud provide you with available resources. This is also known as Software as a Service. Behind the service interface there is usually a grid of computers to provide the resources. The grid may be hosted by one company and consists of a homogeneous environment of hardware and software, making it easier to support and maintain, [or it could be a geographically/administratively distributed array of computers].

Cloud Computing equates to accessing resources and services needed to perform functions with dynamically changing needs. An application or service developer requests access from the cloud rather than a specific endpoint or named resource. What goes on in the cloud manages multiple infrastructures across multiple organizations and consists of one or more frameworks overlaid on top of the infrastructures tying them together. Frameworks provide mechanisms for

- Self-healing
- Self-monitoring
- Resource registration and discovery
- Service level agreement definitions
- Automatic reconfiguration

The cloud is a virtualization of resources that maintains and manages itself. There are, of course, people resources to keep hardware, operation systems, and networking in proper order. But from the perspective of a user or application developer, only the cloud is referenced. There are now frameworks that execute across a heterogeneous environment in a local area network providing a local cloud environment; there are plans for the addition of a network overlay to start providing an infrastructure across the Internet to help achieve the goal of true Cloud Computing.

Grid Computing is a virtualization technology that was talked about in the 1980s and 1990s and entered the scientific computing field in the late-1990s/early-2000s. While Grid Computing is targeted mainly to scientific users, both Cloud and Utility Computing are targeted to enterprises. The common point for both approaches is their reliance on high-speed optical networks to provide advanced and flexibly reconfigurable infrastructure. Optical network researchers are facing challenges in delivering the necessary technologies for supporting Cloud, Grid, and Utility Computing services. Such technologies and services will change the Internet in much the same way as Distributed and Parallel Computing have changed the computation and cyber-infrastructure today [QIA200901].

Cloud Computing per se had its genesis in the successes of enterprise server virtualization and the ensuing IT efficiency improvements achieved through server consolidation in the middle of the 2000 decade. In the past 5 years there has been a lot of press and market activity, and a number of proponents see continued penetration in the future. VMWare (EMC), IBM, and Oracle, among others, are major players in this space. Virtualization was then applied to storage and network hardware, thereby achieving abstraction of the hardware in the entire data center [MIN200501, MIN200601, MIN200701]. Cloud capabilities at press time included, but are not limited to, Amazon Web Services, Google App Engine, and Force.com.

Cloud/Grid Computing providers make the observation that instead of buying servers with preconfigured resources and capacity, a new way to achieve efficiencies is to buy slices of computing resources—such as memory, processor, storage, and network—based on business needs. Obtaining the infrastructure as a utility-based service reduces capital and operating costs, including powering and cooling; businesses do not have to build out and manage their own data centers, and they do not have to pay for under-utilized resources. These Network/Cloud Computing environments provide a secure, virtualized environment to deliver IT infrastructure—including servers, storage, memory, and bandwidth—as a service. Resources can be self-provisioned in real-time to match fluctuating needs, while billing is based on resource consumption (see Table 8.2) [NAV200901]. Services such as Animoto, Gmail, Flickr, and Facebook make use of Cloud/Grid Computing concepts. Press-time headlines indicated that Cisco Systems, Inc., and EMC Corporation were teaming up to sell a new line of networking gear, computers and storage equipment designed for use with Cloud Computing; the line of products, dubbed vBlock, position the two companies to better compete against IBM and Hewlett-Packard Co., which sell a broader array of data center equipment than either Cisco or EMC offer on their own [FIN20901].

Table 8.3 depicts some of the variants of Cloud Computing now available as commercial services [THO200801].

Features of cloud/grid include pay-per-use, instant availability, scalability, hardware abstraction, self-provisioning, virtualization, and service access via the Internet. The Cloud Computing market is typically segmented into public clouds (services offered over the Internet), private clouds (internal enterprise), and hybrid clouds (a mix of both). The public cloud market is often sub-segmented into IaaS (Infrastructure as a Service), PaaS (Platform as a Service), and SaaS (Software as a Service) [LEY200901].

Virtualization solutions (whether at the data center level or via a cloud solution) reduce IT costs at the hardware and management level while increasing the efficiency, utilization, and flexibility of existing computer hardware; in addition, virtualization can clearly reduce wasted power consumption and heat. Virtualization optimizes efficiency by allowing one (or a small set of) (larger) server(s) do the job of multiple (smaller) servers, by sharing the resources of a single infrastructure across multiple environments. Some of the IT administrator-level advantages include [NOR200801]:

Table 8.2   Typical Features of a Cloud Computing Service

On-Demand	Resources can be self-provisioned, automatically, in real-time, to accommodate performance and capacity demands, ensuring that infrastructure is right-sized to match business needs.
Scalable	The IT environment can scale up or down, based on business demands and application performance with no change to the underlying architecture or investment in additional hardware or equipment.
Fully Managed Option	Customers have the ability to transfer the management and support of their self-managed environment to the service provider, leveraging their 24 × 7 hosting services, including server management and various compliance measures such as SAS70[a] and PCI.[b]

[a] Statement on Auditing Standards (SAS) No. 70, Service Organizations, is a widely recognized auditing standard developed by the American Institute of Certified Public Accountants (AICPA). A service auditor's examination performed in accordance with SAS No. 70 ("SAS 70 Audit") is widely recognized, because it represents that a service organization has been through an in-depth audit of its control objectives and control activities, which often include controls over information technology and related processes. In addition, the requirements of Section 404 of the Sarbanes-Oxley Act of 2002 make SAS 70 audit reports even more important to the process of reporting on the effectiveness of internal control over financial reporting. SAS No. 70 is the authoritative guidance that allows service organizations to disclose their control activities and processes to their customers and their customers' auditors in a uniform reporting format [SAS200901].

[b] The Payment Card Industry's Data Security Standard (PCI DSS), a set of comprehensive requirements for enhancing payment account data security, was developed by the founding payment brands of the PCI Security Standards Council, including American Express, Discover Financial Services, JCB International, MasterCard Worldwide and Visa Inc. International, to help facilitate the broad adoption of consistent data security measures on a global basis. The PCI DSS is a multifaceted security standard that includes requirements for security management, policies, procedures, network architecture, software design and other critical protective measures. This comprehensive standard is intended to help organizations proactively protect customer account data. The PCI Security Standards Council is an open global forum for the ongoing development, enhancement, storage, dissemination and implementation of security standards for account data protection. The PCI Security Standards Council's mission is to enhance payment account data security by driving education and awareness of the PCI Security Standards [PCI200901].

■ *Server consolidation and optimization:* Virtualization makes it possible to achieve higher server utilization by pooling common infrastructure resources and avoiding the need for one server for each application model.

Table 8.3 Various Types of Cloud/Grid Services

Applications in the cloud	This service is what people have already used in the form of Gmail, Yahoo mail, the various search engines, Wikipedia, etc. Some company hosts an application in the Internet that many users sign up for and use without any concern for where, how, or by whom the compute cycles and storage bits are provided. The service being sold (or offered in ad-sponsored form) is a complete end-user application. This can be seen as Software as a Service (SaaS).
Platforms in the cloud	In this case, an application platform is offered to developers in the cloud. Developers write their application to a more or less open specification and then upload their code into the cloud where the app is run magically somewhere, typically being able to scale up as usage for the app grows. Examples include Google App Engine, Mosso, and Force.com. The service being sold is the machinery that funnels requests to an application and makes the application work. This is the newest type of cloud service.
Infrastructure in the cloud	This capability is the most general offering; it includes the services that Amazon has pioneered and is where RightScale offers its management platform. Developers and system administrators obtain general compute, storage, queueing, and other resources and run their applications with relatively few limitations. This is the most powerful type of cloud, in that virtually any application and any configuration that is fit for the Internet can be mapped to this type of service. It requires more work on the part of the developers and administrators. This service is indeed similar to the "time-sharing" services offered in the late 1960s and early 1970s, except that now there may be many distributed processors instead of one mainframe, and typically there is an open interface.

■ *Infrastructure cost reduction:* With virtualization, one can reduce the number of servers and related IT hardware in the data center. This leads to reductions in real estate, power, and cooling requirements, thereby resulting in lower IT costs.

■ *Improved operational flexibility and responsiveness:* Virtualization offers a new way of managing IT infrastructure and can help IT administrators spend less time on repetitive tasks such as provisioning, configuration, monitoring, and maintenance.

■ *Improved business continuity:* This allows the IT administrator to eliminate planned downtime and recover quickly from unplanned outages with the ability to securely back up and migrate entire virtual environments with no interruption in service.

To give the reader a sense of what a cloud-enabled enterprise-hosting and application-management services provider offers, we include here for illustrative purposes some service information from two providers: (1) NaviSite and (2) Amazon.

NaviSite is a provider of cloud-enabled IT-as-a-Service solutions. Business applications such as ERP, CRM, and Business Intelligence, and custom applications are typical of many organizations. NaviSite has established NaviCloud, a network-based platform created using enterprise-grade technologies, including Cisco UCS, VMware VSphere, IBM XiV SAN, DoubleTake for Replication, Mezeo for storage abstraction and secondary storage, and 10G Ethernet and Fibre Channel (FC) infrastructure. For a predictable monthly fee, NaviSite provides end-to-end management of typical corporate applications and the infrastructure that supports them. NaviSite services are designed to effectively manage the organization's applications, so that the company's employees can focus on using them. Figures 8.1 and 8.2 provide a pictorial view of the NaviCloud. NaviSite offers a number of options for the Application Management Services:

- *Full Hosting:* NaviSite maintains the organization's application in one of their SAS70 Type II, state-of-the-art data centers. They provide and manage the complete technology stack required to support the organization's critical business application.
- *Remote Hosting:* The organization manages the operating system and physical infrastructure, while NaviSite maintains the availability and support of the

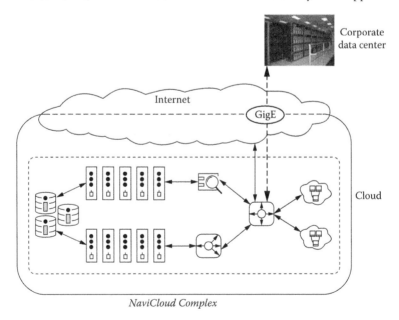

NaviCloud Complex

Figure 8.1 Example of a cloud.

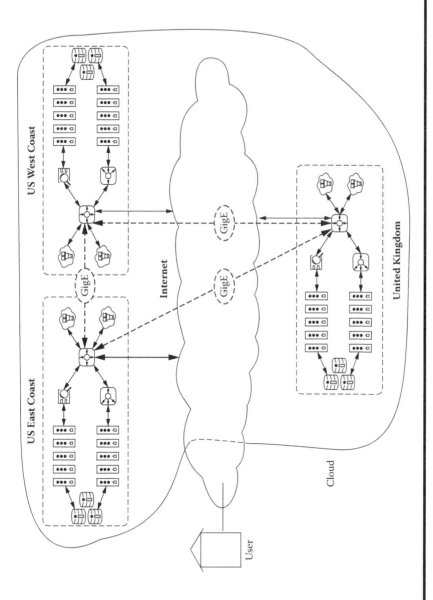

Figure 8.2   Multi-node cloud.

application. This option simplifies day-to-day management, while enabling the organization to keep the physical infrastructure at the organization's facility.

■ *Remote Monitoring:* For companies that do not need complete application management, but would like to take advantage of NaviSite's 24×7 monitoring capabilities, NaviSite offers Remote Application Monitoring. NaviSite monitors the organization's critical business applications 24×7, immediately escalating any issues to minimize the potential for downtime.

■ *Application Support Help Desk:* For organizations that want to provide application support to their end users, NaviSite offers Application Support Help Desk Services. NaviSite's team of application experts will answer a variety of Level I inquiries, including end-user support for user account creation/modification / termination, general application questions, password resets and security administration, application troubleshooting, and batch process verification.

Amazon introduced cloud services a few years ago, specifically EC2 (compute) and S3 (storage); furthermore, Amazon APIs (Application Programming Interfaces) have been used by a large number of developers. All of this allowed customers to deploy and run infrastructure in the cloud. Table 8.4 lists* the Amazon cloud services for illustrative purposes.

Moving along to more recent developments, the first build-your-own-cloud products that were brought to the market came from companies such as Flexiscale (United Kingdom), 3Tera (United States), and Q-layer (Belgium). These products were aimed at the ISPs (Internet Service Providers), who had an urgent market need for innovation. The first new services that these ISPs offered in this space were basically virtual machines—allowing them to run their facilities more efficiently and still charge the same prices to their customers. Soon, companies such as Savvis, GoGrid, and Rackspace added interfaces that enabled end users to control their own infrastructure. In early 2009, Sun Microsystems launched the Virtual Data Center (VDC), a graphical interface with drag-and-drop that enables users to create and manage a full virtual data center in the cloud—the Sun Cloud and its open API has been delayed due to the recent acquisition of Sun by Oracle. In the meantime, action has moved to the private clouds. Enterprises seem ready to "cloud-enable" their infrastructure either in a purely private or a hybrid (enabling cloud-bursting to public clouds for certain services) environment. Many of the leading software providers have announced products. Implementing a private cloud affects the entire business, including not only the entire IT infrastructure (hardware, software, services), but also most business processes (e.g., regulatory compliance). Because few if any of the big software providers have experience in all of these fields, enterprises will have to rely on integrators to build their clouds [LEY200901].

In addition to its applicability to computing, virtualization is also emerging in the area of storage. Storage virtualization is the pooling of physical storage from

---

* Based on Amazon sources.

Table 8.4   Web Services (Cloud Computing Services) from Amazon

| Amazon Elastic Compute Cloud (Amazon EC2) | A Web service that provides resizable compute capacity in the cloud. It is designed to make web-scale computing easier for developers. Amazon EC2's simple Web Service interface allows one to obtain and configure capacity with minimal effort. It provides the firm with complete control of the firm's computing resources and lets the firm run on Amazon's proven computing environment. Amazon EC2 reduces the time required to obtain and boot new server instances to minutes, allowing the firm to quickly scale capacity, both up and down, as the computing requirements change. Amazon EC2 changes the economics of computing by allowing the firm to pay only for capacity that it actually uses. Amazon EC2 provides developers with the tools to build failure-resilient applications and isolate themselves from common failure scenarios.

Amazon EC2 presents a true virtual computing environment, allowing the developer to use web service interfaces to launch instances with a variety of operating systems, load them with the firm's custom application environment, manage the network's access permissions, and run the firm's image using as many or few systems as the firm desires. To use Amazon EC2, one simply:

• Creates an Amazon Machine Image (AMI) containing the firm's applications, libraries, data, and associated configuration settings. Or uses preconfigured, templated images to get up and running immediately.

• Uploads the AMI into Amazon S3. Amazon EC2 provides tools that make storing the AMI simple. Amazon S3 provides a safe, reliable, and fast repository to store the firm's images.

• Uses Amazon EC2 web service to configure security and network access.

• Chooses which instance type(s) and operating system the firm wants, then starts, terminates, and monitors as many instances of the AMI as needed, using the web service APIs or the variety of management tools provided.

• Determines whether the firm wants to run in multiple locations, utilize static IP endpoints, or attach persistent block storage to the firm's instances.

• Pays only for the resources that the firm actually consumes, such as instance-hours or data transfer. |
|---|---|

Continued

Table 8.4   Web Services (Cloud Computing Services) from Amazon (Continued)

Amazon SimpleDB	A Web service providing the core database functions of data indexing and querying in the cloud. By offloading the time and effort associated with building and operating a web-scale database, SimpleDB provides developers with the freedom to focus on application development. A traditional, clustered relational database requires a sizable upfront capital outlay, is complex to design, and often requires extensive and repetitive database administration. Amazon SimpleDB is simpler, requiring no schema, automatically indexing the firm's data, and providing a simple API for storage and access. This approach eliminates the administrative burden of data modeling, index maintenance, and performance tuning. Developers gain access to this functionality within Amazon's proven computing environment, are able to scale as needed, and pay only for what they use.
Amazon S3 is storage for the Internet	A Web service designed to make Web-scale computing easier for developers. Amazon S3 provides a simple web services interface that can be used to store and retrieve any amount of data, at any time, from anywhere on the web. It gives any developer access to the same highly scalable, reliable, fast, inexpensive data storage infrastructure that Amazon uses to run its own global network of web sites. The service aims to maximize benefits of scale and to pass those benefits on to developers. Amazon S3 is intentionally built with a minimal feature set. • Write, read, and delete objects containing from 1 byte to 5 gigabytes of data each. The number of objects one can store is unlimited. • Each object is stored in a bucket and retrieved via a unique, developer-assigned key. • A bucket can be located in the United States or in Europe. All objects within the bucket will be stored in the bucket's location, but the objects can be accessed from anywhere. • Authentication mechanisms are provided to ensure that data is kept secure from unauthorized access. Objects can be made private or public, and rights can be granted to specific users.

Table 8.4 Web Services (Cloud Computing Services) from Amazon (Continued)

Amazon S3 is storage for the Internet	Uses standards-based REpresentational State Transfer (REST) and Simple Object Access Protocol (SOAP) interfaces designed to work with any Internet-development toolkit.
	Built to be flexible so that protocol or functional layers can easily be added. Default download protocol is HTTP. A BitTorrent™ protocol interface is provided to lower costs for high-scale distribution. Additional interfaces will be added in the future.
Amazon CloudFront	A Web service for content delivery. It integrates with other Amazon web services to give developers and businesses an easy way to distribute content to end users with low latency, high data transfer speeds, and no commitments. Amazon CloudFront delivers a firm's content using a global network of edge locations. Requests for the firm's objects are automatically routed to the nearest edge location, so content is delivered with the best possible performance. Amazon CloudFront works seamlessly with Amazon Simple Storage Service (Amazon S3), which durably stores the original, definitive versions of the firm's files. Like other Amazon web services, there are no contracts or monthly commitments for using Amazon CloudFront—one pays only for as much or as little content as one actually delivers through the service.
	Amazon CloudFront has a simple web services interface that lets the developer get started in minutes. In Amazon CloudFront, the firm's objects are organized into distributions. A distribution specifies the location of the original version of the firm's objects. A distribution has a unique CloudFront.net domain name (e.g., abc123. cloudfront.net) that the firm can use to reference the firm's objects through the network of edge locations. One can also map one's own domain name (e.g., images. example.com) to one's distribution. To use Amazon CloudFront, the developer:
	• Stores the original versions of the firm's files in an Amazon S3 bucket.
	• Creates a distribution to register that bucket with Amazon CloudFront through a simple API call.

Continued

**Table 8.4   Web Services (Cloud Computing Services) from Amazon
(Continued)**

Amazon CloudFront	• Uses the firm's distribution's domain name in the firm's web pages or application. When end users request an object using this domain name, they are automatically routed to the nearest edge location for high-performance delivery of the firm's content. • Pay only for the data transfer and requests that the firm actually uses.
Amazon Simple Queue Service (Amazon SQS)	A reliable, highly scalable, hosted queue for storing messages as they travel between computers. By using Amazon SQS, developers can simply move data between distributed components of their applications that perform different tasks, without losing messages or requiring that each component be always available. Amazon SQS makes it easy to build an automated workflow, working in close conjunction with the Amazon Elastic Compute Cloud (Amazon EC2) and the other Amazon Web Services (AWS) infrastructure web services. Amazon SQS works by exposing Amazon's web-scale messaging infrastructure as a web service. Any computer on the Internet can add or read messages without any installed software or special firewall configurations. Components of applications using Amazon SQS can run independently, and do not need to be on the same network, developed with the same technologies, or run at the same time. • Developers can create an unlimited number of Amazon SQS queues with an unlimited number of messages. • A queue can be created in the United States or in Europe. Queue names and message stores are independent of other regions. • The message body can contain up to 8 KB of text in any format. • Messages can be retained in queues for up to 4 days. • Messages can be sent and read simultaneously. • When a message is received, it becomes "locked" while being processed. This keeps other computers from processing the message simultaneously. If the message processing fails, the lock will expire and the message will be available again. In the case where the application needs more time for processing, the "lock" timeout can be changed dynamically via the ChangeMessageVisibility operation.

Table 8.4   Web Services (Cloud Computing Services) from Amazon (Continued)

Amazon Simple Queue Service (Amazon SQS)	• Developers can access Amazon SQS through standards-based Simple Object Access Protocol (SOAP) and Query interfaces. • Developers can securely share Amazon SQS queues with others. Queues can be shared with other AWS accounts and anonymously. Queue sharing can also be restricted by IP address and time-of-day.
Amazon Elastic MapReduce	A Web service that enables businesses, researchers, data analysts, and developers to easily and cost-effectively process vast amounts of data. It utilizes a hosted Hadoop framework running on the web-scale infrastructure of Amazon Elastic Compute Cloud (Amazon EC2) and Amazon Simple Storage Service (Amazon S3). Using Amazon Elastic MapReduce, the firm can instantly provision as much or as little capacity as the firm likes to perform data-intensive tasks for applications such as web indexing, data mining, log file analysis, machine learning, financial analysis, scientific simulation, and bioinformatics research. Amazon Elastic MapReduce lets the firm focus on crunching or analyzing the firm's data without having to worry about time-consuming setup, management, or tuning of Hadoop clusters or the compute capacity upon which they sit. Amazon Elastic MapReduce automatically spins up a Hadoop implementation of the MapReduce framework on Amazon EC2 instances, subdividing the data in a job flow into smaller chunks so that they can be processed (the "map" function) in parallel, and eventually recombining the processed data into the final solution (the "reduce" function). Amazon S3 serves as the source for the data being analyzed, and as the output destination for the end results. To use Amazon Elastic MapReduce, one simply: • Develops the firm's data processing application. Amazon Elastic MapReduce enables job flows to be developed in SQL-like languages, such as Hive and Pig, making it easy to write data analytical scripts without in-depth knowledge of the MapReduce development paradigm. If desired, more sophisticated applications can be authored in the firm's choice of Cascading, Java, Ruby, Perl, Python, PHP, R, or C++.

Continued

Table 8.4   Web Services (Cloud Computing Services) from Amazon (Continued)

Amazon Elastic MapReduce	• Uploads the firm's data and the firm's processing application into Amazon S3. Amazon S3 provides reliable, scalable, easy-to-use storage for the firm's input and output data.
	• Logs into the AWS (Amazon Web Services) Management Console to start an Amazon Elastic MapReduce "job flow." Simply choose the number and type of Amazon EC2 instances the firm wants, specify the location of the firm's data or application on Amazon S3, and then click the "Create Job Flow" button. Alternatively, one can start a job flow by specifying the same information mentioned above via our Command Line Tools or APIs.
	• Monitors the progress of the firm's job flow(s) directly from the AWS Management Console, Command Line Tools, or APIs; and, after the job flow is done, retrieves the output from Amazon S3.
	• Pay only for the resources that the firm actually consumes. Amazon Elastic MapReduce monitors the firm's job flow; and unless the firm specifies otherwise, shuts down the firm's Amazon EC2 instances after the job completes.
Amazon Virtual Private Cloud (VPC) (Amazon VPC)	A secure and seamless bridge between a company's existing IT infrastructure and the AWS cloud. Amazon VPC enables enterprises to connect their existing infrastructure to a set of isolated AWS compute resources via a Virtual Private Network (VPN) connection, and to extend their existing management capabilities such as security services, firewalls, and intrusion-detection systems to include their AWS resources. Amazon VPC integrates today with Amazon EC2, and will integrate with other AWS services in the future. As with all Amazon Web Services, there are no long-term contracts, minimum spend or up-front investments required. With Amazon VPC, one pays only for the resources one uses.
	Amazon VPC enables a firm to use its own isolated resources within the AWS cloud and then connect those resources directly to the firm's own data center using industry-standard encrypted IPSec VPN connections. With Amazon VPC, one can:
	• Create a Virtual Private Cloud on AWS's scalable infrastructure, and specify its private IP address range from any block the firm chooses.

Table 8.4 Web Services (Cloud Computing Services) from Amazon (Continued)

Amazon Virtual Private Cloud (VPC) (Amazon VPC)	• Divide the firm's VPC's private IP address range into one or more subnets in a manner convenient for managing applications and services the firm run in their VPC. • Bridge together the firm's VPC and the IT infrastructure via an encrypted VPN connection. • Add AWS resources, such as Amazon EC2 instances, to the firm's VPC. • Route traffic between the firm's VPC and the Internet over the VPN connection so that it can be examined by the firm's existing security and networking assets before heading to the public Internet. • Extend the firm's existing security and management policies within the firm's IT infrastructure to the firm's VPC as if they were running within the firm's infrastructure. Amazon VPC provides end-to-end network isolation by utilizing an IP address range that the firm specifies, and routing all network traffic between VPC and the firm's data center through an industry-standard encrypted IPSec VPN. This allows the firm to leverage the preexisting security infrastructure, such as firewalls and intrusion-detection systems to inspect network traffic going to and from a VPC.

multiple network storage devices into an asset that appears to be a single storage device manageable from a central console. Storage virtualization is commonly used in a Storage Area Network (SAN). Users can implement storage virtualization with software applications or by using hybrid hardware and software appliances.

We mentioned data de-duplication in Chapter 1. Data de-duplication (also called "intelligent compression" or "single-instance storage") is a method of reducing storage requirements by eliminating redundant data. In effect, this can also be seen as a form of virtualization. With data de-duplication, only one unique instance of the data is actually retained on storage media, redundant data is deleted, and a pointer to the unique data copy is used to support requests for access to that redundant copy. For example, an e-mail system might contain, say, 50 instances of the same file attachment; when the e-mail folder is backed up or archived, all 50 instances of the file are saved, requiring, say, 50 MB storage space. With data de-duplication, only one instance of the attachment is actually stored; each subsequent instance is referenced back to the one saved copy. In this example, a 50 MB storage requirement is reduced to only 1 MB. Data de-duplication can operate at the file, the block, and even the bit level. File de-duplication eliminates duplicate files (as in the example above), but block and bit de-duplication are more efficient by looking within a file and saving unique iterations of each block or bit. Data de-duplication is often used alongside conventional compression and delta differencing. Data de-duplication lowers storage space requirements, which will save money on the need for additional expansion racks of equipment, with ensuing requirements for powering and then cooling.

8.2 Cloud/Grid/Network Computing Technology

This section provides a more extensive discussion of Grid/Cloud Computing. Virtualization is a well-known concept in networking, from Virtual Channels in Asynchronous Transfer Mode, to Virtual Private Networks, to Virtual LANs, and Virtual IP Addresses. However, an even more fundamental type of virtualization is achievable with today's ubiquitous networks: machine cycle and storage virtualization through the auspices of Cloud/Grid Computing and IP storage. As already noted, Cloud/Grid Computing is also known as Utility Computing, On-Demand Computing, and Distributed Computing.

Cloud/Grid Computing is intrinsically network-based: Resources are distributed all over an intranet, an extranet, or the Internet. Cloud/Grid Computing cannot exist without networks (the "grid"), because the user is requesting computing or storage resources that are located miles or continents away. Users need not be concerned about the specific technology used in delivering the computing or storage power: All the user wants and gets is the requisite "service." One can think of Grid Computing as middleware that shields the user from the raw technology itself. The network delivers the job requests anywhere in the world and returns the results, based on an established service level agreement. The advantages of Cloud/Grid Computing are the fact that there can be a mix-and-match of different hardware in the network; the cost is lower because there is a better, statistically averaged, utilization of the underlying resources; also, there is higher availability because if a processor were to fail, another processor is automatically switched into service. Think of an environment of a Redundant Array of Inexpensive Computers (RAIC), similar to the concept of Redundant Array of Inexpensive Disks (RAID).

At the enterprise level, server consolidation and infrastructure optimization help organizations consolidate servers and increase utilization rates, greatly reduce power and cooling costs, and manage and automate IT processes for maximum availability, performance, and scalability. Virtualization offers cost savings when users address the biggest and most obvious inefficiencies by grouping similar workloads, availability needs, security, and agility to hosted virtualized platforms; this represents consolidating the "low-hanging fruit." The longer-term goal relates to moving workloads around networks, within blades, and across heterogeneous infrastructures in a portable way; that movement may take longer to realize, however, because virtualization vendors have generally used proprietary tools to manage their specific platforms as silos [WEI200801]. A virtual infrastructure is a dynamic mapping of physical resources to business needs. Users can get locally based virtualization by using middleware such as VMWare that allows a multitude of servers right in the corporate data center to be utilized more efficiently. Typically, corporate servers are utilized for less than 30% to 40% of their available computing power. Using a virtualization mechanism, the firm can improve utilization, increase availability, reduce costs, and make use of a plethora of mix-and-match processors; as a minimum, this drives to server consolidation. The possibility exists, according to the industry, that with Cloud/Grid Computing, companies can save as much as 30% of certain key line items of the operations budget (in an ideal situation), which is typically a large fraction of the total IT budget [SUN200301, COH200301, MIN200501]. While a virtual machine represents the physical resources of a single computer, a virtual infrastructure represents the physical resources of the entire IT environment, aggregating x86 computers and their attached network and storage into a unified pool of IT resources. For example, a typical approach to virtualization (e.g., used by VMWare) inserts a thin layer of software directly on the computer hardware or on a host operating system. This software layer creates virtual machines and contains a virtual machine monitor that allocates hardware resources dynamically and transparently so that multiple operating systems can run concurrently on a single physical computer without even knowing it [VMW200701]. A press-time survey by Blade. org found that 82% of large IT organizations have implemented or are planning to implement virtualization on blade servers in the data center; for medium-sized businesses (SMB) organizations, 28% have already virtualized on blade servers, with an additional 36% in the planning stage. Another 2008 survey found that organizations tend to implement virtualization on new servers rather than upgrading older servers with virtualization capabilities; more than 50% of all new servers were shipped as virtualized servers in 2009. High numbers of virtual machines (VMs) were found, with the average virtualized x86 server hosting eight or more VMs compared with two to four VMs per physical server in 2005 [EAS200901].

Moving beyond own-data-center virtualization, the concept of Cloud/Grid Computing is straightforward: With Cloud/Grid Computing, an organization can transparently integrate, streamline, and share dispersed, heterogeneous pools of hosts, servers, storage systems, data, and networks that they may own into one synergistic system, in order to deliver agreed-upon service at specified levels of application efficiency and processing performance. Additionally, or, alternatively, an organization can simply secure commoditized "machine cycles" or storage capacity from a remote provider, "on-demand," without having to own the "heavy iron" to do the "number crunching." Either way, to an end user or application, this arrangement (ensemble) looks like one large, cohesive, virtual, transparent computing system [DEV200301, MCC200301]. Broadband networks play a fundamental enabling role in making Cloud Computing possible, and this is the motivation for looking at this technology from the perspective of communication.

According to IBM's definition [ZHA200201, HAW200301],

> A grid is a collection of distributed computing resources available over a local or wide area network that appear to an end user or application as one large virtual computing system. The vision is to create virtual dynamic organizations through secure, coordinated resource-sharing among individuals, institutions, and resources. Grid Computing is an approach to distributed computing that spans not only locations but also organizations, machine architectures, and software boundaries to provide unlimited power, collaboration, and information access to everyone connected to a grid... The Internet is about getting computers to talk together; Grid Computing is about getting computers to work together. Grid will help elevate the Internet to a true computing platform, combining the qualities of service of enterprise computing with the ability to share heterogeneous distributed resources—everything from applications, data, storage and servers.

Another definition, this one from The Globus Alliance (a research and development initiative focused on enabling the application of grid concepts to scientific and engineering computing), is as follows [GLO200301]:

> The grid refers to an infrastructure that enables the integrated, collaborative use of high-end computers, networks, databases, and scientific instruments owned and managed by multiple organizations. Grid applications often involve large amounts of data and/or computing and often require secure resource sharing across organizational boundaries, and are thus not easily handled by today's Internet and Web infrastructures.

Yet another industry-formulated definition of Grid Computing is as follows [FOS199901] [FOS200201]:

> A computational grid is a hardware and software infrastructure that provides dependable, consistent, pervasive, and inexpensive access to high-end computational capabilities. A grid is concerned with coordinated resource sharing and problem solving in dynamic, multi-institutional virtual organizations. The key concept is the ability to negotiate resource-sharing arrangements among a set of participating parties (providers and consumers) and then to use the resulting resource pool for some purpose. The sharing that we are concerned with is not primarily file exchange but rather direct access to computers, software, data, and other resources, as is required by a range of collaborative problem-solving and resource-brokering strategies emerging in industry, science, and engineering. This sharing is, necessarily, highly controlled, with resource providers and consumers defining clearly and carefully just what is shared, who is allowed to share, and the conditions under which

sharing occurs. A set of individuals and/or institutions defined by such sharing rules form what we call a virtual organization (VO).

Whereas the Internet is a network of communication, Cloud/Grid Computing is seen as a network of computation: The field provides tools and protocols for resource sharing of a variety of IT resources. Grid Computing approaches are based on coordinated resource sharing and problem solving in dynamic, multi-institutional VOs. A (short) list of examples of possible VOs includes application service providers, storage service providers, machine-cycle providers, and members of industry-specific consortia. These examples, among others, represent an approach to computing and problem solving based on collaboration in data-rich and computation-rich environments [FOS200101, MYE200301]. The enabling factors in the creation of Cloud Computing systems in recent years have been the proliferation of broadband (optical-based) communications, the Internet, and the World Wide Web infrastructure, along with the availability of low-cost, high-performance computers using standardized (open) Operating Systems [CHE200201, FOS199901, FOS200101]. This kind of computing is also known by a number of other names (although some of these terms have slightly different connotations), such as just "cloud computing" (the most current term), "grid" (the term "the grid" was coined in the mid-1990s to denote a proposed Distributed Computing infrastructure for advanced science and engineering), "computational grid," "computing-on-demand," "on-demand computing," "just-in-time computing," "platform computing," "network computing," "computing utility" (the term used by this author in the late 1980s [MIN198701]), "utility computing," "cluster computing," and "high-performance distributed computing."

Prior to the deployment of Cloud/Grid Computing, a typical business application had a dedicated server platform of servers and an anchored storage device assigned to each individual server. Applications developed for such platforms were not able to share resources, and from an individual server's perspective it was not possible, in general, to predict, even statistically, what the processing load would be at different times. Consequently, each instance of an application needed to have its own excess capacity to handle peak usage loads. This predicament typically resulted in higher overall costs than would otherwise need to be the case [HAN200301]. To address these lacunae, Cloud/Grid Computing aims at exploiting the opportunities afforded by the synergies, the economies of scale, and the load smoothing that result from the ability to share and aggregate distributed computational capabilities, and deliver these hardware-based capabilities as a transparent service to the end user.* To reinforce the point, the term "synergistic" implies "working together so that the total effect is greater than the sum of the individual constituent elements."

---

* As implied in the opening paragraphs, a number of solutions in addition to Grid Computing (e.g., virtualization) can be employed to address this and other computational issues—Grid Computing is just one approach.

From a service provider perspective, Cloud/Grid Computing is somewhat akin to an ASP environment, but with a much higher level of performance and assurance [BUY200301]. The following points describe some of the concepts embodied in Cloud/Grid Computing and other related technologies.

- ◾ Cloud/Grid Computing:
    - (Virtualized) Distributed Computing environment that enables the dynamic "runtime" selection, sharing, and aggregation of (geographically) distributed autonomous (autonomic) resources based on the availability, capability, performance, and cost of these computing resources, and, simultaneously, also based on an organization's specific baseline and/or burst processing requirements.
    - Enables organizations to transparently integrate, streamline, and share dispersed, heterogeneous pools of hosts, servers, storage systems, data, and networks into one synergistic system, in order to deliver agreed-upon service at specified levels of application efficiency and processing performance.
    - An approach to Distributed Computing that spans multiple locations and/or multiple organizations, machine architectures, and software boundaries to provide power, collaboration, and information access.
    - Infrastructure that enables the integrated, collaborative use of computers, supercomputers, networks, databases, and scientific instruments owned and managed by multiple organizations.
    - A network of computation, namely, tools and protocols for coordinated resource sharing and problem solving among pooled assets ... allows coordinated resource sharing and problem solving in dynamic, multi-institutional virtual organizations.
    - Simultaneous application of the resources of many networked computers to a single problem ... concerned with coordinated resource sharing and problem solving in dynamic, multi-institutional virtual organizations.
    - Decentralized architecture for resource management, and a layered hierarchical architecture for implementation of various constituent services.
    - Combines elements such as Distributed Computing, high-performance computing, and disposable computing, depending on the application.
    - Local, metropolitan, regional, national, or international footprint. Systems may be in the same room, or may be distributed across the globe; they may running on homogenous or heterogeneous hardware platforms; they may be running on similar or dissimilar operating systems; and they may owned by one or more organizations.
    - Types (for classical "grids"): (1) computational grids: machines with set-aside resources stand by to "number-crunch" data or provide coverage for other intensive workloads; (2) scavenging grids: commonly used to locate and exploit machine cycles on idle servers and desktop machines for use in resource-intensive tasks; and (3) data grids: a unified interface

for all data repositories in an organization, and through which data can be queried, managed, and secured.
- Computational grids can be local enterprise grids (also called Private clouds), and Internet-based grids. Enterprise grids are middleware-based environments to harvest unused "machine cycles," thereby displacing otherwise-needed growth costs.

■ Virtualization:
- An approach that allows several operating systems to run simultaneously on one (large) computer (e.g., IBM's z/VM operating system lets multiple instances of Linux coexist on the same mainframe computer).
- More generally, it is the practice of making resources from diverse devices accessible to a user as if they were a single, larger, homogenous, appear-to-be-locally-available resource.
- Dynamically shifting resources across platforms to match computing demands with available resources: the computing environment can become dynamic, enabling autonomic shifting applications between servers to match demand.
- The capability to divide a single physical device into multiple logical devices, thereby making a single physical server operate like multiple servers [EAS200901].
- The abstraction of server, storage, and network resources in order to make them available dynamically for sharing by IT services, both internal to and external to an organization. In combination with other server, storage, and networking capabilities, virtualization offers customers the opportunity to build more efficient IT infrastructures. Virtualization is seen by some as a step on the road to Utility Computing.
- A proven software technology that is rapidly transforming the IT landscape and fundamentally changing the way that people compute [VMW200701].

■ Clusters:
- Aggregating of processors in parallel-based configurations, typically in local environment (within a data center); all nodes work cooperatively as a single unified resource.
- Resource allocation is performed by a centralized resource manager and scheduling system.
- Comprised of multiple interconnected independent nodes that cooperatively work together as a single unified resource; unlike clouds/grids, cluster resources are typically owned by a single organization.
- All users of clusters have to go through a centralized system that manages allocation of resources to application jobs. Cluster management systems have centralized control, complete knowledge of system state and user requests, and complete control over individual components.

■ (Basic) Web Services (WSs):

- Web Services provide standard infrastructure for data exchange between two different distributed applications (grids provide an infrastructure for aggregation of high-end resources for solving large-scale problems)>
- Web Services are expected to play a key constituent role in the standardized definition of Grid Computing, because Web Services have emerged as a standards-based approach for accessing network applications.

■ Peer-to-Peer (P2P):
- P2P is concerned with the same general problem as Cloud/Grid Computing, namely, the organization of resource sharing within virtual communities.
- Grid community focuses on aggregating distributed high-end machines such as clusters, whereas the P2P community concentrates on sharing low-end systems such as PCs connected to the Internet,
- Like P2P, Cloud/Grid Computing allows users to share files (many-to-many sharing). With grid, the sharing is not only in reference to files, but also other IT resources.

Table 8.5 provides a basic glossary of Cloud Computing terms, synthesized from a number of industry sources.

Grid Computing proper started out as the simultaneous application of the resources of many networked computers to a single (scientific) problem [FOS199901]. Grid Computing has been characterized as the "massive integration of computer systems" [WAL200201]. Computational grids have been used for a number of years to solve large-scale problems in science and engineering. The noteworthy fact is that, at this juncture, the approach can already be applied to a mix of mainstream business problems. Specifically, Grid Computing is now beginning to make inroads into the commercial world, including financial services operations, making the leap forward from such scientific venues as research labs and academic settings [HAN200301].

To deploy a cloud/grid, a commercial organization needs to assign computing resources to the shared environment and deploy appropriate grid middleware on these resources, enabling them to play various roles that need to be supported in the grid (e.g., scheduler, broker). Some (minor) application re-tuning or parallelization may, in some instances, be required; data accessibility will also have to be taken into consideration. A security framework will also be required. If the organization subscribes to the service provider model, then grid deployment would mean establishing adequate access bandwidth to the provider, some possible application re-tuning, and the establishment of security policies (the assumption being that the provider will itself have a reliable security framework).

The concept of providing computing power as a utility-based function is generally attractive to end users requiring fast transactional processing and "scenario modeling" capabilities. The concept may also be attractive to IT planners looking to control costs and reduce data center complexity. The ability to have a cluster, an entire data center, or other resources spread across a geography connected by

Table 8.5   Basic Glossary of Cloud Computing

Cloud Computing	The latest term to describe a grid/utility computing service. Such service is provided in the network. From the perspective of the user, the service is virtualized. In turn, the service provider will most likely use virtualization technologies (virtualized computing, virtualized storage, etc.) to provide the service to the user.
Clusters	Aggregating of processors in parallel-based configurations, typically in local environment (within a data center); all nodes work cooperatively as a single unified resource. Resource allocation is performed by a centralized resource manager and scheduling system. Comprised of multiple interconnected independent nodes that cooperatively work together as a single unified resource.
Common Object Request Broker Architecture (CORBA)	Object Management Group's open, vendor-independent architecture and infrastructure that computer applications use to work together over networks. A CORBA-based program from any vendor, on almost any computer, operating system, programming language, and network, can interoperate with a CORBA-based program from the same or another vendor, on almost any other computer, operating system, programming language, and network. Because of the easy way that CORBA integrates machines from so many vendors, with sizes ranging from mainframes through minis and desktops to handhelds and embedded systems, it is the middleware for large (and even not-so-large) enterprises. One of its most important, as well as most frequent, uses is in servers that must handle a large number of clients, at high hit rates, with high reliability [OMG200701].
Core Architecture Data Model (CADM)	A formal model defining the data organization for a repository of C4ISR/DoDAF-compliant architecture products (artifacts). The CADM provides a common schema for repositories of architecture information. Tool builders or vendors providing support for DoDAF-style architecture descriptions typically implement the CADM with a database [SYS200501].
Data De-duplication	Also called "intelligent compression" or "single-instance storage" is a method of reducing storage requirements by eliminating redundant data. In effect, this can also be seen as a form of virtualization. With data de-duplication, only one unique instance of the data is actually retained on storage media; redundant data is deleted, and a pointer to the unique data copy is used to support requests for access to that redundant copy.

Continued

Table 8.5   Basic Glossary of Cloud Computing (Continued)

Data Grid	A kind of Grid Computing grid used for housing and providing access to data across multiple organizations; users are not focused on where this data is located as long as they have access to the data [MIN200501].
Developer	A high-level role responsible for the implementation of the solution. Again, more specialized actors within this role might actually carry out the work, such as a database programmer, a Java developer, a Web developer, and a business process choreographer, to name a few. Developers work on specific layers of the application stack and each requires specialized skills for that layer [MIT200601].
Enterprise	A(ny) collection of corporate or institutional task-supporting functional entities that has a common set of goals or a single mandate. In this context, an enterprise is, but is not limited to, an entire corporation, a division or department of a corporation, a group of geographically dispersed organizations linked together by common administrative ownership, a government agency (or set of agencies) at any level of jurisdiction, and so on. This also encompasses the concept on an "extended enterprise," which is a logical aggregation that includes internal business units of a firm along with partners and suppliers [TOG200501].
Enterprise Service Bus (ESB)	A connectivity infrastructure for integrating applications and services by performing the following actions between services and requestors: ROUTING messages between services, CONVERTING transport protocols between requestor and service, TRANSFORMING message formats between requestor and service, and HANDLING business events from disparate sources [IBM200701].
eXtensible Markup Language (XML)	A structured language that was published as a W3C Recommendation in 1998. It is a meta language because it is used to describe other languages, the elements they can contain, and how those elements can be used. These standardized specifications for specific types of information make them, along with the information that they describe, portable across platforms.

Table 8.5   Basic Glossary of Cloud Computing (Continued)

Fibre Channel over IP (FCIP)	A protocol for transmitting Fibre Channel (FC) data over an IP network. It allows the encapsulation/tunneling of FC packets and transport via Transmission Control Protocol/Internet Protocol (TCP/IP) networks (gateways are used to interconnect FC Storage Area Networks (SANs) to the IP network and to set up connections between SANs). Protocol enables applications developed to run over FC SANs to be supported under IP, enabling organizations to leverage their current IP infrastructure and management resources to interconnect and extend FC SANs.
Grid Computing	Aka Cloud Computing, Utility Computing, etc., an environment that can be built at the local (data center), regional, or global level, where individual users can access computers, databases, and scientific tools in a transparent manner, without having to directly take into account where the underlying facilities are located.
	(Virtualized) Distributed Computing environment that enables the dynamic "runtime" selection, sharing, and aggregation of (geographically) distributed autonomous (autonomic) resources based on the availability, capability, performance, and cost of these computing resources, and simultaneously, also based on an organization's specific baseline or burst processing requirements. Grid Computing enables organizations to transparently integrate, streamline, and share dispersed, heterogeneous pools of hosts, servers, storage systems, data, and networks into one synergistic system, in order to deliver agreed-upon service at specified levels of application efficiency and processing performance. Grid Computing is an approach to Distributed Computing that spans multiple locations or multiple organizations, machine architectures, and software boundaries to provide power, collaboration, and information access. Grid Computing is infrastructure that enables the integrated, collaborative use of computers, supercomputers, networks, databases, and scientific instruments owned and managed by multiple organizations. Grid Computing is a network of computation: namely, tools and protocols for coordinated resource sharing and problem solving among pooled assets… allow coordinated resource sharing and problem solving in dynamic, multi-institutional virtual organizations [MIN200501].

Continued

Table 8.5   Basic Glossary of Cloud Computing (Continued)

Grid Computing Topologies	Local, metropolitan, regional, national, or international footprint. Systems may be in the same room, or may be distributed across the globe; they may running on homogenous or heterogeneous hardware platforms; they may be running on similar or dissimilar operating systems; and they may owned by one or more organizations [MIN200501].
Grid Computing Types	Computational grids: machines with set-aside resources stand by to number-crunch data or provide coverage for other intensive workloads; (2) Scavenging grids: commonly used to locate and exploit CPU cycles on idle servers and desktop machines for use in resource-intensive tasks; and (3) Data grids: a unified interface for all data repositories in an organization, through which data can be queried, managed, and secured. Computational grids can be local, Enterprise grids (also called Intragrids), and Internet-based grids (also called Intergrids). Enterprise grids are middleware-based environments to harvest unused machine cycles, thereby displacing otherwise-needed growth costs [MIN200501].
Grid Service	A network-based service, possibly a Web Service, that conforms to a set of conventions (interfaces and behaviors) that define how a client interacts with a grid (cloud) capability [MIN200501].
Integration Centric - Business Process Management Suites (IC-BPMS)	Integration capabilities products that support process improvement and that have evolved out of the enterprise application integration (EAI). Originally this space was dominated by proprietary, closed-framework solutions; at this time these products are based on SOA and on standards-based integration technology. Vendors have added embedded Enterprise Service Bus (ESB) and Business Process Management (BPM) capabilities.
Internet FCP (iFCP)	A protocol that converts FC frames into TCP enabling native FC devices to be connected via an IP network. Encapsulation protocols for IP storage solutions where the lower-layer FC transport is replaced with TCP/IP and Gigabit Ethernet. The protocol enables existing FC storage devices or Storage Area Networks (SANs) to attach to an

Table 8.5   Basic Glossary of Cloud Computing (Continued)

Internet FCP (iFCP)	IP network. The operation is as follows: FC devices, such as disk arrays, connect to an iFCP gateway or switch. Each FC session is terminated at the local gateway and converted to a TCP/IP session via iFCP. A second gateway or switch receives the iFCP session and initiates an FC session. In iFCP, TCP/IP switching and routing elements complement and enhance, or replace, FC SAN fabric components.
Internet Small Computer System Interface (iSCSI)	A protocol that serializes SCSI commands and converts them to TCP/IP. Encapsulation protocols for IP storage solutions for the support of Direct Attached Storage (DAS) (specifically SCSI-3 commands) over IP network infrastructures (at the physical layer, iSCSI supports a Gigabit Ethernet interface so that systems supporting iSCSI interfaces can be directly connected to standard Gigabit Ethernet switches or IP routers; the iSCSI protocol sits above the physical and data-link layers).
IP Storage	Using IP and Gigabit Ethernet to build SANs. Traditional SANs were developed using the FC transport, because it provided gigabit speeds compared to 10 and 100 Mbps Ethernet used to build messaging networks at that time. FC equipment was costly, and interoperability between different vendors' switches was not completely standardized. Since Gigabit Ethernet and IP have become commonplace, IP storage enables familiar network protocols to be used, and IP allows SANs to be extended throughout the world. Network management software and experienced professionals in IP networks are also widely available.

The following protocols are applicable:

• iFCP is a gateway-to-gateway protocol that allows the replacement of FC fabric components, allowing attachment of existing FC-enabled storage products to an IP network.

• Metro Fibre Channel Protocol (mFCP) is another proposal for handling "IP storage." It is identical to iFCP, except that TCP is replaced by User Datagram Protocol (UDP). |

Continued

Table 8.5 Basic Glossary of Cloud Computing (Continued)

IP Storage	• iSCSI is a transport protocol for SCSI that operates on top of TCP. It provides a new mechanism for encapsulating SCSI commands on an IP network. iSCSI is a protocol for a new generation of storage end-nodes that natively use Transmission Control Protocol/Internet Protocol (TCP/IP) and replaces FCP with a pure TCP/IP implementation. iSCSI has broad industry support.
	• FCIP is FC over TCP/IP. Here FC uses IP-based network services to provide the connectivity between the SAN islands over Local Area Networks (LANs), Metropolitan Area Networks (MANs), or Wide Area Networks (WANs). FCIP relies on TCP for congestion control and management, and upon both TCP and FC for data error and data loss recovery. FCIP treats all classes of FC frames the same as datagrams.
Message-Oriented Middleware (MOM)	A client/server infrastructure that increases the interoperability, portability, and flexibility of an application by allowing the application to be distributed over multiple heterogeneous platforms. It reduces the complexity of developing applications that span multiple operating systems and network protocols by insulating the application developer from the details of the various operating system and network interfaces. Application Programming Interfaces (APIs) that extend across diverse platforms and networks are typically provided by the MOM. Applications exchange messages that can contain formatted data, requests for action, or both.
Peer-to-Peer (P2P)	P2P is concerned with same general problem as Grid Computing, namely, the organization of resource sharing within virtual communities.
	The Grid community focuses on aggregating distributed high-end machines such as clusters, whereas the P2P community concentrates on sharing low-end systems such as PCs connected to the Internet.
	Like P2P, Grid Computing allows users to share files (many-to-many sharing). With Grid, the sharing is not only in reference to files, but also other IT resources.

Table 8.5   Basic Glossary of Cloud Computing (Continued)

Service Orientation	A way of thinking about business processes as linked, loosely coupled tasks supported by services. A new service can be created by composing a primitive group of services. This recursive definition is important because it enables the construction of more complex services above a set of existent ones [SOU200601].
Service-Oriented Architecture (SOA)	SOA describes an IT architecture based around the concept of delivering reusable, business services that are underpinned by IT components in such a way that the providers and the consumers of the business services are loosely coupled, with no knowledge of the technology, platform, location, or environment choices of each other [STA200501]. A business-driven IT architectural approach that supports integrating business as linked, repeatable business tasks, or services. SOA helps businesses innovate by ensuring that IT systems can adapt quickly, easily, and economically to support rapidly changing business needs [IBM200701]. In an SOA, resources are made available to other participants in the network as independent services that are accessed in a standardized way.
Service-Oriented Computing (SOC)	The computing paradigm that utilizes services as fundamental elements for developing applications [PAP200301].
Service-Oriented Network (SON)	A service-oriented architecture for the development, deployment, and management of network services [SOU200601]. The application paradigm that utilizes services distributed across a network as fundamental functional elements.
Service-Oriented Network Architecture (SONA)	Cisco Systems' architectural framework that aims at delivering business solutions to unify network-based services such as security, mobility, and location with the virtualization of IT resources.
Simple Object Access Protocol (SOAP)	A standard of the W3C that provides a framework for exchanging XML-based information. A lightweight protocol for exchange of XML-based information in a distributed environment.
SOA Infrastructure	A simplified, virtualized, and distributed application framework that supports SOA.

Continued

Table 8.5   Basic Glossary of Cloud Computing (Continued)

Storage	Infrastructure (typically in the form of appliances) that is used for the permanent or semi-permanent online retention of structured (e.g., databases) and unstructured (e.g., business/e-mail files) corporate information. Typically includes (1) a controller that manages incoming and outgoing communications as well as the data steering onto the physical storage medium (e.g., RAIDs [Redundant Arrays of Independent Disks], semiconductor memory, etc.); and (2) the physical storage medium itself. The communications mechanism could be a network interface (i.e., Gigabit Ethernet), a channel interface (i.e., SCSI), or a SAN Interface (i.e., FC).
Storage Appliance	A storage platform designed to perform a specific task, such as NAS, routers, virtualization, etc.
Storage Virtualization	Software (sub)systems (typically middleware) that abstract the physical and logical storage assets from the host systems.
Tiered Storage	A process for the assignment of different categories of data to different types of storage media. The purpose is to reduce total storage cost and optimize accessibility. Organizations are reportedly finding cost savings and improved data management with a tiered storage approach. In practice the assignment of data to particular media tends to be an evolutionary and complex activity. Storage categories may be based on a variety of design/architectural factors, including levels of protection required for the application or organization, performance requirements, and frequency of use. Software exists for automatically managing the process based on a company-defined policy. Tiered storage generally introduces more vendors into the environment and interoperability is important.
	An example of tiered storage is as follows: Tier-1 data (e.g., mission-critical files) could be effectively stored high-quality DAS (but relatively expensive) media such as double-parity RAIDs. Tier-2 data (e.g., quarterly financial records) could be stored on media affiliated with a SAN; this media tends to be less expensive than DAS drives, but there may be network latencies associated with the

Table 8.5   Basic Glossary of Cloud Computing (Continued)

Tiered Storage	access. Tier-3 data (e.g., e-mail backup files) could be stored on recordable compact discs (CD-Rs) or tapes. (Clearly there could be more than three tiers, but the management of the multiple tiers then becomes fairly complex.) Another example (in the medical field) is as follows: Real-time medical imaging information may be temporarily stored on DAS disks as a Tier 1, say, for a couple of weeks. Recent medical images and patient data may be kept on FC drives (tier-2) for about a year. After that, less-frequently accessed images and patient records are stored on AT Attachment (ATA) drives (tier-3) for 18 months or more. Tier-4 consists of a tape library for archiving.
Universal Discovery, Description and Integration (UDDI)	A standardized method for publishing and discovering information about Web Services. UDDI is an industry initiative that attempts to create a platform-independent, open framework for describing services, discovering businesses, and integrating business services. UDDI deals with the process of discovery in the SOA (WSDL is often used for service description, and SOAP for service invocation.) Being a Web Service itself, UDDI is invoked using SOAP. In addition, UDDI also defines how to operate servers and how to manage replication among several servers.
Virtual Infrastructure	A dynamic mapping of physical resources to business needs [VMW200701].
Virtualization	An approach that allows several operating systems to run simultaneously on one (large) computer (e.g., IBM's z/VM operating system lets multiple instances of Linux coexist on the same mainframe computer). More generally, it is the practice of making resources from diverse devices accessible to a user as if they were a single, larger, homogenous, appear-to-be-locally-available resource. Dynamically shifting resources across platforms to match computing demands with available resources: the computing environment can become dynamic, enabling autonomic shifting applications between servers to match demand.

Continued

Table 8.5   Basic Glossary of Cloud Computing (Continued)

Virtualization	The abstraction of server, storage, and network resources in order to make them available dynamically for sharing by IT services, both internal and external to an organization. In combination with other server, storage, and networking capabilities, virtualization offers customers the opportunity to build more efficient IT infrastructures. Virtualization is seen by some as a step on the road to Utility Computing.
	Virtualization is a proven software technology that is rapidly transforming the IT landscape and fundamentally changing the way that people compute [VMW200701].
Web Services (WSs)	A software system designed to support interoperable machine-to-machine interaction over a network. It has an interface described in a machine-processable format (specifically WSDL). Other systems interact with the Web Service in a manner prescribed by its description using SOAP messages, typically conveyed using HTTP with an XML serialization in conjunction with other Web-related standards [IBM200701].
	Web Services provide standard infrastructure for data exchange between two different distributed applications (grids provide an infrastructure for aggregation of high-end resources for solving large-scale problems). Web Services are expected to play a key constituent role in the standardized definition of Grid Computing, because Web Services have emerged as a standards-based approach for accessing network applications.
Web Services Description Language (WSDL)	An XML-based language used to describe Web Services and how to locate them; it provides information on what the service is about, where it resides, and how it can be invoked.
Web Services Networking	Assembly of a more complex service from service modules that reside on different nodes connected to a network.

the Internet (or, alternatively, connected by an intranet or extranet), operating as a single transparent virtualized system that can be managed as a service, rather than as individual constituent components, likely will, over time, increase business agility, reduce complexity, streamline management processes, and lower operational costs [HAN200301]. Grid technology allows organizations to utilize numerous computers to solve problems by sharing computing resources. The problems to be solved might involve data processing, network bandwidth, or data storage, or a combination thereof.

In a grid environment, the ensemble of resources is able to work together cohesively because of defined protocols that control connectivity, coordination, resource allocation, resource management, security, and chargeback. Generally, the protocols are implemented in the middleware. The systems "glued" together by a computational grid may be in the same room, or may be distributed across the globe; they may be running on homogenous or heterogeneous hardware platforms; they may be running on similar or dissimilar operating systems; and they may owned by one or more organizations. The goal of Grid Computing is to provide users with a single view or single mechanism that can be utilized to support any number of computing tasks: The grid leverages its extensive informatics capabilities to support the "number crunching" needed to complete the task, and all the user perceives is, essentially, a large virtual computer undertaking his work [DEV200301].

In recent years one has seen an increasing roster of published articles, conferences, tutorials, resources, and tools related to the topic of Cloud Computing. However, as already implied, a number of the basic concepts of Grid Computing go back as far as the mid-1960s and early 1970s. Recent advances, such as ubiquitous high-speed networking in both private and public venues (e.g., high-speed intranets/high-speed Internet), make the technology more deployable at the practical level, particularly when looking at corporate environments.

As far back as 1987, this researcher was advocating the concept of Grid Computing in internal Bell Communications Research White Papers (e.g., in Special Reports SR-NPL-000790—an extensive plan written by the author listing progressive data services that could be offered by local telcos and Regional Bell Operating Systems (RBOSs), entitled "A Collection of Potential Network-Based Data Services" [MIN198701]). In a section called "Network for a Computing Utility" it was stated that

> The proposed service provides the entire apparatus to make the concept of the Computing Utility possible. This includes as follows: (1) the physical network over which the information can travel, and the interface through which a guest PC/workstation can participate in the provision of machine cycles and through which the service requesters submit jobs; (2) a load sharing mechanism to invoke the necessary servers to complete a job; (3) a reliable security mechanism; (4) an effective accounting mechanism to invoke the billing system; and (5) a detailed directory of servers. … Security is one of the major issues for this service, particularly if the PC is not fully dedicated to this function, but also used for other local activities. Virus threats, infiltration and corruption of data, and other damage must be appropriately addressed and managed by the service; multi-task and robust operating systems are also needed for the servers to assist in this security process … The

Computing Utility service is beginning to be approached by the Client/ Server paradigm now available within a Local Area Network (LAN) environment… This service involves capabilities that span multiple 7-layer stacks. For example, one stack may handle administrative tasks, another may invoke the service (e.g., Remote Operations), still another may return the results (possibly a file), and so on… Currently no such service exists in the public domain. Three existing analogues exist, as follows: (1) timesharing service with a centralized computer; (2) highly-parallel computer systems with hundreds or thousands of nodes (what people now call cluster computing), and (3) gateways or other processors connected as servers on a LAN. The distinction between these and the proposed service is the security and accounting arenas, which are much more complex in the distributed, public (grid) environment… This service is basically feasible once a transport and switching network with strong security and accounting (chargeback) capabilities is deployed,… A high degree of intelligence in the network is required… a physical network is required…security and accounting software is needed…protocols and standards will be needed to connect servers and users, as well as for accounting and billing. These protocols will have to be developed before the service can be established…

Security is a key consideration in Grid Computing. The user wants to get its services in a trustworthy and confidential manner. Then there is the desire for guaranteed levels of service and predictable, reduced costs. Finally, there is the need for standardization, so that a user with the appropriate middleware client software can transparently reach any registered resource in the network. Grid Computing supports the concept of the Service Oriented Architecture, where clients obtain services from loosely coupled service-provider resources in the network. Web Services based on the Simple Object Access Protocol (SOAP) and Universal Description, Discovery and Integration (UDDI) protocols are now key building blocks of a cloud/grid environment [MIN200601].

8.3 Potential Applications and Financial Benefits of Cloud/Grid Computing

Proponents take the position that Cloud/Grid Computing represents a "next step" in the world of computing, and that Grid Computing promises to move the Internet evolution to the next logical level. According to some (see [YAN200401, YAN200301, CHU200201], among others), Utility Computing is a positive, fundamental shift in computing architecture, and many businesses will be completely transformed over the next decade by using grid-enabled services, as these businesses integrate not only applications across the Internet, but, also raw computer power and storage. Furthermore, proponents prognosticate that infrastructure will appear that will be able to connect multiple regional and national computational grids, creating a universal source of pervasive and dependable computing power that will support new classes of applications [BER200301].

The benefits gained from Grid Computing can translate into competitive advantages in the marketplace. For example, the potential exists for grids to [IBM200301, CHE200201]:

- Enable resource sharing
- Provide transparent access to remote resources
- Makes effective use of computing resources, including platforms and data sets
- Reduce significantly the number of servers needed (25% to 75%)
- Allow on-demand aggregation of resources at multiple sites
- Reduce execution time for large-scale data processing applications
- Provide access to remote databases and software
- Provide load smoothing across a set of platforms
- Provide fault tolerance
- Take advantage of time zone and random diversity (in peak hours, users can access resources in off-peak zones)
- Provide the flexibility to meet unforeseen emergency demands by renting external resources for a required period instead of owning them
- Enable the realization of a Virtual Data Center

As an example, VMware customers that have adopted our virtual infrastructure solutions have reported dramatic results, including [VMW200701]

- 60% to 80% utilization rates for x86 servers (up from 5% to 15% in non-virtualized PCs)
- Cost savings of more than $3,000 annually for every workload virtualized
- Ability to provision new applications in minutes instead of days or weeks
- 85% improvement in recovery time from unplanned downtime

There are well-known advantages in sharing resources, as a routine assessment of the behavior of the M/M/1 queue (memoryless/memoryless/1 server queue) versus the M/M/m queue (memoryless/memoryless/n servers queue) demonstrates: a single, more powerful queue is more efficient than a group of discrete queues of comparable aggregate power. Grid Computing represents a development in virtualization: As we have stated, it enables the abstraction of Distributed Computing and data resources such as processing, network bandwidth, and data storage to create a single system image; this grants users and applications seamless access (when properly implemented) to a large pool of IT capabilities. Just as an Internet user views a unified instance of content via the Web, a Grid Computing user essentially sees a single, large virtual computer [IBM200301]. "Virtualization"—the driving force behind Grid Computing—has been a key factor since the earliest days of electronic business computing.

Studies have shown that when problems can be parallelized, such as in the case of data mining, records analysis, and billing (as may be the case in a bank, securities company, financial services company, insurance company, etc.), then significant savings are achievable. Specifically, when a classical model may require, say, $100K to process 100K records, a grid-enabled environment may take as little as $20K to process the same number of records. Hence, the bottom line is that Fortune 500 companies have the potential to save 30% or more in the Run-the-Engine costs on the appropriate line item of their IT budgets.

Grid Computing can also be seen as part of a larger re-hosting initiative and underlying IT trend at many companies (where alternatives such as Linux® or possibly Windows Operating Systems could, in the future, be the preferred choice over the highly reliable, but fairly costly UNIX solutions). While each organization is different and the results vary, the directional cost trend is believable. Vendors engaged in this space include (but are not limited to) IBM, Hewlett-Packard (HP), Sun, and Oracle. IBM uses "on-demand" to describe its initiative; HP has its Utility Data Center (UDC) products; Sun has its N1 Data-center Architecture; and Oracle has the 10g family of "grid-aware" products. Several software vendors also have a stake in Grid Computing, including, but not limited to, Microsoft, VMWare, Computer Associates, Veritas Software, and Platform Computing [BED200301]. VMWare was one of the leading suppliers of virtualization software at press time.*

8.4 Cloud/Grid Types, Topologies, and Components

Cloud/Grid Computing embodies a combination of a decentralized architecture for resource management, and a layered hierarchical architecture for implementation of various constituent services [GRI200301]. A Cloud/Grid Computing system can have local, metropolitan, regional, national, or international footprints. In turn, the autonomous resources in the constituent ensemble can span a single organization, multiple organizations, or a service provider space. Clouds/Grids can focus on the pooled assets of one organization or span virtual organizations that use a common suite of protocols to enable grid users and applications to run services in a secure, controlled manner [MYE200301]. Furthermore, resources can be

---

* VMware Infrastructure 3 has been positioned by the vendor as a tool for creating a self-optimizing IT infrastructure today with the most widely deployed software suite for optimizing and managing industry-standard IT environments through virtualization. VMware Infrastructure 3 is the next generation of industry-leading infrastructure virtualization software that virtualizes servers, storage, and networking, allowing multiple unmodified operating systems and their applications to run independently in virtual machines while sharing physical resources. The suite delivers virtualization, management, resource optimization, application availability, and operational automation capabilities [VMW200701].

logically aggregated for a long period of time (say, months or years), or for a temporary period of time (say, minutes, days, or weeks).

Traditional Grid Computing proper often combines elements such as Distributed Computing, high-performance computing, and disposable computing, depending on the application of the technology and the scale of the operation. Grids can, in practical terms, create a virtual supercomputer out of existing servers, workstations, and even PCs, to deliver processing power not only to a company's own stakeholders and employees, but also to its partners and customers. This metacomputing environment is achieved by treating such IT resources as processing power, memory, storage, and network bandwidth as pure commodities. Like an electricity or water network, computational power can be delivered to any department or any application where it is needed most at any given time, based on specified business goals and priorities. Furthermore, Grid Computing allows chargeback on a per-usage basis rather than for a fixed infrastructure cost [HAN200301]. Grids have historically encompassed the following types [DEV200301]:

- Computational grids, where machines with set-aside resources stand by to "number-crunch" data or provide coverage for other intensive workloads
- Scavenging grids, commonly used to find and harvest machine cycles from idle servers and desktop machines for use in resource-intensive tasks (scavenging is usually implemented in a way that is unobtrusive to the owner/user of the processor)
- Data grids that provide a unified interface for all data repositories in an organization, and through which data can be queried, managed, and secured

As already noted, no claim is made herewith that there is a single solution to a given enterprise IT problem; Cloud/Grid Computing is one of the available solutions. For example, while some of the machine-cycle inefficiencies can be addressed by virtual servers/re-hosting (e.g., VMWare, MS VirtualPC and VirtualServer, LPARs from IBM, partitions from Sun and HP, which do not require a grid infrastructure), one of the possible approaches to this inefficiency issue is, indeed, Cloud/Grid Computing. Grid Computing does have an emphasis on geographically distributed, multi-organization, utility-based (outsourced), networking-reliant methods, while clustering and re-hosting have more (but not exclusively) of a data-center-focused, single-organization-oriented approach. Organizations will need to perform appropriate functional, economic, and strategic assessments to determine which approach is, in final analysis, best for their specific environment.

Figures 8.3, 8.4, and 8.5 provide a pictorial view of some Grid Computing environments. Figure 8.3 depicts the traditional computing environment where a multitude of often-underutilized servers support a disjoint set of applications and data sets. As implied by this figure, the typical IT environment prior to Cloud/Grid Computing, operated as follows: A business-critical application runs on a designated server. While the average utilization may be relatively low, during peak cycles

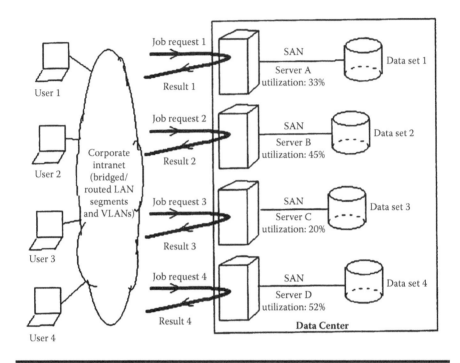

Figure 8.3 Standard computing environment.

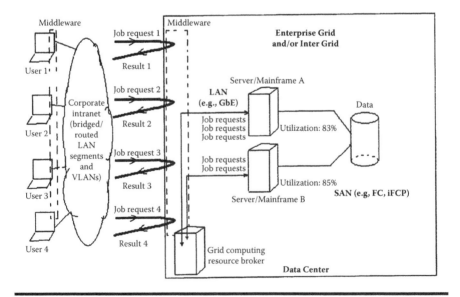

Figure 8.4 Grid computing environment (local implementation).

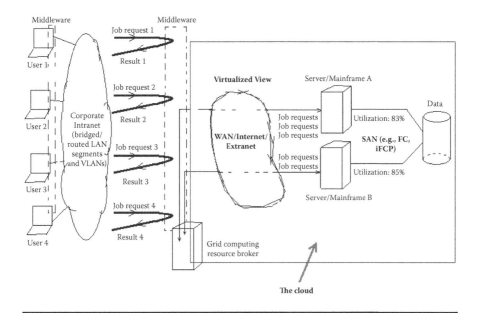

Figure 8.5   Cloud/grid computing environment (remote implementation).

the server in question can get overtaxed. As a consequence of this instantaneous overtaxation, the application can slow down, experience a halt, or even stall. In this traditional instance, the large data set that this application is analyzing exists only in a single data store (note that while multiple copies of the data could exist, it would not be easy with the traditional model to synchronize the databases if two programs independently operated aggressively on the data at the same time). Server capacity and access to the data store place limitations on how quickly desired results can be returned. Machine cycles on other servers are unable to be constructively utilized, and available disk capacity remains unused [IBM200301].

Figure 8.4 depicts an organization-owned computational grid; here, a middleware application running on a Cloud/Grid Computing Broker manages a small(er) set of processors and an integrated data store. A computational grid is a hardware and software infrastructure that provides dependable, consistent, pervasive, and inexpensive access to high-end computational capabilities. In a grid environment, workload can be broken up and sent in manageable pieces to idle server cycles. Not all applications are necessarily instantly migratable to a grid environment, without at least some re-design. While legacy business applications may, a priori, fit such a class of applications, a number of Fortune 500 companies are indeed looking into how such legacy applications can be modified or re-tooled such that they can be made to run on grid-based infrastructures. A scheduler sets rules and priorities for routing jobs on a cloud/grid-based infrastructure. When servers and storage are enabled for Grid Computing, copies of the data can be stored in formerly unused space and easily made available [IBM200301]. A grid also provides mechanisms for managing the distributed data in a seamless way [PAD200301, FOS199901]. Grid middleware provides facilities to allow use of the grid for applications and users. Middleware such as Globus [FOS199701], Legion [GRI199701], and UNICORE (UNiform Interface to COmputer Resources) [HUB200101] provide software infrastructure to handle the various challenges of computational and data grids [PAD200301].

Figure 8.5 depicts the utility-oriented implementation of a computational grid. This concept is analogous to electric power network (grid) where power generators are distributed, but the users are able to access electric power without concerning themselves about the source of energy and its pedestrian operational management [GRI200301, HAN200301]. As suggested by these figures, Cloud/Grid Computing aims to provide seamless and scalable access to distributed resources. Computational grids enable the sharing, selection, and aggregation of a wide variety of geographically distributed computational resources (such as supercomputers, computing clusters, storage systems, data sources, instruments, and developers), and presents them as a single, unified resource for solving large-scale computing- and data-intensive applications (e.g., molecular modeling for drug design, brain activity analysis, and high-energy physics). An initial grid deployment at a company can be scaled over time to bring in additional applications and new data. This allows gains in speed and accuracy without significant cost increases.

Some grids focus on data federation and availability; other grids focus on computing power and speed. Many grids involve a combination of the two. For end users, all infrastructure complexity stays hidden [IBM200301]. Data (database) federation makes disparate corporate databases look like the constituent data is all in the same database. Significant gains can be secured if one can work on all the different databases, including selects, inserts, updates, and deletes as if all the tables existed in a single database.* Almost every organization has significant unused computing capacity, widely distributed among a tribal arrangement of PCs, midrange platforms, mainframes, and supercomputers. For example, if a company has 10,000 PCs, at an average computing power of 333 MIPS, this equates to an aggregate 3 Tera (1012) floating-point operations per second (TFLOPS) of potential computing power. As another example, in the United States there are an estimated 300 million computers; at an average computing power of 333 MIPS, this equates to a raw computing power of 100,000 TFLOPS. Mainframes are generally idle 40% of the time; UNIX servers are actually "serving" something less than 10% of

---

* The federator system operates on the tables in the remote systems, the "federatees." The remote tables appear as virtual tables in the federator database. Client application programs can perform operations on the virtual tables in the federator database, but the real persistent storage is in the remote database. Each federatee views the federator as just another database client connection. The federatee is simply servicing client requests for database operations. The federator needs client software to access each remote database. Client software for IBM Informix®, Sybase, Oracle, etc. would need to be installed to access each type of "federatee" [IBM200301].

the time; and most PCs do nothing for 95% of a typical day [IBM200301]. This is an inefficient situation for customers. TFLOPS speeds that are possible with Grid Computing enable scientists to address some of the most computationally intensive scientific tasks, from problems in protein analysis that will form the basis for new drug designs, to climate modeling, to deducing the content and behavior of the cosmos from astronomical data [WAL200201].

The key components of Cloud/Grid Computing include the following [DEV200301]:

- *Resource management:* The grid must be aware of what resources are available for different tasks.
- *Security management:* The grid needs to take care that only authorized users can access and use the available resources.
- *Data management:* Data must be transported, cleansed, parceled, and processed.
- *Services management:* Users and applications must be able to query the grid in an effective and efficient manner.

8.5 Comparison with Other Approaches

As we saw earlier, certain IT computing constructs are not clouds/grids; we revisit these distinctions here. In some instances, these technologies are the optimal solution for an organization's problem; in other cases, Cloud/Grid Computing is the best solution, particularly if in the long term, one is especially interested in supplier-provided Utility Computing.

The distinction between clusters and clouds/grids relates to the way resources are managed. In case of clusters (aggregating of processors in parallel-based configurations), resource allocation is performed by a centralized resource manager and scheduling system; also, nodes cooperatively work together as a single unified resource. In case of grids, each node has its own resource manager and such a node does not aim at providing a single system view [BUY200301]. A cluster is comprised of multiple interconnected independent nodes that cooperatively work together as a single unified resource. This means that all users of clusters have to go through a centralized system that manages the allocation of resources to application jobs. Unlike grids, cluster resources are almost always owned by a single organization. Actually, many clouds/grids are constructed using clusters or traditional parallel systems as their nodes, although this is not a requirement.

Cloud/Grid Computing also differs from basic Web Services, although it now makes use of these services. Web Services have become an important component of Distributed Computing applications over the Internet [GRI200401]. The World Wide Web as a whole is not (yet, in itself) a grid: Its open, general-purpose protocols support access to distributed resources but not the coordinated use of those resources to deliver negotiated qualities of service [FOS200201]. So, while the Web is mainly focused on communication, Grid Computing enables resource sharing and collaborative resource interplay toward common business goals. Web Services provide standard infrastructure for data exchange between two different distributed applications, while grids provide an infrastructure for aggregation of high-end resources for solving large-scale problems in science, engineering, and commerce. However, there are similarities as well as dependencies: (1) similar to the case of the World Wide Web, Grid Computing keeps complexity hidden: multiple users experience a single, unified experience; and (2) Web Services are utilized to support Grid Computing mechanisms: these Web Services, will play a key constituent role in the standardized definition of Grid Computing, because Web Services have emerged in the past few years as a standards-based approach for accessing network applications. The recent trend is to implement grid solutions using Web Services technologies; for example, the Globus Toolkit 3.0 middleware is being implemented using Web Services technologies. In this context, low-level Grid Services are instances of Web Services (a Grid Service is a Web Service that conforms to a set of conventions that provide for controlled, fault-resilient, and secure management of stateful services) [GRI200301, FOX200201].

Both Peer-to-Peer (P2P) and Cloud/Grid Computing are concerned with the same general problem, namely, the organization of resource-sharing within VOs. As is the case with P2P environments, Grid Computing allows users to share files; but unlike P2P, Grid Computing allows many-to-many sharing. Furthermore, with grid, the sharing is in reference not only to files, but other resources as well. The grid community generally focuses on aggregating distributed high-end machines such as clusters, while the P2P community concentrates on sharing low-end systems such as PCs connected to the Internet [CHE200201]. Both disciplines take the same general approach to solving this problem, namely, the creation of overlay structures that coexist with, but need not correspond in structure to underlying organizational structures. Each discipline has made technical advances in recent years, but each also has—in current instantiations—a number of limitations: There are complementary aspects regarding the strengths and weaknesses of the two approaches that suggest that the interests of the two communities are likely to grow closer over time [IAM200301, CHE200201].

Cloud/Grid Computing also differs from virtualization. Resource virtualization is the abstraction of server, storage, and network resources in order to make them available dynamically for sharing by IT services, both inside and outside an organization. Virtualization is a step along the way on the road to Utility Computing (Grid Computing) and, in combination with other server, storage, and networking capabilities, offers customers the opportunity to build, according to advocates, an IT infrastructure "without" hard boundaries or fixed constraints [HEP200201]. Virtualization has somewhat more of an emphasis on local resources, while Grid Computing has more of an emphasis on geographically distributed inter-organizational resources (see Section 8.6).

Cloud/Grid Computing deployment, although potentially related to a re-hosting initiative, is not just re-hosting. As Figure 8.6 depicts, re-hosting implies the reduction of a large number of servers (possibly using some older or proprietary OS) to a smaller set of more powerful and more modern servers (possibly running on open source OSs). This is certainly advantageous from an operations, physical maintenance, and power and space perspective. There are savings associated with re-hosting. However, applications are still assigned specific servers. Grid Computing, on the other hand, permits the true virtualization of the computing function, as seen in Figure 8.6. Here, applications are not preassigned a server, but the "run-time" assignment is made based on real-time considerations. (Note: In the bottom diagram, the hosts could be collocated or spread all over the world. When local hosts are aggregated in tightly coupled configurations, they generally tend to be of the cluster parallel-based computing type; such processors, however, can also be non-parallel-computing-based grids [e.g., by running the Globus Toolkit]. When geographically dispersed hosts are aggregated in Distributed Computing configurations, they generally tend to be of the Grid Computing type and not running in a clustered arrangement. Figure 8.6 does not show geography, and the reader should conclude that the hosts are arranged in a Grid Computing arrangement.)

In summary, like clusters and Distributed Computing, grids bring computing resources together. Unlike clusters and Distributed Computing, which need physical proximity and operating homogeneity, Clouds and Grids can be geographically distributed and can be heterogeneous. Like virtualization technologies, Grid Computing enables the virtualization of IT resources. Unlike virtualization technologies that virtualize a single system, Cloud/Grid Computing enables the virtualization of broad-scale and disparate IT resources [IBM200301].

8.6 Enterprise-Level Virtualization

As if one needed confirmation, power and cooling are pressing issues for IT managers according to a press-time study by market research company IDC, which found that nearly 21% of data center managers feel that power and cooling capacity counts as their number-1 concern, as shown in Figure 8.7. The survey found that the majority of data center managers had experienced some operational and business impacts from power and cooling issues: 43% of managers highlighted "increased operational cost" as an issue and 33% identified "server downtime." Figure 8.8 shows that the growing operating costs—management and administration—continue to take a higher percentage (versus new server spending) of the IT budget [EAS200901].

Data center strategies that have been used in the recent past include the following [EAS200901]:

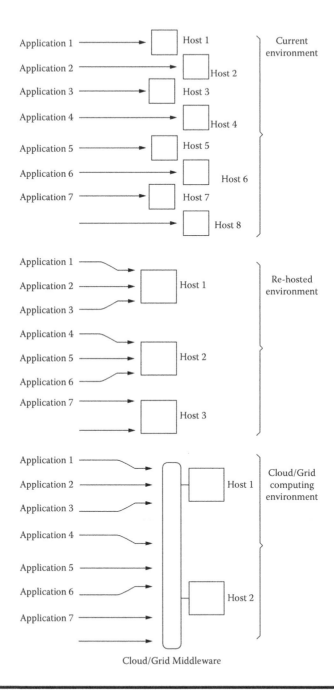

Figure 8.6 A comparison with re-hosting.

Q. *What is the number 1 challenge that your datacenter faces today?*

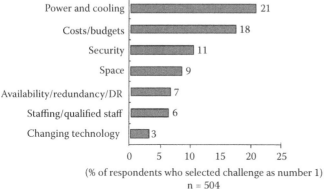

(% of respondents who selected challenge as number 1)
n = 504

Q. *Has your organization experienced any of the following business impacts from issues related to power and cooling?*

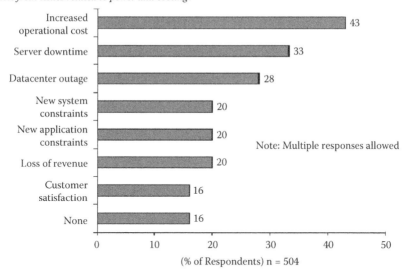

(% of Respondents) n = 504

Figure 8.7   Power-related concerns by data center administrators.

- Consolidating to fewer systems employing basic design techniques to decrease server count, thereby reducing all the expenses associated with each physical server
- Consolidating to fewer systems using virtualization techniques
- Consolidating to fewer, more energy-efficient servers to reduce power and cooling costs; this reduction in power and cooling demand can also impact

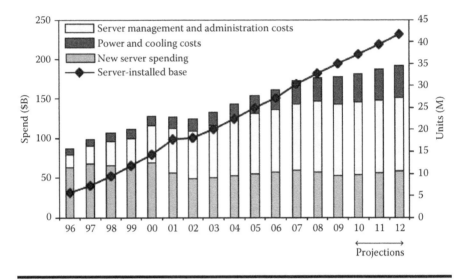

Figure 8.8 Global data center operating costs, including power and cooling.

the physical operations of the data center, reducing systems downtime and limiting the risk of potential data center outages
■ Deploying newer generations of servers that incorporate throughput and performance capabilities that are significantly higher than those of legacy servers; the performance gains of new, current-generation servers lead to fewer servers providing more and better application performance for a larger number of legacy servers

These techniques allow IT applications to run on fewer, newer servers, making acquisition costs drop (lower CAPEX); furthermore, because these fewer, newer, and more energy-efficient servers consume less power and cooling resources, operating expenses fall as well (lower OPEX).

At the turn of the 2000s decade, a high percentage of servers in a data center were stand-alone, pedestal-based servers. Soon thereafter, the mix of servers being installed changed, with large numbers of small, rack-optimized servers appearing in the data center, to be followed several years later by blade servers, managed in groups, within a blade chassis. The deployment of high-density servers brought about reduced CAPEX but did little to address operational expenses (OPEX) within the data center. In fact, the OPEX costs continue to rise—especially power/cooling costs (growing eight times as fast as the spending rate for the servers themselves) and maintenance/management costs associated with IT staff costs (growing at four times the rate of new server spending). At the same time, compute density and power density have increased significantly, with the advent of multicore processors in the x86 server space (servers based on x86 processors made by Intel and

AMD). As a result, CAPEX has been controlled in recent years, also because of the declining average sales prices for servers: Since 2004, the average system cost per processor has dropped 18% to $2,000, and the average core cost has dropped 70% to $715. While servers have become much more powerful over time, acquisition costs and energy requirements have dropped; however, the older-generation servers have fewer cores per processor—and throw more heat—while generating increasing power and cooling, maintenance, and management costs over time, affecting IT operational costs and IT productivity. The IDC study found that the typical x86 server is in place for 4.5 years, and other surveys indicate that almost 40% of deployed servers have been operating in place for 4 years or longer (this represents over 12 million single core-based servers still in use) [EAS200901].

Server virtualization has been used in recent years to address some of these issues. Virtualization allows IT to use a single server as if it were multiple servers; firms initially focused on virtualization primarily to consolidate multiple workloads on a single machine, but at this time virtualization plays a more direct design criterion in new applications and deployments.

Three representative products available in recent years are EMC's VMware, HP's Utility Data Center, and Platform Computing's Platform LFS. With virtualization, the logical functions of the server, storage, and network elements are separated from their physical functions (e.g., processor, memory, I/O, controllers, disks, switches). In other words, all servers, storage, and network devices can be aggregated into independent pools of resources. Some elements may even be further subdivided (server partitions, storage Logical Unit Numbers (LUNs)) to provide an even more granular level of control. Elements from these pools can then be allocated, provisioned, and managed—manually or automatically—to meet the changing needs and priorities of one's business. Virtualization can span the following domains [HEP200201]:

1. *Server virtualization for horizontally and vertically scaled server environments.* Server virtualization enables optimized utilization, improved service levels, and reduced management overhead.
2. *Network virtualization, enabled by intelligent routers, switches, and other networking elements supporting virtual LANs.* Virtualized networks are more secure and more able to support unforeseen spikes in customer and user demand.
3. *Storage virtualization (server, network, and array-based).* Storage virtualization technologies improve the utilization of current storage subsystems, reduce administrative costs, and protect vital data in a secure and automated fashion.
4. *Application virtualization enables programs and services to be executed on multiple systems simultaneously.* This computing approach is related to horizontal scaling, clusters, and Grid Computing, where a single application is able to cooperatively execute concurrently on a number of servers.

5. *Data center virtualization*, whereby groups of servers, storage, and network resources can be provisioned or reallocated on-the-fly to meet the needs of a new IT service or to handle dynamically changing workloads [HEP200201].

Vendor advertisements such as the one from Hewlett-Packard that follows were circulating at press time:

Next-generation HP ProLiant servers, such as HP ProLiant DL380 G6 Server, help improve efficiency, reduce costs, and deliver returns in as little as 3 months. Powered by Intel® Xeon® 5500 Series Processors, the newest line of HP ProLiant servers provides 11:1 consolidation ratio:

Reduce costs … comparison of HP-measured performance, power and cooling, and estimated software license fee data for HP ProLiant DL380 G4 (single-core Intel® Xeon® processors) at 30% load with HP DL 380 G6 (quad-core Intel® Xeon® Processor 5500 series) at 40% load

Reduce power and cooling costs up to 95%… Comparisons of the public SPECpower_ssj2008 submission results of the two HP ProLiant DL380 servers above on www. spec.org shows > 20x improvement in performance/watt between the HP ProLiant DL380 G4 and G6 servers. As a result, the same amount of compute capacity can be accomplished with fewer HP ProLiant G6 servers, yielding 95% less power consumed related to server energy and cooling costs.

Lower software license fees up to 90%… Using an estimated $800 annual Linux OS license cost for the existing ProLiant G4 servers and the target ProLiant G6 server, with 11:1 consolidation ratio, software licensing is reduced from $8800 annually to $800, a reduction of 90.9%.

Decrease servers managed by 90%… Consolidating from 11 servers to 1 = 90.0% reduction

---

The universal problem that virtualization is solving in a data center is that of dedicated resources. While this approach does address performance, this method lacks fine granularity. Typically, IT managers make an educated guess as to how many dedicated servers they will need to handle peaks, by purchasing extra servers and then later finding out that significant portions of these servers were grossly underutilized. A typical data center has a large amount of idle infrastructure, bought and set up online to handle peak traffic for different applications. Virtualization offers a way of moving resources dynamically from one application to another. However, specifics of the desired virtualizing effect depend on the specific application deployed [SME200301].

In summary, enterprise virtualization is a proven software technology that is rapidly transforming the IT landscape and fundamentally changing the way that people compute. Multiple virtual machines share hardware resources without interfering with each other so that the IT planner can safely run several operating systems and applications at the same time on a single computer. The VMware approach to virtualization inserts a thin layer of software directly on the computer hardware or on a host operating system. Virtualization software offers a virtualization platform that can scale across hundreds of interconnected physical computers and storage devices to form an entire virtual infrastructure.

8.7 Service-Oriented Network (SON) Concepts

To deliver the services discussed above for Cloud/Grid Computing, the network has to evolve from being just a traditional Layer 1 infrastructure (an assembly of communications channels), or a just a traditional Layer 2 infrastructure (an assembly of cell/frame switches and supporting channels), or a just a traditional Layer 3 infrastructure (an assembly of routers and supporting channels), to higher layers of the protocol model, that is, to encompass Layers 4 to 7. This gives rise to a Service-Oriented Network (SON). A SON is a service-based architecture for the development, deployment, and management of network-provided services. We provide a brief discussion of this topic here. The interested reader should consult reference [MIN200801] if interested in additional information on this topic.

While it may not be practical to take a statically standardized approach to all business applications, architects can benefit their organizations by taking a uniform approach to the network, since it is the common connectivity fabric for all applications, computing resources, and video and voice communication. The network is pervasive, affecting all fixed and mobile devices, end users, and even entities outside the enterprise, such as supply-chain partners. The network enables resources across enterprise, even those in discrete silos, to deliver business agility by integrating IT assets with critical business processes, by creating a platform for application optimization, and by achieving process improvements. The network also has the ability to consolidate security and identity services, so that the integrity of business applications, as they migrate, can be maintained. Furthermore, the network can support the virtualization of resources such as storage, firewalls, or policy enforcement. An intelligent network creates the kind of dynamic, application- and service-aware infrastructure that a nonintegrated infrastructure is unable to provide [MIL200701]. The question is: How does one build a SON? The discussion that follows provides some insight. The tools of the Service-Oriented Architecture (SOA) can be applied to this end.

The emergence of Web Services (WSs) technologies has triggered a major paradigm shift in Distributed Computing. Distributed Computing can be supported by any number of models and approaches; however, it is useful when some kind of (de facto) industry standard is used. One such standard is Distributed Object Architecture (DOA). DOA makes use of technologies such as Common Object Request Broker Architecture (CORBA), Distributed Component Object Model (DCOM), Distributed Communications Environment (DCE), and Remote Method Invocation (RMI). However, at this time, one is seeing a general transition from DOA to SOAs. The emergence of approaches based on WSs has triggered a major paradigm shift in Distributed Computing.

In an SOA environment, a set of network-accessible operations and associated resources are abstracted as a "service." The SOA vision allows for a service to be described in a standardized fashion, published to a service registry, discovered, and invoked by a service consumer. SOAs make use of technologies such Web Services Description Language (WSDL), SOAP, and UDDI. Like all its predecessors, SOA promises to provide ubiquitous application integration; SOA also promises to provide platform-independent standards that enable application integration and dynamic any-to-any real-time connectivity, locally or even across distributed networks. Although the nature of SOA is different from previous architecture approaches, traditional architecture modeling approaches can be built upon and used to describe SOA.

Original applications of SOA were related to (application) software development. However, at this juncture, SOA is finding direct applications in networking, and applications are making use of networks to reach SOA business-logic functional modules. Two major trends are affecting SOA networking as vendors iteratively build and deploy SOA solutions:

1. Many SOA platforms have been built on a core message-oriented middleware (MOM) or Enterprise Service Buses (ESB) and these SOA platforms are constantly evolving as feature/functionality is added. At the high end, SOA platforms will likely evolve into Integration Centric–Business Process Management Suites (IC-BPMS), and at the low end SOA platform functionality (mainly ESB functionality) is being commoditized from application servers into embedded network appliances.
2. The focus of SOA adoption is moving from Enterprise SOA to Business-to-Business (B2B) SOA (also known as Extended Enterprise SOA) requiring the management of services beyond the immediate enterprise where these services are being deployed over a more distributed network infrastructure between more partners.

These two trends present challenges in the productizing SOA platforms and associated SOA infrastructure in general, but also, specifically, for SOA network infrastructure. The objective of SOA networking is to support the deployment of network-centric SOA infrastructure solutions to enable reliable, consistent, and predictable communications between WSs deployed across a distributed enterprise or between enterprises.

To support the most sophisticated enterprise application integration and business processes, SOA networks must solve a number of problems:

- *Ubiquitous, secure, and reliable messaging:* As SOA evolves, the IT environment is transforming from a few large applications to a network of many shared services, and there are a number of factors such as reliable messaging, security, policy management, and ubiquity that become more important. SOA platforms for internal enterprise integration are challenging in their own right; however, managing distributed services beyond the enterprise is even more of a challenge when a network is introduced.
- *Enterprise SOA interoperability and mediation:* An enterprise IT environment is characterized by many incompatibilities: in software platforms, standards adoption, service sophistication, invocation patterns, developer capabilities, and so on. These factors must be mitigated to enable seamless service sharing within and between enterprises. Sharing services through loose coupling lies at the foundation of the benefits of SOA, enabling business agility, lowering IT cost, and reducing time to market for new applications. These incompatibilities between heterogeneous SOA environments create impedances to service sharing, and therefore must be removed as we move from Enterprise SOA to B2B SOA if we want to achieve the goals of SOA.

Middleware is a traditional way of making disparate applications "communicate" with each other without actually being integrated. But middleware itself is an application that introduces its own layer of maintenance and management; middleware also has limitations in terms of extensibility and adaptability. However, as SOA and Web 2.0 evolve, the network has the capability to complement and enhance this evolution as desktop environments embrace WSs directly. By embedding eXtensible Markup Language (XML) translation, message routing, and event notification within the network, the service-oriented environments are optimized with efficiency gains in file transfers, database synchronizations, e-mail, and Web acceleration, among other processes [MIL200701].

The evolution of SOA can be described as a continuum—from XML to WS to SOA; and this evolution will likely continue to evolve to Business Process Management (BPM). And, likewise the evolution of SOA Networking has followed this continuum from XML Networking to Web Services Networking to SOA Networking; it will continue to evolve over time. This trend is driven by business' desire to integrate with partners and to have more intelligent exchanges with these partners, which inherently means exchanging more data and metadata. While the evolution continues, it does not mean that SOA Networking is the solution for all applications and all environments. For example, XML Networking may be adequate for environments where the simple exchange of documents is needed. And Web Services Networking may be adequate for the simple exchange of structured data between databases. SOA Networking is used to describe the management of services within and between homogenous and heterogeneous SOA environments.

Because this is a nascent and evolving field, a number of concepts have evolved (not always completely nonoverlapping or fully defined). Some of these terms are discussed next. In general, however, the fundamental concepts are similar.

■ *XML Networking:* In a Web environment, HyperText Markup Language (HTML) is used to display information and XML is used to structure data in a representation that can be easily understood and agreed upon between a number of parties. Agreeing on the same data representation allows more readily for the exchange of documents and data. XML is a foundation for SOAP messaging, which in turn is the foundation of Web Services and SOA Networking. XML Networking is usually used for the simple exchange of XML documents over a network and focuses on the lower level of the stack.

■ *Web Services Networking:* As noted, WSs build upon XML and SOAP, and allow for the exchange of structured data, using a commonly agreed-upon, standards-based communications framework. Exposing WSs to the network requires a common interface called WSDL that is built upon the SOAP messaging standard. UDDI is also part of this first generation of WS standards and allows for discovery of services beyond the immediate enterprise; UDDI, however, has not yet gained in popularity for discovering services even though many vendors have adopted support of service registries (UDDI v3.0). WSs Networking is based upon standards and has been adopted primarily as B2B solutions for the exchange of structured data. These exchanges presently tend to be very limited in scope and are not usually implemented as part of an overall SOA architecture; and therefore are not to be used for orchestrating, choreographing, or communicating services between SOA environments. It must be noted that building a bunch of Web Services is not equivalent to architecting an SOA environment and therefore networking a bunch of Web Services is not the same as SOA Networking.

■ *SOA Networking:* The evolution of XML and WSs continues to move beyond simple exchange of documents and database records to support distributed SOA environments. XML and WSs have become building blocks for an architectural platform—SOA. While standards bodies such as OASIS (Organization for the Advancement of Structured Information Standards) are working on providing a common framework (SOA–RM; Service-Oriented Architecture–Reference Model) for understanding SOA and related concepts, these concepts are still at a very high level of abstraction and do not specifically address the network as part of an overall SOA solution.

Because standards bodies have not defined the terminology around networking in SOA environments, the following definitions were put together to describe networking terminology used in the industry, although they may not be commonly accepted. The following definitions will help clarify what SOA networking is and is not.

SON: The term "SON" has been used by the IT Industry to describe networking of "distributed (Web) services" in networked environments. The term "SON" has also been used by the Telecom Industry to describe delivery and the composition of "network services," that is, Voice, Video, IPTV network-based applications. These network services and the Service Delivery Platform (SDP) used to deliver these solutions in telecom environments may or may not be based on SOA principles. And finally, the term "SON" is also being used in the context of service virtualization, where "service" is a generic term describing various network services such as security, identity, presence. To add to the confusion, there may be instances where a network service is equivalent to a Web Service, and vice versa, but this may not always be the case. There is some ambiguity surrounding the term "SON" because it is not clear what is meant by a service and whether services are "Web Services" or "Network Services." And finally, it is also unclear whether these services are being used in an SOA environment(s).

SOA Networking (SOAN): The term "SOA Networking" relates to the networking of services in Service-Oriented environments (which can be homogeneous or heterogeneous) and are based on SOA principles. Web services are used in these environments to support the deployment of SOAs. Presently, most deployments focus primarily on homogenous environments. However, as we move to networks between heterogeneous environments, there are requirements to mediate between these environments. SOAN is presently being used interchangeably with Service-Oriented Networking, but is not as ambiguous as Service-Oriented Networking because it clearly includes the term SOA in the definition. SOA Networking is based on XML, SOAP, and Web Services but does not encompass XML Networking and Web Services Networking; because it is possible to have the exchange of XML documents or the exchange of Web services without using an SOA. SOA Networking is used to describe the use of Web Services and the associated messaging used to support an SOA. In SOA Networking, more of the business logic starts to become distributed in the network itself, and the differentiation between what an application is and what a network is, is replaced with the notion of services and business processes. Ideally, these business processes can be composed of services irrespective of distance, location, and platforms. Figure 8.9 depicts an example of an SOA-based network [SOU200601].

Service-Oriented Network Architecture (SONA): A vendor-specific architectural framework that aims at delivering business solutions to unify network-based

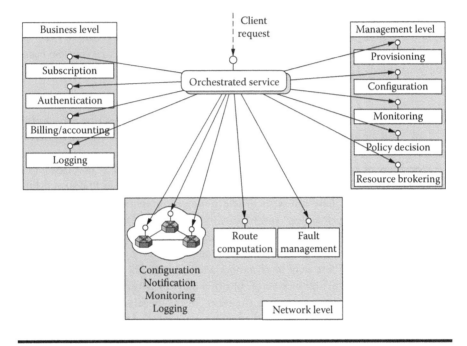

Figure 8.9   A service-oriented Architecture for developing network services.

services such as security, mobility, and location with the virtualization of IT resources. SONA is comprised of layers; so, as applications migrate to Web-based services, this architectural approach enables each layer to complement the functionality of the other layers, thereby presenting a loosely coupled framework. SONA's layers are [MIL200701]:

Applications layer: includes all software used for business purposes (e.g., enterprise resource planning) or collaboration (e.g., conferencing). As Web-based applications rely on XML schema and become tightly interwoven with routed messages, they become capable of supporting greater collaboration and more effective communications across an integrated networked environment.

Integrated network services layer: a layer that optimizes communications between applications and services by taking advantage of distributed network functions such as continuous data protection, multiprotocol message routing, embedded Quality-of-Service (QoS), Input/Output (I/O) virtualization, server load balancing, Secure Socket Layer (SSL) VPN, identity, location and IPv6-based services. These intelligence network-centric services can be used by the application layer through either transparent or exposed interfaces presented by the network.

Network systems layer: a layer that provides the corporate campus, the data center, and remote branches with a broad suite of collaborative connectivity functions, including peer-to-peer, client-to-server and storage-to-storage connectivity.

Application-Oriented Networking (AON): A vendor-specific concept defined before SOA was commonly adopted in the industry that primarily focuses on routing, performance, and managing QoS associated with different types of applications. Because services are actually smaller compose-able units than an application, AON typically does not address applications that are built upon services in an SOA environment. AON focuses on the networking of applications and associated acceleration, performance, QoS, and managing of applications over networks. It preceded SOA and therefore is not built upon Web Services and SOA principles, which call for the reuse and compose-ability of services. Managing applications built upon these types of services by their very nature are much more dynamic than traditional applications because services have different characteristics or principles associated with their use—the two major differences being compose-ability and reuse. Supposedly, AON is part of SONA.

8.8 Conclusion

Virtualization of processing and storage capabilities, whether strictly within an enterprise or via a network-provided service such as Cloud Computing, not only intrinsically saves the organization money because of improved ICT efficiencies, but also results in lower power and cooling expenditure, thus supporting a greening agenda.

References

[ATI200901]   ATIS, ATIS Report on Environmental Sustainability, A Report by the ATIS Exploratory Group on Green, ATIS, March 2009, 1200 G Street, NW, Suite 500, Washington, DC 20005.

[BED200301]   A. Bednarz and D. Dubie, How to: How to get to utility computing, *Network World*, 1 December 2003.

[BER200301]   F. Berman, G. Fox, and A.J. Hey, *Grid Computing: Making the Global Infrastructure a Reality*, Wiley, Chichester, U.K., 2003.

[BUY200301]   R. Buyya, Frequently Asked Questions, Grid Computing Info Centre, *GridComputing Magazine*, 2003.

[CHE200201]   M. Chetty and R. Buyya, Weaving computational grids: How analogous are they with electrical grids?, *Computing in Science & Engineering*, July/August 2002.

[CHU200201]   L-J. Zhang, J.-Y. Chung, and Q. Zhou, Developing grid computing applications. 1: Introduction of a Grid Architecture and Toolkit for Building Grid Solutions, 1 October 2002, Updated 20 November 2002, IBM Corporation, 1133 Westchester Avenue, White Plains, NY 10604.

[COH200301]   R.B. Cohen and E. Feser, Grid Computing, Projected Impact in North Carolina's Economy & Broadband Use through 2010, Rural Internet Access Authority, September 2003.

[DEV200301]   Developerworks Staff, Start Here to Learn about Grid Computing, August 2003, IBM Corporation, 1133 Westchester Avenue, White Plains, NY 10604.

[EAS200901]   M. Eastwood, J.S. Bozman, J.C. Pucciarelli, and R. Perry, The Business Value of Consolidating on Energy-Efficient Servers: Customer Findings, Hewlett-Packard/IDC White paper, May 2009.

[FIN20901]   J. Finkle, Cisco, EMC Team up on Cloud Computing: Sources, Reuters, 30 October 2009.

[FOS199701]   I. Foster and C. Kesselman, Globus: A metacomputing infrastructure toolkit. *The International Journal of Supercomputer Applications and High Performance Computing*, 11(2): 115–128, 1997.

[FOS199901]   I. Foster and C. Kesselman, Editors. *The Grid: Blueprint for a Future Computing Infrastructure*. Morgan Kaufmann Publishers, San Francisco, CA 1999.

[FOS200101]   I. Foster, C. Kesselman, and S. Tuecke, The anatomy of the grid: Enabling scalable virtual organizations, *International Journal of High-Performance Computing Applications,* 15(3), 200, 2001.

[FOS200201]   I. Foster, What Is the Grid? A Three Point Checklist, Argonne National Laboratory and University of Chicago, 20 July 2002, Argonne National Laboratory, 9700 Cass Ave, Argonne, IL 60439.

[FOX200201]   G. Fox, M. Pierce, R.G.D. Gannon, and M. Thomas, Overview of Grid Computing Environments, GFD-I.9, February 2003, Copyright © Global Grid Forum (2002).

[GLO200301]   Globus Alliance, Press Releases, c/o Carl Kesselman, USC/Information Sciences Institute, 4676 Admiralty Way, Suite 1001, Marina del Rey, CA 90292-6695, carl@isi.edu, http://www.globus.org, info@globus.org.

[GRI199701]   A.S. Grimshaw, W.A. Wulf, and the Legion Team, The legion vision of a worldwide virtual computer, *Communications of the ACM*, 40(1): 39–45, January 1997.

[GRI200301]   Grid Computing Info Centre (GRID Infoware), *Enterprise Architect Magazine: Grid Computing*, Answers to the *Enterprise Architect Magazine* Query, http://www.cs.mu.oz.au/~raj/GridInfoware/gridfaq.html.

[GRI200301]   http://www.gridcomputing.com/.

[GRI200401]   Grid Computing using .NET and WSRF.NET Tutorial, GGF11, Honolulu, HI, 6 June 2004.

[HAN200301]   M. Haney, Grid Computing: Making inroads into financial services, 4(5): 24 April 2003, IBM's Developerworks Grid Library, IBM Corporation, White Plains, NY. www.ibm.com.

[HAR200901]   K. Hartig, What is cloud computing?, *Cloud Computing Journal*, SYS-CON Media, Inc., 15 April 2009.

[HAW200301]   T. Hawk, IBM Grid Computing General Manager, Grid Computing Planet Conference and Expo, San Jose, CA, 17 June 2002. Also as quoted by Globus Alliance, Press Release, 1 July 2003.

[HEP200201]   Hewlett-Packard Company, HP Virtualization: Computing without Boundaries or Constraints, Enabling an Adaptive Enterprise, HP White paper, 2002, Hewlett-Packard Company, 3000 Hanover St., Palo Alto, CA 94304-1185.

[HUB200101]   V. Huber. UNICORE: A Grid Computing environment for distributed and parallel computing, *Lecture Notes in Computer Science,* 2127: 258–266, 2001.

[IAM200301]   I. Foster and A. Iamnitchi, On death, taxes, and the convergence of peer-to-peer and grid computing, *2nd International Workshop on Peer-to-Peer Systems (IPTPS'03)*, February 2003, Berkeley, CA.

[IBM200301]   IBM Press Releases, IBM Corporation, 1133 Westchester Ave., White Plains, NY 10604, www.ibm.com.

[IBM200701]   IBM, Service Oriented Architecture—SOA, *Service Oriented Architecture Glossary*, IBM Corporation, 1 New Orchard Rd., Armonk, NY 10504-1722.

[LEY200901]   T. Leyden, A brief history of cloud computing, *Cloud Computing Journal*, 20 October 2009.

[MCC200301]   M. McCommon, Letter from the Grid Computing Editor: Welcome to the new developerWorks Grid Computing resource, Editor, Grid Computing resource, IBM, 7 April 2003.

[MIL200701]   M. Milinkovich, Agility Principle: Service-Oriented Network Architecture, 7 December 2007, Service-Oriented Network Architecture, http://www.ebizq.net, Cisco Systems.

[MIN198701]   D. Minoli, A Collection of Potential Network-Based Data Services, Bellcore/Telcordia Special Report, SR-NPL-000790, 1987, Piscataway, NJ.

[MIN200501]   D. Minoli, *A Networking Approach to Grid Computing* (Wiley, 2005).

[MIN200601]   D. Minoli, The virtue of virtualization, *Networkworld*, 27 February 2006.

[MIN200701]   D. Minoli, Grid computing for commercial enterprise environments, in Handbook on Information Technology in Finance (International Handbooks on Information Systems), Edited by F. Schlottmann, D. Seese, and C. Weinhardt (Springer Verlag, October 2007).

[MIN200801]   D. Minoli, *Enterprise Architecture A to Z: Frameworks, Business Process Modeling, SOA, and Infrastructure Technology* (Auerbach, 2008).

[MIT200601]   T. Mitra, Business-driven development, 13 January 2006, *IBM developerWorks*, On-line magazine, http://www-128.ibm.com/developerworks/webservices/library/ws-bdd/.

[MYE200301]   T. Myer, Grid Computing: Conceptual Flyover for Developers, May 2003, IBM Corporation, 1133 Westchester Ave., White Plains, NY 10604.

[NAV200901]   NaviSite, Inc., Promotional Materials, Andover, MA, www.navisite.com.

[NOR200801]   K. Normandeau, Top 5 Reasons to Adopt Virtualization Strategies, Virtualization Strategies, 7 October 2008, www.VirtulizationStrategies.com.

[OMG200701]   OMG, Object Management Group, 140 Kendrick Street, Building A Suite 300, Needham, MA 02494.

[PAD200301]   P. Padala, A Survey of Grid File Systems, Editor, GFS-WG (Grid File Systems Working Group), Global Grid Forum, 19 September 2003. The Global Grid Forum, 9700 South Cass Ave., Bldg. 221/A142, Lemont, IL 60439.

[PAP200301]   M.P. Papazoglou and D. Georgakopoulos, Service-oriented computing: Introduction, *Communications of the ACM,* 46(10): 24–28, October 2003.

[PCI200901]   The PCI Security Standards Council, 401 Edgewater Place, Suite 600, Wakefield, MA 01880.

[QIA200901]   C. Qiao, D. Simeonidou, B. St. Arnaud, and P. Tomsu, Grid vs cloud computing and why this should concern the optical networking community, *OFC/NFOEC 2009* Presentation, 21–25 March 2009, San Diego, CA.

[SAS200901]   Statement on Auditing Standards (SAS) No. 70 http://www.sas70.com/.

[SME200301]   M. Smetanikov, HP Virtualization a Step toward Planetary Network, Web Host Industry Review (theWHIR.com), 15 April 2003. Web Host Industry Review, Inc. 552 Church Street, Suite 89, Toronto, Ontario, Canada M4Y 2E3.

[SOU200601]   V.A.S.M. de Souza and E. Cardozo, SOANet — A service oriented architecture for building compositional network services, *Journal of Software*, 1(2): 465–477, August 2006.

[STA200501]   Staff, Enterprise-Class SOA Management with webMethods Servicenet, Leveraging a Web Services Infrastructure Platform to Deliver on the Promise of SOA, January 2006, White Paper, December 2005, webMethods, Inc., 3877 Fairfax Ridge Road, South Tower, Fairfax, VA 22030.

[SUN200301]   Sun Networks Press Release, Network Computing Made More Secure, Less Complex with New Reference Architectures, Sun Infrastructure Solution, 17 September 2003. Sun Microsystems, Inc., 4150 Network Circle, Santa Clara, CA 95054.

[SYS200501]   Systems and Software Consortium, Glossary, http://www.software.org/pub/architecture/pzdefinitions.asp.

[THO200801]   T. von Eicken, The three levels of cloud computing, *Cloud Computing Journal,* 31 July 2008, http://cloudcomputing.sys-con.com/node/581961.

[TOG200501]   TOGAF, Preliminary Phase: Framework and Principles, http://www.opengroup.org/architecture/togaf8-doc/arch/p2/p2_prelim.htm.

[VMW200701]   VMWare Promotional Materials, VMware, Inc. World Headquarters, 3401 Hillview Ave., Palo Alto, CA 94304, http://www.vmware.com.

[WAL200201]   M. Waldrop, Hook enough computers together and what do you get? A new kind of utility that offers supercomputer processing on tap, *MIT Enterprise Technology Review*, May 2002, Technology Review, One Main Street, 7th Floor, Cambridge, MA 02142.

[WEI200801]   G.J. Weiss and C. Haight, Will Data Center Virtualization Spiral Out of Control? A Model to Deal with It, Gartner, Publication Date: 24 July 2008, ID Number: G00159966.

[YAN200301]   Enterprise Computing & Networking Report on Utility Computing in Next-Gen IT Architectures, Yankee Group, August 2003 Yankee Group, 31 St. James Avenue (aka the Park Square Building), Boston, MA 02116.

[YAN200401]   Enterprise Computing & Networking Report on Performance Management Road Map for Utility Computing, Yankee Group, February 2004. Yankee Group, 31 St. James Avenue (aka the Park Square Building), Boston, MA 02116.

[ZHA200201]   L.-J. Zhang, J.-Y. Chung, and Q. Zhou, Developing Grid Computing Applications. Part 1: Introduction of a Grid Architecture and Toolkit for Building Grid Solutions, October 1, 2002, Updated November 20, 2002, IBM's Developerworks Grid Library, IBM Corporation, 1133 Westchester Avenue, White Plains, NY 10604, www.ibm.com.

Glossary

Abatement: The act of eliminating or reducing greenhouse gases from being emitted into the atmosphere [ATI200902].

Absolute Humidity: The total amount of moisture contained in a cubic foot of air, measured in grains per cubic foot.

Absolute Pressure: The total pressure on a surface, measured in pounds per square inch. Equal to gauge pressure plus 14.7 (atmospheric pressure).

Absorbent: Any substance that can absorb another substance, changing it physically or chemically in the process. Lithium bromide is the absorbent used in absorption chillers [TRA199901].

Absorber: The low-pressure section of the absorption machine that has a lithium bromide solution that is cooled to attract refrigerant vapor from the evaporator [TRA199901].

Absorption System: Refrigeration system that uses heat to produce a cooling effect.

Air Conditioning Airflow Efficiency (ACAE): The amount of heat removed per standard cubic foot of airflow per minute [42U201001].

Air Handling Unit (AHU): Device used to condition and circulate air as part of an HVAC system. Typically, the air handler is a metal box housing a blower, heating and/or cooling elements, filter racks or chambers, sound attenuators, and dampers. Air handlers connect to ductwork that distributes the conditioned air throughout the building.

Air Mixing: The unintended mixing of cold and hot air.

Air Space: Space where the air space below a raised floor or above a suspended ceiling is used to recirculate data center/telecom room environmental air.

Air Vent Valve: Valve connected to the top of the water box or connecting pipe used to vent trapped air.

Air, Bypass: Air diverted around a cooling coil in a controlled manner for the purpose of avoiding saturated discharge air. On an equipment room scale, bypass air can also refer to the supply air that "short-cycles" around the load and returns to the air handler without producing effective cooling at the load [GRE201001].

Air, Cabinet: Air that passes through a cabinet housing telecom/datacom equipment for the purpose of cooling it.

Air, Conditioned: Air treated to control its temperature, relative humidity, purity, pressure, and movement.

Air, Equipment: In this context, airflow that passes through the IT, datacom, or telecom equipment.

Air, Return: Air extracted from a space and totally or partially returned to an air conditioner, furnace, or other heat source.

Air/Liquid Cooling: removal of heat.

>**Air cooling:** direct removal of heat at its source using air.

>**Liquid cooling:** direct removal of heat at its source using a liquid (usually water, water/glycol mixture, Fluorinert™ or refrigerant).

>**Air-cooled rack or cabinet:** system conditioned by removal of heat using air.

>**Liquid-cooled rack or cabinet:** system conditioned by removal of heat using a liquid.

>**Air-cooled equipment:** equipment conditioned by removal of heat using air.

>**Liquid-cooled equipment:** equipment conditioned by removal of heat using a liquid.

>**Air-cooled server:** server conditioned by removal of heat using air.

>**Liquid-cooled server:** server conditioned by removal of heat using a liquid.

>**Air-cooled blade:** blade conditioned by removal of heat using air.

>**Liquid-cooled blade:** blade conditioned by removal of heat using a liquid.

>**Air-cooled board:** circuit board conditioned by removal of heat using air.

>**Liquid-cooled board:** circuit board conditioned by removal of heat using a liquid.

>**Air-cooled chip:** chip conditioned by removal of heat from the chip using air.

>**Liquid-cooled chip:** chip conditioned by removal of heat using a liquid.

>(terms from [GRE201001])

Air-Cooled System: System where conditioned air is supplied to the inlets of the rack/cabinet for convective cooling of the heat rejected by the components of the electronic equipment within the rack. Within the rack, the transport of heat from the actual source component within the rack itself can be either liquid or air based, but the heat rejection media from the rack to the terminal cooling device outside of the rack is air [GRE201001].

Airside Economizer: An economizer that directs exterior air into the data center when the air temperature is at or below the cooling setpoint [42U201001].

Aisle: The open space between rows of racks. Best practice dictates that racks should be arranged with consistent orientation of front and back to create "cold" and "hot" aisles [42U201001].

Alternate Fuel for Essential Agricultural Use: (gas utility industry) Any fuel that is economically practicable and reasonably available as determined by the Federal Energy Regulatory Commission (FERC) [MGE200902].

American Society of Heating, Refrigerating and Air-Conditioning Engineers (ASHRAE): An international technical society organized to advance the arts and sciences of air management.

Annual Charge Adjustment (ACA): (gas utility industry) A surcharge as permitted by Section 154.38 (d) (6) of the FERC's Regulations to permit an interstate pipeline company to recover from its shippers all Total Annual Charges assessed it by the FERC under Part 382 of the FERC's Regulations [MGE200902].

Atmospheric Pressure: The pressure exerted upon the Earth's surface by the weight of atmosphere above it.

Backup Generation Service: (electric utility industry) An optional service for customers with demands greater than or equal to 75 kW who wish to enhance their distribution system reliability through contracting with the company for the use of portable diesel or gas-fired backup generators. The service provides for backup generation if customers should ever experience a distribution-related outage [MGE200901].

Balancing Service Charge: (gas utility industry) The charge applicable to a customer's adjusted imbalance volume when Daily Balancing Service (DBS-1) is being utilized [MGE200902].

Balancing Services: (gas utility industry) The company's managing of a customer's natural gas supply to enable the customer to match the customer's daily usage with the customer's confirmed pipeline delivery of third-party natural gas supplies. Customers who wish to receive third-party natural gas supplies must obtain balancing service through the balancing provisions in the Daily Balancing Service (DBS-1) or Comprehensive Balancing Service (CBS-1) [MGE200902].

Balancing Valve: A device to balance the water flow.

Base Load Generation: (electric utility industry) Those generating facilities within a utility system that are operated to the greatest extent possible to maximize system mechanical and thermal efficiency and minimize system operating costs [MGE200901].

Base Load Unit/Station: (electric utility industry) Units or plants that are designed for nearly continuous operation at or near full capacity to provide all or part of the base load. An electric generation station normally operated to meet all, or part, of the minimum load demand of a power company's system over a given amount of time [MGE200901].

Base Rate: (electric utility industry) That part of the total electric rate covering the general costs of doing business unrelated to fuel expenses [MGE200901].

Benchmark: A reference point.

Billing Cycle: The regular periodic interval used by a utility for reading the meters of customers for billing purposes [MGE200902].

Blade Server: A modular structured server, containing one, two, or more microprocessors and memory, that is intended for a single, dedicated application (such as serving Web pages) and that can be easily inserted into a space-saving rack with many similar servers. One product offering, for example, makes it possible to install up to 280 blade server modules vertically in a single floor-standing cabinet. Blade servers, which share a common high-speed bus, are designed to create less heat and thus save energy costs as well as space [GRE201001].

Blanking Panel: A device mounted in unused U spaces in a rack that restricts bypass airflow, also called blanking or filler plates [42U201001].

Boiler Fuel: (gas utility industry) Natural gas used as a fuel for the generation of steam or hot water (including natural gas used as a fuel for externally fired pressure vessels using heat transfer fluids other than water) or electricity, including the utilization of gas turbines for the generation of electricity [MGE200902].

Boiling Point: The temperature at which the addition of any heat will begin a change of state from a liquid to a vapor.

Btu (BTU): One British Thermal Unit, the amount of heat required to raise the temperature of one pound of water one degree Fahrenheit at 60 degrees Fahrenheit [MGE200902].

Building Automation System (BAS): Centralized building control. Usually utilized for the purpose of monitoring and controlling environment, lighting, power, security, fire safety, and elevators.

Building Management System (BMS): Computer-based tools used to manage data center assets [42U201001].

Busy Hour Drain: The actual average current drawn by a circuit or group of circuits during the busy hour of the season (excluding power fail/battery discharge events). Busy hour drains for DC power plants are measured by power monitor devices [ATI200902].

Bypass: A piping detour around a component.

Bypass Airflow: Conditioned air that does not reach computer equipment, escaping through cable cutouts, holes under cabinets, misplaced perforated tiles, or holes in the computer room perimeter walls [42U201001].

Cabinet: Device for holding IT equipment, also called a rack.

CADE: Corporate Average Data Center Efficiency.

Calibrate: To adjust a control or device in order to maintain its accuracy.

Capacity: (electric utility industry) The load for which a generating unit, generating plant, or other electrical apparatus is rated either by the user or by the manufacturer [MGE200901]. The maximum usable output of a machine.

Carbon Footprint: A (currently in-vogue) term to describe a calculation of total carbon emissions of some system, namely, to describe energy consumption. It is the amount of greenhouse gases (GHG) produced to support a given activity (or piece of equipment), typically expressed in equivalent tons of carbon dioxide (CO_2).

Carbon footprints are normally calculated for a period of a year. One way to calculate carbon footprints is to use a Life Cycle Assessment; this takes into account the emissions required to produce, use, and dispose of the object. Another way is to analyze the emissions resulting from the use of fossil fuels. Several carbon footprint calculators are available. One well-known calculator is offered by the U.S. Environmental Protection Agency (EPA); it is a personal emissions calculator. This calculator allows individuals to estimate their own carbon footprint by asking the individual to supply various fields relating to transportation, home energy, and waste. The calculator shows how many pounds of carbon dioxide are formed per year. Safe Climate is another carbon footprint calculator; to determine the footprint; this calculator requires information regarding energy use in gallons, liters, therms, thousand cubic feet, and kilowatt-hours per month (or year).

Carbon neutral: Calculating total carbon emissions, reducing them where possible, and balancing remaining emissions with the purchase of carbon offsets [GAL200801].

Central Air Conditioner: A central air-conditioner model consists of one or more factory-made assemblies that normally include an evaporator or cooling coil(s), compressor(s), and condenser(s). Central air conditioners provide the function of air-cooling, and may include the functions of air circulation, air cleaning, dehumidifying, or humidifying. A split system is a system with components located both inside and outside a building. A "single package unit" is a system that has all components completely contained in one unit [ENE200901].

Certified Wood Products: Products made from lumber harvested in a sustainable manner and certified by a reliable third party, such as the Forest Stewardship Council (FSC) [GAL200801].

Charge: The refrigerant and lithium bromide contained in a sealed system; the total of refrigerant and lithium bromide required.

Chilled-Water System: A type of air-conditioning system that has no refrigerant in the unit itself. The refrigerant is contained in a chiller, which is located remotely. The chiller cools water, which is piped to the air conditioner to cool the space. An air or process conditioning system containing chiller(s), water pump(s), a water piping distribution system, chilled-water cooling coil(s), and associated controls. The refrigerant cycle is contained in a remotely located water chiller. The chiller cools the water, which is pumped through the piping system to the cooling coils [GRE201001].

Circuit (electric): (electric utility industry) A conductor or a system of conductors through which electric current flows or is intended to flow.

City Gate: (gas utility industry) The points of delivery between the interstate pipelines providing service to the facilities of a gas utility [MGE200902].

Close-Coupled Cooling: Cooling technology that is installed adjacent to server racks and enclosed to direct airflow directly to the rack without mixing with DC air [42U201001].

Cloud Computing: The latest term to describe a grid/utility computing service. Such service is provided in the network. From the perspective of the user, the service is virtualized. In turn, the service provider will most likely use virtualization technologies (virtualized computing, virtualized storage, etc.) to provide the service to the user.

Cloud Computing: *Applications in the cloud:* This is what people have already used in the form of gmail, yahoo mail, the various search engines, Wikipedia, etc. Some company hosts an application in the Internet that many users sign up for and use without any concern about where, how, or by whom the compute cycles and storage bits are provided. The service being sold (or offered in ad-sponsored form) is a complete end-user application. This can be seen as Software as a Service (SaaS) [THO200801].

Cloud Computing: *Infrastructure in the cloud:* This is the most general offering; it includes the services that Amazon has pioneered and where RightScale offers its management platform. Developers and system administrators obtain general compute, storage, queuing, and other resources and run their applications with the fewest limitations. This is the most powerful type of cloud in that virtually any application and any configuration that is fit for the Internet can be mapped to this type of service. It requires more work on the part of the developers and administrators [THO200801].

Platforms in the cloud: Here an application platform is offered to developers in the cloud. Developers write their application to a more or less open specification and then upload their code into the cloud where the app is run magically somewhere, typically being able to scale up as usage for the app grows. Examples include Google App Engine, Mosso, and Force. com. The service being sold is the machinery that funnels requests to an application and makes the application work. This is the newest type of cloud service [THO200801].

Cloud/Grid Computing Types: (1) Computational grids: machines with set-aside resources stand by to "number-crunch" data or provide coverage for other intensive workloads; (2) Scavenging grids: commonly used to locate and exploit CPU cycles on idle servers and desktop machines for use in resource-intensive tasks; and (3) Data grids: a unified interface for all data repositories in an organization, and through which data can be queried, managed, and secured. Computational grids can be local, enterprise grids (also private clouds), and Internet-based grids. Enterprise grids are middleware-based environments to harvest unused "machine cycles," thereby displacing otherwise-needed growth costs [MIN200501].

Clusters: Aggregating of processors in parallel-based configurations, typically in local environments (within a data center); all nodes work cooperatively as a single unified resource. Resource allocation is performed by a centralized resource manager and scheduling system.

Comprised of multiple interconnected independent nodes that cooperatively work together as a single unified resource; unlike grids, cluster resources are typically owned by a single organization. All users of clusters have to go through a centralized system that manages allocation of resources to application jobs. Cluster management systems have centralized control, complete knowledge of system state and user requests, and complete control over individual components.

CO_2 Equivalent: A metric used to equate the impact of various greenhouse gas emissions upon the environment's global warming potential based on CO_2 [ATI200902].

Coefficient of Effectiveness (CoE): Uptime Institute metric based on the Nash-Sutcliffe model efficiency coefficient.

Coefficient of Performance: The ratio of the amount of work obtained from a machine to the amount of energy supplied.

Coefficient of Performance (COP): A measure of efficiency in the heating mode that represents the ratio of total heating capacity (Btu) to electrical input (also in Btu) [ENE200901]. The ratio of the cooling load (in kW) to power input at the compressor (in kW). The ratio of the rate of heat removal to the rate of energy input (in consistent units) for a complete cooling system or factory-assembled equipment, as tested under a nationally recognized standard or designated operating conditions [GRE201001]. The definition is similar to EER, but the units are different.

Cold Aisle: An aisle where rack fronts face into the aisle. Chilled airflow is directed into this aisle so that it can then enter the fronts of the racks in a highly efficient manner [42U201001].

Cold Aisle Containment (CAC): A system that directs cooled air from air-conditioning equipment to the inlet side of racks in an efficient manner [42U201001].

Cold Computing: Refrigeration of electronics and opto-electronics using thermoelectric refrigeration; these technologies typically require only small-scale or localized spot cooling of small components, which do not impose a large heat load [TRI200201].

Cold Spot: An area where the ambient air temperature is below acceptable levels. Typically caused by cooling equipment capacity exceeding heat generation [42U201001].

Cold Supply Infiltration Index (CSI): Term that quantifies the amount of hot air mixing with cold inlet air prior to entering the rack.

Commissioning: A quality control process required by most green building certification programs. This process incorporates verification and documentation to ensure that building systems and assemblies are planned, designed, installed, tested, operated, and maintained to meet specified requirements [GAL200801].

Common Object Request Broker Architecture (CORBA): Object Management Group's open, vendor-independent architecture and infrastructure that computer applications use to work together over networks. A CORBA-based program from any vendor, on almost any computer, operating system, programming language, and network, that can interoperate with a CORBA-based program from the same or another vendor, on almost any other computer, operating system, programming language, and network. Because of the easy way that CORBA integrates machines from so many vendors, with sizes ranging from mainframes through minis and desktops to handheld devices and embedded systems, it is the middleware for large (and even not-so-large) enterprises. One of its most important, as well as most frequent, uses is in servers that must handle large numbers of clients, at high hit request rates, with high reliability [OMG200701].

Company Administered Balancing Pool: (gas utility industry) A pool of aggregated customer volumes administered by a utility company under Daily Balancing Service for customers who are not members of a Third-Party Pool [MGE200902].

Compound Gauge: A gauge used to measure pressures above and below atmospheric pressure (0 psig).

Computational Fluid Dynamics (CFD): Scientific calculations applied to airflow analysis.

Computer Room Air Handler (CRAH): Computer room air handler that uses chilled water to cool air.

Condensable: A gas that can be easily converted to liquid form, usually by lowering the temperature and/or increasing pressure.

Condensate: Water vapor that liquefies due to its temperature being lowered to the saturation point.

Condensation Point: The temperature at which the removal of any heat will begin a change of state from a vapor to a liquid.

Condenser: A device in which the superheat and latent heat of condensation is removed to effect a change of state from a vapor to a liquid. Some subcooling is also usually accomplished.

Condenser Water: Water that removes the heat from the lithium bromide in the absorber and from condenser vapor. The heat is rejected to atmosphere by a cooling tower.

Conditioned Air: Air treated to control its temperature, relative humidity, purity, and movement.

Conductor: A substance or body that allows an electric current to pass continuously along it.

Conduit: A duct designed to contain underground cables, conductors, and wires.

Conjunctive Billing: (electric utility industry) The combination of the quantities of energy, demand, or other items of two or more meters or services into respective single quantities for the purpose of billing, as if the bill were for a single meter or service [MGE200901].

Connection Charge: (electric utility industry) An amount to be paid by a customer in a lump sum or in installments for connecting the customer's facilities to the supplier's facilities [MGE200901].

Constant Volume (CV) AC: An air conditioner (AC) that supplies a constant rate of airflow to a space at reasonably high velocity and remove an equal amount of air from the space (return air).

Contactor: A switch that can repeatedly cycle, making and breaking an electrical circuit: a circuit control. When sufficient current flows through the coil built into the contactor, the resulting magnetic field causes the contacts to be pulled in or closed [TRA199901].

Control: Any component that regulates the flow of fluid, or electricity.

Control Device: Any device that changes the energy input to the chiller/heater when the building load changes and shuts it down when the chiller is not needed.

Cool Roofs: Energy-efficient roofing systems that reflect the sun's radiant energy before it penetrates the interior of the building. These systems can reduce the building's energy requirements for air conditioning [GAL200801].

Cooling Capacity: The quantity of heat in Btu (British Thermal Units) that an air conditioner or heat pump is able to remove from an enclosed space during a 1-hour period [ENE200901].

Cooling Tower: Heat-transfer device where atmospheric air cools warm water, generally by direct contact (heat transfer and evaporation).

Cooling, Air: Conditioned air is supplied to the inlets of the rack/cabinet/server for convective cooling of the heat rejected by the components of the electronic equipment within the rack. Within the rack, the transport of heat from the actual source component within the rack itself can be either liquid or air based, but the heat rejection medium from the rack to the building cooling device outside the rack is air. The use of heat pipes or pumped loops inside a server or rack where the liquid remains is still considered air cooling [GRE201001].

Cooling, Liquid: Conditioned liquid is supplied to the inlets of the rack/cabinet/server for thermal cooling of the heat rejected by the components within the rack.

Cooperative (Cooperatively Owned Electric Utility): A group of persons organized in a joint venture for the purpose of supplying electricity to a specified area. Such ventures are generally exempt from federal income tax laws. The Rural Electric Service (RES, formerly the Rural Electrification Administration or REA) finances most cooperatives [MGE200901].

Cooperative, Rural Electric (Coop): A consumer-owned utility established to provide electric service in rural parts of the United States. Consumer cooperatives are incorporated under the laws of the 46 states in which they operate. A consumer cooperative is a nonprofit enterprise, owned and controlled by the people it serves. These systems obtain most of their financing through insured and guaranteed loans administered by the Rural Electric Service and from their own financing institution, the National Rural Utilities Cooperative Financing Corporation [MGE200901].

Core Architecture Data Model (CADM): A formal model defining the data organization for a repository of C4ISR/DoDAF-compliant architecture products (artifacts). The CADM provides a common schema for repositories of architecture information. Tool builders or vendors providing support for DoDAF-style architecture descriptions typically implement the CADM with a database [SYS200501].

CRAC (Computer Room Air Conditioning): Computer room air conditioner (pronounced "crack") that uses a compressor to mechanically cool air. A modular packaged environmental control unit designed specifically to maintain the ambient air temperature and humidity of spaces that typically contain telecom/datacom/data center equipment. These products can typically perform all (or a subset) of the following functions: cool, reheat, humidify, dehumidify. CRAC units designed for data and communications equipment room applications typically meet the requirements of ANSI/ASHRAE Standard 127-2001, Method of Testing for Rating Computer and Data Processing Room Unitary Air-Conditioners.

Cradle-to-Cradle Design: Designing objects that are not just resource efficient, but essentially waste-free. This is accomplished using building materials that can be either recycled or reused, or composted or consumed [GAL200801].

Critical Load: Computer equipment load delivered by PDU output.

Cubic Feet per Minute (CFM): An airflow volume measurement.

Curtailment: (gas utility industry) A reduction in gas deliveries or gas sales necessitated by a shortage of supply [MGE200902].

Customer Charge: An amount to be paid periodically by a customer for electric service based upon costs incurred for metering, meter reading, billings, etc., exclusive of demand or energy consumption [MGE200901].

Customer's Gas Commodity Account: (gas utility industry) The amount of natural gas, under Comprehensive Balancing Service and measured in therms, that (a) has been delivered to the utility for redelivery to the customer but has not been consumed by the customer or (b) has been consumed by the customer in excess of the amount that has been delivered to the utility by the customer. A positive balance in the Customer's Gas Commodity Account means that the customer has used less natural gas than has been delivered to the utility for redelivery to the customer, and that the utility is holding such amount of the customer's gas in storage. A negative balance in the Customer's Gas Commodity Account means that the customer has used more natural gas than has been delivered to the utility for redelivery to the customer [MGE200902].

Cutout: An open area in a raised floor that allows airflow or cable feeds.

Daily Transportation Quantity: (gas utility industry) The average amount of natural gas that the utility estimates a customer served under Comprehensive Balancing Service will use for each day of a calendar month [MGE200902].

Data Center infrastructure Efficiency (DCiE): An efficiency measure that is calculated by dividing the IT equipment power consumption by the power consumption of the entire data center. This measure is the inverse of PUE.

Data De-duplication: Also called "intelligent compression" or "single-instance storage," it is a method of reducing storage requirements by eliminating redundant data. In effect, this can also be seen as a form of virtualization. With data de-duplication, only one unique instance of the data is actually retained on storage media; redundant data is deleted and a pointer to the unique data copy is used to support requests for access to that redundant copy.

Data Grid: A kind of Grid Computing grid used for housing and providing access to data across multiple organizations; users are not focused on where this data is located as long as they have access to the data [MIN200501].

Daylighting: Using various design methods, such as windows and skylights, to reduce the building's reliance on electric lighting. Numerous studies have highlighted the productivity benefits of natural lighting for building occupants.

Dead Band: An HVAC energy-saving technique whereby sensitivity setpoints of equipment are set more broadly to improve coordination of the equipment and avoid offsetting behaviors; also dead band control strategy [42U201001].

De-duplication: *See* Data De-duplication.

Default Balancing Service: (gas utility industry) A Third-Party Pool has selected pipeline balancing services, other than the utility's balancing services, and the pipeline balancing services do not cover the entire imbalance of the pool. The utility will provide Company Balancing Service for the remainder of the imbalance at Maximum Rates, as specified in the Daily Balancing Service rate schedule [MGE200902].

Degree-Day: A unit measuring the extent to which the outdoor mean (average of maximum and minimum) daily dry-bulb temperature falls below or rises above an assumed base. The base is normally taken as 65°F for heating and for cooling unless otherwise designated. One degree-day is counted for each degree below (deficiency heating) or above (excess cooling) the assumed base, for each calendar day on which such deficiency or excess occurs [MGE200901].

Delta T (ΔT, Delta temperature): The spread between the inlet and outlet air temperatures of air conditioning equipment, measured as the maximum achievable difference between inlet (return) and outlet (supply) temperatures [42U201001].

Demand: (electric utility industry) The rate at which electric energy is delivered to or by a system, part of a system, or a piece of equipment. It is expressed usually in kilowatts at a given instant or averaged over any designated period of time. The primary source of "demand" is the power-consuming equipment of customers [MGE200901].

Demand Charge: (electric utility industry) That part of the charge for electric service based upon the electric capacity (kW) consumed and billed on the basis of billing demand under an applicable rate schedule [MGE200901].

Demand Costs: (electric utility industry) Costs related to and vary with power demand (kW), such as fixed production costs, transmission costs, and a part of the distribution costs [MGE200901].

Demand Interval: (electric utility industry) The period of time during which the electric energy flow is averaged in determining demand, such as 60 minute, 30 minute, 15 minute, or instantaneous. Electric utilities typically use 15-minute demands for most demand-billed rate classes [MGE200901].

Demand Reading: (electric utility industry) Highest or maximum demand for electricity an individual customer registers in a given interval (15 minutes) during the month or billing period. The metered demand or billing demand reading sets the demand charge for the month or billing period [MGE200901].

Demand, Annual Maximum: (electric utility industry) The greatest demand that occurred during a prescribed demand interval (15 minutes) in a calendar year [MGE200901].

Demand, Annual System Maximum: (electric utility industry) The greatest demand on an electric system during a prescribed demand interval in a calendar year [MGE200901].

Demand, Average: (electric utility industry) The demand on, or the power output of, an electric system or any of its parts over any interval of time, as determined by dividing the total number of kilowatt hours by the number of units of time in the interval [MGE200901].

Demand, Billing: (electric utility industry) The demand upon which billing to a customer is based, as specified in a rate schedule or contract. It may be based on the contract year, a contract minimum or a previous maximum and, therefore, does not necessarily coincide with the actual measured demand of the billing period [MGE200901].

Demand, Coincident: (electric utility industry) The sum of two or more demands that occur in the same demand interval [MGE200901].

Demand, Customer Maximum 15 Minutes: (electric utility industry) The greatest rate at which electrical energy has been used during any period of 15 consecutive minutes in the current or preceding 11 billing months [MGE200901].

Demand, Instantaneous Peak: (electric utility industry) The demand at the instant of greatest load, usually determined from the readings of indicating or graphic meters [MGE200901].

Demand, Maximum: (electric utility industry) The greatest demand that occurred during a specified period of time such as a billing period [MGE200901].

Demand, Maximum Monthly 15 Minutes: (electric utility industry) The greatest rate at which electrical energy has been used during any period of 15 consecutive minutes in the billing month [MGE200901].

Demand, Maximum On-Peak 15 Minutes: (electric utility industry) The greatest rate at which electrical energy has been used during any on-peak period of 15 consecutive minutes in the billing month [MGE200901].

Demand, Noncoincident: (electric utility industry) The sum of two or more individual demand that do not occur in the same demand interval [MGE200901].

De-materialization: The replacement of goods and services with virtual equivalents; for example music, video, and book downloads in place of CDs, DVDs, and physical books or magazines.

Developer: A high-level role responsible for the implementation of the solution. More specialized actors within this role might actually carry out the work, such as a database programmer, a Java developer, a Web developer, and a business process choreographer, to name a few. Developers work on specific layers of the application stack and each requires specialized skills for that layer [MIT200601].

Dew-Point Temperature (DPT): The temperature at which a moist air sample at the same pressure would reach water vapor saturation. At this saturation point, water vapor begins to condense into liquid water fog or solid frost, as heat is removed. The temperature at which air reaches water vapor saturation, typically used when examining environmental conditions to ensure they support optimum hardware reliability.

Direct Expansion (DX) System: An air-conditioning system where the cooling effect is obtained directly from the refrigerant. It typically incorporates a compressor; almost invariably, the refrigerant undergoes a change of state in the system.

Dispatch, Dispatching: (electric utility industry) The operating control of an integrated electric system to [MGE200901]

- Assign generation to specific generating plants and other sources of supply to effect the most reliable and economical supply as the total of the significant area loads rises or falls.
- Control operations and maintenance of high-voltage lines, substations, and equipment, including administration of safety procedures.
- Operate the interconnection.
- Schedule energy transactions with other interconnected electric utilities.

Distribution: (electric utility industry) The act or process of delivering electric energy from convenient points on the transmission system (usually a substation) to consumers. The network of wires and equipment that distributes, transports, or delivers electricity to customers. The delivery of electric energy to customers on the distribution service. Electric energy is carried at high voltages along the transmission lines. For consumers needing lower voltages, it is reduced in voltage at a substation and delivered over primary distribution lines extending throughout the area where the electricity is distributed. For users needing even lower voltages, the voltage is reduced once more by a distribution transformer or line transformer. At this point, it changes from primary to secondary distribution [MGE200901].

Distribution: (gas utility industry) The process of transporting natural gas through the utility's facilities to the customer's facilities.

Distribution Line: (electric utility industry) One or more circuits of a distribution system either direct-buried, in conduit, or on the same line of poles or supporting structures, operating at relatively low voltage as compared with transmission lines [MGE200901].

Distribution Service: (electric utility industry) The network of wires and equipment that carries electric energy from the transmission system to the customer's premises. The costs to support, operate, and maintain this local delivery system are included in the rates and are usually priced in cents per kilowatt-hour for energy-only customers and in dollars per kilowatt for demand-billed customers [MGE200901].

Dry-Bulb Temperature (DBT): The temperature of an air sample, as determined by an ordinary thermometer, the thermometer's bulb being dry. The SI unit is Kelvin; in the United States, the unit is degrees Fahrenheit.

Economic Dispatch: The start-up, shutdown, and allocation of load to individual generating units to effect the most economical production of electricity for customers [MGE200901].

Economizer, Air: A ducting arrangement and automatic control system that allow a cooling supply fan system to supply outdoor (outside) air to reduce or eliminate the need for mechanical refrigeration during mild or cold weather [GRE201001].

Economizer, Water: A system by which the supply air of a cooling system is cooled directly or indirectly or both by evaporation of water or by other appropriate fluid (in order to reduce or eliminate the need for mechanical refrigeration) [GRE201001].

Efficiency: The amount of usable energy produced by a machine, divided by the amount of energy supplied to it; that is, the ratio of the output useful energy to the input energy of any system.

Efficiency, HVAC System: The ratio of the useful energy output to the energy input, in consistent units and expressed in percent.

Electricity Service: The network of generating plants, wires, and equipment needed to produce or purchase electricity (generation) and to deliver it to the local distribution system (transmission). Priced in cents per kilowatt-hour for energy-only customers and in dollars per kilowatt and in cents per kilowatt-hour for demand-billed customers [MGE200901].

Energy Charge: That part of the charge for electric service based upon the electric energy (kWh) consumed or billed [MGE200901].

Energy Costs: Costs, such as fuel, related to and varying with energy production or consumption [MGE200901].

Energy Efficiency: Designing buildings to use less energy for the same or higher performance as conventional buildings. All buildings systems can contribute to higher energy efficiency [GAL200801].

Energy Efficiency Ratio (EER): A measure of efficiency in the cooling mode that represents the ratio of total cooling capacity (Btu/hour) to electrical energy input (watts) [ENE200901]. It is measured by ratio of Btu/hour of cooling or heating load to watts of electrical power input (Btu/hr/W). (When consistent units are used, this ratio is equal to COP.)

Energy, Electric: As commonly used in the electric utility industry, electric energy means kilowatt hours (see [MGE200901]).

Energy, Off-Peak: Energy supplied during periods of relatively low system demand as specified by the supplier. For MGE, this is from 9 p.m. to 10 a.m., Monday through Friday, all holidays, and all weekends (see [MGE200901]).

Energy, On-Peak: Energy supplied during periods of relatively high system demand, as specified by the supplier.

Energy, Primary: Energy available from firm power.

Energy, Secondary or Supplemental: Energy available from nonfirm power.

Enhanced Transportation Service (ETS): (gas utility industry) ANR's Rate Schedule ETS, Enhanced Transportation Service as specified in ANR's FERC Gas Tariff (see [MGE200902]).

Enterprise: A(ny) collection of corporate or institutional task-supporting functional entities that has a common set of goals and/or a single mandate. In this context, an enterprise is, but is not limited to, an entire corporation, a division or department of a corporation, or a group of geographically dispersed organizations linked together by common administrative ownership, a government agency (or set of agencies) at any level of jurisdiction, and so on. This also encompasses the concept of an "extended enterprise," which is a logical aggregation that includes internal business units of a firm, along with partners and suppliers (see [TOG200501]).

Enterprise Service Bus (ESB): A connectivity infrastructure for integrating applications and services by performing the following actions between services and requestors: ROUTING messages between services, CONVERTING transport protocols between requestor and service, TRANSFORMING message formats between requestor and service, and HANDLING business events from disparate sources [IBM200701].

Equilibrium: When refrigerant (water) molecules leave the solution at the same rate that they are being absorbed, the solution is said to be in equilibrium.

Equilibrium Chart: A pressure-temperature concentration chart that can be used to plot solution equilibrium at any point in the absorption cycle.

Equipment Room: Data center or networking/telecom rooms—including carrier central office room and points-of-presence—that houses computer and/or telecom equipment.

Equivalent Full Cabinet (EFC): The number of full cabinets that would exist if all the equipment in the data center were concentrated in full cabinets [42U201001].

Essential Company Use: (gas utility industry) Any use of natural gas that is necessary to maintain service to the balance of customers who are not curtailed (see [MGE200902]).

Example 1: Cooling degree-day. Assume the maximum and minimum outdoor temperatures were 80°F and 60°F. The outdoor mean would be 70°F. The base is 65°. Since 70° is 5° greater than (in excess of) 65°, there are 5 cooling degree-days for this day.

Example 2: Heating degree-day. Assume the high was 32°F and the low was 10°F. The mean is 21°F. Since 21° is 44° lower (deficient) than the 65° base, there are 44 heating degree-days for this day.

Evacuate: To remove, through the use of the vacuum pump, all noncondensables from a machine [TRA199901].

Evaporative Condenser: Condenser where the removal of heat from the refrigerant is achieved by the evaporation of water from the exterior of the condensing surface, induced by the forced circulation of air and cooling by the air [GRE201001].

Evaporator: Heats and vaporizes refrigerant liquid from the condenser, using building system water.

eXtensible Markup Language (XML): A structured language that was published as a World Wide Web Consortium (W3C) Recommendation in 1998. It is a meta language because it is used to describe other languages, the elements they can contain, and how those elements can be used. These standardized specifications for specific types of information make them, along with the information that they describe, portable across platforms.

Fahrenheit, degree (°F): The common scale of temperature measurement in the English system of units. It is based on the freezing point of water being 32°F and the boiling point of water being 212°F at standard pressure conditions.

Fan: Device for moving air. Various systems include:

- Airfoil fan: Shaped blade in a fan assembly to optimize flow with less turbulence.
- Axial fan: Fan that moves air in the general direction of the axis about which it rotates.
- Centrifugal fan: Fan in which the air enters the impeller axially and leaves it substantially in a radial direction.
- Propeller fan: Fan in which the air enters and leaves the impeller in a direction substantially parallel to its axis.

(terms from [GRE201001])

Federal Energy Regulatory Commission (FERC): An independent agency created within the Department of Energy (1 October 1977), FERC is vested with broad regulatory authority. Virtually every facet of electric and natural gas production, transmission, and sales conducted by private, investor-owned utilities, corporations, or public marketing agencies was placed under FERC purview through either direct or indirect jurisdiction if any aspect of their operations were conducted in interstate commerce. As successor to the former Federal Power Commission (FPC), the FERC inherited practically all the FPC's interstate regulatory functions over the electric power and natural gas industries [MGE200901].

Feedstock: (gas utility industry) Natural gas used as a raw material for its chemical properties in creating an end product [MGE200902].

Fibre Channel over IP (FCIP): A protocol for transmitting Fibre Channel (FC) data over an IP network. It allows the encapsulation/tunneling of FC packets and transport via Transmission Control Protocol/Internet Protocol (TCP/IP) networks (gateways are used to interconnect FC Storage Area Networks (SANs) to the IP network and to set up connections between SANs. Protocol enables applications developed to run over FC SANs to be supported under IP, enabling organizations to leverage their current IP infrastructure and management resources to interconnect and extend FC SANs.

Firm Customer: (gas utility industry) A customer receiving service under Rate Schedules or contracts designed to provide customer's gas supply and distribution needs on a continuous basis [MGE200902].

Firm Obligation: (electric utility industry) A commitment to supply electric energy or to make capacity available at any time specified during the period covered by the commitment [MGE200901].

Firm Service: (gas utility industry) A service offered to customers under schedules or contracts that anticipate no interruptions [MGE200902].

Firm Storage Service (FSS): (gas utility industry) ANR's Rate Schedule FSS, Firm Storage Service as specified in ANR's FERC Gas Tariff [MGE200902].

Firm Transportation Service (FTS): (gas utility industry) Transportation services for which facilities have been designed, installed, and dedicated to a certified quantity [MGE200902].

Fixed Costs: Costs that do not change or vary with usage, output, or production.

Force majeure: Acts of God, strikes, lockouts, or other industrial disturbances, acts of a public enemy, wars, blockades, insurrections, riots, epidemics, landslides, lighting, earthquakes, fires, storms, storm warnings, floods, washouts, arrests, and restraints of governments and people, present or future, valid orders, decisions or rulings of any governmental authority having jurisdiction, civil disturbances, explosions, breakage or accident to machinery or lines of pipe, freezing of wells or lines of pipe, and any other cause, whether of the kind herein enumerated or otherwise, not within the control of the power or gas company that, by the exercise of due diligence, the power or gas company is unable to prevent or overcome. The settlement of strikes or lockouts shall be entirely within the discretion of the power or gas company and the above requirement that any inability to carry out obligations hereunder due to force majeure shall be remedied with all reasonable dispatch and shall not require the settlement of strikes or lockouts by conceding to the demand of the opposing party when such course is inadvisable in the discretion of the power or gas company [MGE200902].

Freezing Point: The temperature at which the removal of any heat will begin a change of state from a liquid to a solid.

Fuel Cost Adjustments: (electric utility industry) A provision in a rate schedule that provides for an adjustment to the customer's bill if the cost of fuel at the supplier's generating stations varies from a specified unit cost [MGE200901].

Gas: Natural gas, manufactured gas, propane-air gas, or any mixture of hydrocarbons or of hydrocarbons and noncombustible gases, in a gaseous state, consisting predominately of methane determined on a Btu basis [MGE200902].

Gas/Electric Package Unit: A single package unit with gas heating and electric air conditioning that is often installed on a slab or a roof [ENE200901].

Gauge Pressure: A fluid pressure scale in which atmospheric pressure equals zero pounds and a perfect vacuum equals 30 inches mercury (Hg).

Generation, Generating Plant Electric Power: The large-scale production of electricity in a central plant. A power plant consists of one or more units. Each unit includes an individual turbine generator. Turbine generators (turbines directly connected to electric generators) use steam, wind, hot gas, or falling water to generate power [MGE200901].

Gigawatt (gW, GW): One gigawatt equals 1 billion (1,000,000,000) watts, 1 million (1,000,000) kilowatts, or 1 thousand (1,000) megawatts.

Gigawatt-hour (gWh, GWh): One gigawatt-hour equals 1 billion (1,000,000,000) watt-hours, 1 million (1,000,000) kilowatt-hours, or 1 thousand (1,000) megawatt-hours.

Graywater: Wastewater that does not contain contaminants and can be reused for productive purposes such as irrigation or nonpotable purposes such as toilet flushing. Graywater reuse is restricted in many jurisdictions; check with local health and building officials [GAL200801].

Green Power, Green Pricing: Optional service choices that feature renewable fuels such as wind or solar, usually priced at some form of premium.

Greenhouse Gas (GHG): Specific gases that absorb terrestrial radiation and contribute to the global warming or greenhouse effect. Six greenhouse gases are listed in the Kyoto Protocol: carbon dioxide, methane, nitrous oxide, hydrofluorocarbons, perfluorocarbons, and sulfur hexafluoride [ATI200902].

Grid Computing: (aka Cloud Computing, Utility Computing, etc.) An environment that can be built at the local (data center), regional, or global level, where individual users can access computers, databases, and scientific tools in a transparent manner, without having to directly take into account where the underlying facilities are located.

(Virtualized) distributed computing environment that enables the dynamic "runtime" selection, sharing, and aggregation of (geographically) distributed autonomous (autonomic) resources based on the availability, capability, performance, and cost of these computing resources, and, simultaneously, also based on an organization's specific baseline and/or burst processing requirements. Grid Computing enables organizations to transparently integrate, streamline, and share dispersed, heterogeneous pools of hosts, servers, storage systems, data, and networks into one synergistic system, in order to deliver agreed-upon service at specified levels of application efficiency and processing performance. Grid Computing is an approach to distributed computing that spans multiple locations and/or multiple organizations, machine architectures, and software boundaries to provide power, collaboration, and information access. Grid Computing is infrastructure that enables the integrated, collaborative use of computers, supercomputers, networks, databases, and scientific instruments owned and managed by multiple organizations. Grid Computing is a network of computation: namely, tools and protocols for coordinated resource sharing and problem solving among pooled assets ... allows coordinated resource sharing and problem solving in dynamic, multi-institutional virtual organizations [MIN200501].

Grid Computing Topologies: Local, metropolitan, regional, national, or international footprint. Systems may be in the same room or distributed across the globe; they may running on homogenous or heterogeneous hardware platforms; they may be running on similar or dissimilar operating systems; and they may owned by one or more organizations [MIN200501].

Grid Service: A Web Service that conforms to a set of conventions (interfaces and behaviors) that define how a client interacts with a grid capability [MIN200501].

Harmonic Distortion: Multiples of power frequency superimposed on the power waveform that causes excess heating in wiring and fuses.

Heat Exchanger: Any device for transferring heat from one fluid to another. A device used to transfer heat energy, typically used for removing heat from a chilled liquid system, namely, a device to transfer heat between two physically separated fluids.

Heat Exchanger, Counterflow: Heat exchanger in which fluids flow in opposite directions approximately parallel to each other [GRE201001].

Heat Exchanger, Cross-flow: Heat exchanger in which fluids flow perpendicular to each other [GRE201001].

Heat Exchanger, Parallel-flow: Heat exchanger in which fluids flow approximately parallel to each other and in the same direction [GRE201001].

Heat Exchanger, Plate: (aka plate liquid cooler) Thin plates formed so that liquid to be cooled flows through passages between the plates and the cooling fluid flows through alternate passages [GRE201001].

Heat of Condensation: The latent heat energy liberated in the transition from a gaseous to a liquid state.

Heat Pump: A heat pump model consists of one or more factory-made assemblies that normally include an indoor conditioning coil(s), compressor(s), and outdoor coil(s), including means to provide a heating function. Heat pumps shall provide the function of air heating with controlled temperature, and may include the functions of air-cooling, air circulation, air cleaning, dehumidifying, or humidifying [ENE200901].

Heat Transfer: The three methods of heat transfer are conduction, convection and radiation.

Heat, Total (Enthalpy): A thermodynamic quantity equal to the sum of the internal energy of a system plus the product of the pressure-volume work done on the system. Namely, $H = E + PV$ where H = enthalpy or total heat content, E = internal energy of the system, P = pressure, and V = volume. Specifically, one has H = sensible heat + latent heat. Sensible heat is the heat that causes a change in temperature while latent heat is the change of enthalpy during a change of state.

Heating Seasonal Performance Factor (HSPF): A measure of a heat pump's energy efficiency over one heating season. It represents the total heating output of a heat pump (including supplementary electric heat) during the normal heating season (in Btu) as compared to the total electricity consumed (in watt-hours) during the same period [ENE200901].

High-Flow Constraint Day: (gas utility industry) A day when the utility expects natural gas demand to exceed its available deliverable supply of gas for gas sales service needs. On such a day, the utility will interrupt interruptible customers and/or require customers using third-party natural gas supplies to use no more than their daily confirmed pipeline deliveries to avoid incurring pipeline penalties and assure that adequate supplies are available for Firm Sales Service needs [MGE200902].

High-Performance Building: A building that uses significantly less energy than a conventional building. Such a building also features high water efficiency and superior indoor air quality [GAL200801].

High-Performance Data Center (HPDC): A data center with above average kilowatt loading, typically greater than 10 kW per rack.

Horizontal Displacement (HDP): An air-distribution system used predominantly in telecommunications central offices in Europe and Asia; typically, this system introduces air horizontally from one end of a cold aisle [GRE201001].

Horizontal Overhead (HOH): An air-distribution system that is used by some long-distance carriers in North America. This system introduces the supply air horizontally above the cold aisles and is generally utilized in raised-floor environments where the raised floor is used for cabling [GRE201001].

Hot Aisle: An aisle where rack backs face into the aisle. Heated exhaust air from the equipment in the racks enters this aisle and is then directed to the CRAC return vents [42U201001].

Hot Aisle Containment (HAC): A system that directs heated air from the outlet side of racks to air-conditioning equipment return ducts in a highly efficient manner [42U201001].

Hot Aisle/Cold Aisle: A common means of providing cooling to datacom rooms in which IT equipment is arranged in rows and cold supply air is supplied to the cold aisle, pulled through the inlets of the IT equipment, and exhausted to a hot aisle to minimize recirculation of the hot exhaust air with the cold supply air. Supply air is introduced into a region called the cold aisle. On each side of the cold aisle, equipment racks are placed with their intake sides facing the cold aisle. A hot aisle is the region between the backs of two rows of racks. The cooling air delivered is drawn into the intake side of the racks. This air heats up inside the racks and is exhausted from the back of the racks into the hot aisle [GRE201001].

Hotspot: An area, typically related to a rack or set of racks, where ambient air temperature is above acceptable levels. Typically caused by heat generation in excess of cooling equipment capacity [42U201001].

Humidity: Water vapor within a given space or volume.

Humidity, Absolute: The mass of water vapor in a specific volume of a mixture of water vapor and dry air.

Humidity, Relative (RH): Ratio of the partial pressure or density of water vapor to the saturation pressure or density, respectively, at the same dry-bulb temperature and barometric pressure of the ambient air.

Humidity Ratio: The proportion of mass of water vapor per unit mass of dry air at the given conditions (DBT, WBT, DPT, RH, etc.). The humidity ratio is also known as moisture content or mixing ratio.

HVAC (Heating, Ventilation and Air Conditioning): The set of components used to condition interior air, including heating and cooling equipment as well as ducting and related airflow devices. HVAC systems provide heating, cooling, humidity control, filtration, fresh air makeup, building pressure control, and comfort control. CRACs are HVAC systems specifically designed for data centers applications.

HVAC&R: HVAC and Refrigeration.

Hydrocarbon: Any of a number of compounds composed of carbon and hydrogen.

Imbalance: (gas utility industry) When a party receives or delivers a quantity of natural gas, then redelivers or uses a larger or smaller quantity of natural gas to another party [MGE200902].

Inches of Mercury Column: A unit used in measuring pressures. One inch of mercury column equals a pressure of 0.491 pounds per square inch.

Inches of Water Column: A unit used in measuring pressures. One inch of water column equals a pressure of 0.578 ounces per square inch. One-inch mercury column equals about 13.6-inches water column (see [TRA199901]).

Independent Power Facility: A facility, or portion thereof, that is not in a utility's rate base. In the past, such a facility could sell only to electric utilities for resale to ultimate customers as a wholesale transaction. Today, laws are changing to allow these plants to sell directly to ultimate customers as retail transactions or retail wheeling [MGE200901].

Independent Power Producer (IPP): Any person who owns or operates, in whole or in part, one or more new independent power production facilities [MGE200901].

Inhibitor: A chemical in the lithium bromide solution used to protect the metal shell and tubes from corrosive attack by the absorbent solution.

Inlet Air: The air entering the referenced equipment. For air conditioning equipment this is the heated air returning to be cooled, also called return air. For racks and servers this is the cooled air entering the equipment [42U201001].

Input Rate: The quantity of heat or fuel supplied to an appliance, expressed in volume or heat units per unit time, such as cubic feet per hour or Btu per hour [TRA199901].

In-Row Cooling: Cooling technology installed between server racks in a row that delivers cooled air to equipment more efficiently.

Integrated Design: The main method used by green builders to design high-performance buildings on conventional budgets. This is accomplished by incorporating efficient building system design that reduces the anticipated energy use of the building so that smaller building systems can be installed [GAL200801].

Integrated Energy Efficiency Ratio (IEER): A measure that expresses cooling part-load EER efficiency for commercial unitary air-conditioning and heat pump equipment on the basis of weighted operation at various load capacities [ENE200901].

Integration Centric Business Process Management Suites (IC-BPMS): Integration capabilities products that support process improvement and that have evolved out of the Enterprise Application Integration (EAI). Originally this space was dominated by proprietary, closed-framework solutions; at this time these products are based on Simple Object Access (SOA) and on standards-based integration technology. Vendors have added embedded ESB and Business Process Management (BPM) capabilities.

Internet FCP (iFCP): A protocol that converts Fibre Channel (FC) frames into Transmission Control Protocol (TCP), enabling native FC devices to be connected via an IP network. Encapsulation protocols for IP storage solutions where the lower-layer FC transport is replaced with TCP/IP and Gigabit Ethernet. The protocol enables existing FC storage devices or Storage Area Networks (SANs) to attach to an IP network. The operation is as follows: FC devices, such as disk arrays, connect to an iFCP gateway or switch. Each FC session is terminated at the local gateway and converted to a TCP/IP session via iFCP. A second gateway or switch receives the iFCP session and initiates an FC session. In iFCP, TCP/IP switching and routing elements complement and enhance, or replace, FC SAN fabric components.

Internet Small Computer System Interface (iSCSI): A protocol that serializes Small Computer System Interface (SCSI) commands and converts them to Transmission Control Protocol/Internet Protocol (TCP/IP). Encapsulation protocols for IP storage solutions for the support of Directly Attached Storage (DAS) (specifically SCSI-3 commands) over IP network infrastructures (at the physical layer, iSCSI supports a Gigabit Ethernet interface so that systems supporting iSCSI interfaces can be directly connected to standard Gigabit Ethernet switches or IP routers; the iSCSI protocol sits above the physical and data-link layers).

Interruptible Customer: (gas utility industry) A customer receiving service under rate schedules or contracts that permit interruption of service on short notice due to insufficient gas supply or capacity to deliver that supply [MGE200902].

Interruptible Service: (gas utility industry) Low-priority service offered to customers under schedules or contracts that anticipate and permit interruption on short notice, generally in peak-load seasons, by reason of the claim of firm service customers and higher-priority users. Gas is available at any time of the year if the supply is sufficient and the supply system is adequate [MGE200902].

Inverted Block Rate Design: A rate design for a customer class for which the unit charge for electricity increases as usage increases [MGE200901].

IP Storage (Internet Protocol Storage): Using IP and Gigabit Ethernet to build Storage Area Networks (SANs). Traditional SANs were developed using the Fibre Channel (FC) transport because it provided gigabit speeds compared to 10 and 100 Mbps Ethernet used to build messaging networks at that time. FC equipment has been costly, and interoperability between different vendors' switches was not completely standardized. Since Gigabit Ethernet and IP have become commonplace, IP Storage enables familiar network protocols to be used, and IP allows SANs to be extended throughout the world. Network management software and experienced professionals in IP networks are also widely available.

■ *Internet FCP (iFCP)*: A gateway-to-gateway protocol that allows the replacement of FC fabric components, allowing attachment of existing FC enabled storage products to an IP network.

■ *Metro Fibre Channel Protocol (mFCP):* Another proposal for handling IP storage, it is identical to iFCP, except that Transmission Control Protocol (TCP) is replaced by User Datagram Protocol (UDP).

■ *Internet Small Computer System Interface (iSCSI):* A transport protocol for SCSI that operates on top of TCP. It provides a new mechanism for encapsulating SCSI commands on an IP network. iSCSI is a protocol for a new generation of storage end-nodes that natively use TCP/IP, and replaces FCP with a pure TCP/IP implementation. iSCSI has broad industry support.

■ *Fiber Channel over Internet Protocol (FCIP):* Is FC over TCP/IP, where FC uses IP-based network services to provide the connectivity between the SAN islands over Local Area Networks (LANs), Metropolitan Area Networks (MANs), or Wide Area Networks (WANs). FCIP relies on TCP for congestion control and management, and on both TCP and FC for data error and data loss recovery. FCIP treats all classes of FC frames the same as datagrams.

ITC (Information and Communications Technology): The field of data processing (information technologies) and networking (telecom).

Kilowatt (kW): One kilowatt equals 1,000 watts.

Kilowatt-hour (kWh): This is the basic unit of electric energy equal to one kilowatt of power supplied to or taken from an electric circuit steadily for 1 hour; 1 kilowatt-hour equals 1,000 watt-hours.

kVA: Kilovolt amperes = Voltage × Current (amperage).

kWc: Kilowatts of cooling, alternate unit of measurement for the cooling capacity of a CRAH.

Latent Cooling Capacity: Cooling capacity related to wet-bulb temperature and objects that produce condensation.

Latent Heat: Heat that produces a change of state without a change in temperature; that is, ice to water at 32°F or water to steam at 212°F.

Latent Heat of Condensation: The amount of heat energy (in Btu) that must be removed to change the state of 1 pound of vapor to 1 pound of liquid at the same temperature [TRA199901].

Latent Heat of Vaporization: The amount of heat energy (in Btu) required for changing the state of 1 pound of a liquid to 1 pound of vapor at the same temperature.

LCCA (Life-Cycle Cost Analysis): A method of evaluating energy and water conservation technologies that saves money and resources for the long term but may cost more money initially than conventional technologies. Identical to the total cost of ownership concept used in many other industries [GAL200801].

LED (Light-Emitting Diode): A lighting technology that reduces energy consumption, allows lighting to be programmed by computer, and permits wide variations in lighting color.

LEED™ (Leadership in Energy and Environmental Design): A third-party certification program operated by the U.S. Green Building Council (USGBC). LEED™ is the primary U.S. benchmark for the design, construction, and operation of high-performance Green buildings [GAL200801].

Line Noise: Distortions superimposed on the power waveform that causes electromagnetic interference.

Liquid-Cooled System: Conditioned liquid (e.g., water, usually above the dew point) is channeled to the actual heat-producing electronic equipment components and used to transport heat from that component where it is rejected via a heat exchanger (air to liquid or liquid to liquid) or extended to the cooling terminal device outside of the rack [GRE201001].

Liquid Cooling: A general term used to refer to cooling technology that uses a liquid circulation system to evacuate heat as opposed to a condenser, most commonly used in reference to specific types of in-row or close-coupled cooling technologies [42U201001].

Lithium Bromide: The absorbent used in the absorption machine.

Load: The kilowatt (kW) consumption of equipment, typically installed in a rack. Also, the heat level a cooling system is required to remove from the data center environment [42U201001].

Load Curve: A curve on a chart showing power (kilowatts) supplied, plotted against time of occurrence, and illustrating the varying magnitude of the load during the period covered [MGE200901].

Load Factor: The ratio of the average load in kilowatts supplied during a designated period to the peak or maximum load in kilowatts occurring in that period. Load factor, in percent, also can be derived by multiplying the kilowatt-hours (kWh) in the period by 100 and dividing by the product of the maximum demand in kilowatts and the number of hours in the period. Example: Load Factor calculation: Load Factor = Kilowatt-hours/Hours in period/Kilowatts.

Assume a 30-day billing period or 30 times 24 hours for a total of 720 hours. Assume a customer used 10,000 kWh and had a maximum demand of 21 kW. The customer's load factor would be 66% ((10,000 kWh/720 hours/21 kW)×100) [MGE200901].

Load Level: For any benchmark that submits various amounts of work to a System Under Test (SUT), a load level is one such amount of work.

Load Management: (electric utility industry) Economic reduction of electric energy demand during a utility's peak generating periods. Load management differs from conservation in that load-management strategies are designed to either reduce or shift demand from on-peak to off-peak times, while conservation strategies may primarily reduce usage over the entire 24-hour period. Motivations for initiating load management include the reduction of capital expenditure (for new power plants), circumvention of capacity limitations, provision for economic dispatch, cost of service reductions, system efficiency improvements, or system reliability improvements. Actions may take the form of normal or emergency procedures. Many utilities encourage load management by offering customers a choice of service options with various price incentives [MGE200901].

Load Shifting: (electric utility industry) Involves moving load from on-peak to off-peak periods. Popular applications include use of storage water heating, storage space heating, cool storage, and customer load shifts to take advantage of time-of-use or other special rates [MGE200901].

Locally Sourced Materials: Building products manufactured and/or extracted within a defined radius of the building site. The U.S. Green Building Council defines local materials as those that are manufactured, processed, and extracted within a 500-mile radius of the building site. Locally sourced materials require less transport, reducing transportation-related costs and environmental impacts [GAL200801].

Loop/Looped: (electric utility industry) An electrical circuit that provides two sources of power to a load or to a substation so that if one source is de-energized, the remaining source continues to provide power [MGE200901].

Loss (Losses): The general term applied to energy (kilowatt-hours) and power (kilowatts) lost or unaccounted for in the operation of an electric system. Losses occur primarily as energy transformations from kilowatt-hours to waste heat in electric conductors and apparatus [MGE200901].

Low-Flow Constraint Day: (gas utility industry) Any day that the utility anticipates that it may be subject to pipeline or supplier penalties if natural gas supplies delivered to the utility exceed demand. On such a day, the utility will require customers using third-party natural gas supplies to use no less than their daily confirmed pipeline deliveries to avoid incurring pipeline penalties [MGE200902].

Low-Flush Toilets: A toilet that uses less water than a traditional unit, thereby lowering costs. Dual-flush toilets are an example of a low-flush technology.

Makeup Air Handler (MAH): Device used to manage humidity control for the entire data center. An MAH is a larger air handler that conditions 100%

outside air. Synonymous with MAU. The MAH filters and conditions the makeup air, either adding or removing water vapor.

Makeup Air Unit (MAU): A large air handler that conditions 100% outside air. Synonymous with MAH.

Marginal Price: The payment required to obtain one additional unit of a good.

Market Value: The current or prevailing price of a security or commodity as indicated by current market quotations, and therefore the price at which additional amounts presumably can be purchased or sold [MGE200901].

Maximum Day Requirements: (gas utility industry) Will be determined from actual telemetered maximum day usage or by dividing the maximum monthly deliveries (customers shall have their maximum daily requirement calculated on the most recent maximum monthly consumption) by the number of billing days in that month. Volumes specified in the Curtailment Priority Categories of this tariff shall apply in the aggregate rather than on a unit-of-equipment basis [MGE200902].

Maximum Temperature Rate of Change: An ASHRAE standard established to ensure stable air temperatures. The standard is 9°F/hour.

Megawatt (MW): One megawatt equals 1 million (1,000,000) watts.

Megawatt-hour (MWh): One megawatt-hour equals 1 million (1,000,000) watt-hours.

MERV (Minimum Efficiency Reporting Value): ASHRAE 52.2, for air filtration measured in particulate size.

Message-Oriented Middleware (MOM): A client/server infrastructure that increases the interoperability, portability, and flexibility of an application by allowing the application to be distributed over multiple heterogeneous platforms. It reduces the complexity of developing applications that span multiple operating systems and network protocols by insulating the application developer from the details of the various operating system and network interfaces—Application Programming Interfaces (APIs)—that extend across diverse platforms and networks are typically provided by the MOM. Applications exchange messages that can contain formatted data, requests for action, or both.

Minimum Charge: A provision in a rate schedule stating that a customer's bill cannot fall below a specified level. For example, the electric energy charge may be 6.4 cents per kilowatt-hour (kWh), the customer charge may be $4, with a minimum monthly charge equal to the customer charge. In this case, a customer would be billed for $4 if usage were anywhere between 0 and 62 kWh. However, even though the minimum charge may be stated in terms of a customer charge, a minimum charge differs from a customer charge in that charges for energy consumed are added to a customer charge, whereas a minimum charge ensures that the bill for energy consumed does not fall below a certain amount, even if little or no energy is consumed. A minimum charge is similar to a customer charge because it is designed to recover fixed costs of services such as meter reading, billing, and facilities maintenance. Although this charge does not generally recover the full cost of these services, it does give the customer a price signal that these costs do exist [MGE200901].

Monthly Delivered Quantities: (gas utility industry) The amount of natural gas that the utility estimates that a customer receiving service under Comprehensive Balancing Service will use in a calendar month [MGE200902].

NEBS™: A trademark of Telcordia Technologies, Inc., it defines a set of physical, environmental, and electrical requirements for a Central Office (CO) of a carrier; specs can be used by other entities.

Network: (electric utility industry) A system of transmission or distribution lines so cross-connected and operated as to permit multiple power supply to any principal point on it.

Nominal Cooling Capacity: The total cooling capacity of air-conditioning equipment; includes both latent and useful capacities. Due to humidity control in data centers, the latent capacity should be deducted from the nominal capacity to determine the useful capacity [42U201001].

Nomination: (gas utility industry) A request for a physical quantity of gas under a specific purchase, sales, or transportation agreement or for all contracts at a specific point [MGE200902].

Noncondensable Gas: Air or any gas in the machine that will not liquefy under operating pressures and temperatures.

Non-raised Floor: Facilities without a raised floor utilize overhead ducted supply air to cool equipment. Such ducted overhead supply systems are typically limited to a cooling capacity of 1000 W/m^2.

North American Electric Reliability Council (NERC): Formed by the electric utility industry in 1968 to promote the reliability of its generation and transmission systems. NERC consists of nine Regional Reliability Councils and one affiliate encompassing virtually all the electric systems in the United States, Canada, and Mexico.

Overcooling: A situation where air is cooled below optimum levels. Typically used in reference to rack inlet temperatures.

Overnomination: (gas utility industry) A nomination that is more than the amount of gas used by the customer during the period of time covered by the nomination (e.g., a Gas Day or a Gas Month). For balancing purposes, the difference between an overnomination and the usage volumes will be divided by the nomination to determine the overnomination percentage. Under Daily Balancing Service, the absolute difference between the nomination and the usage volumes will be subject to the appropriate Balancing Service Charges or Penalties. Usage volumes will be adjusted as necessary to be on a comparable basis with nomination volumes before an overnomination or undernomination is calculated [MGE200902].

Passive solar design: Design strategies that reduce or eliminate the use of fossil fuels and electricity for heating, cooling, and building lighting. This is accomplished by incorporating sunlight and natural ventilation into the basic design of a building, thereby minimizing the need for mechanical system capacity [GAL200801].

Peak Day: (gas utility industry) The maximum daily quantity of gas distributed through the utility's system.

Peer-to-Peer (P2P): P2P is concerned with same general problem as Grid Computing, namely, the organization of resource sharing within virtual communities. Grid community focuses on aggregating distributed high-end machines such as clusters, whereas the P2P community concentrates on sharing low-end systems such as PCs connected to the Internet. Like P2P, Grid Computing allows users to share files (many-to-many sharing). With grid, the sharing is not only in reference to files, but also other IT resources.

Photovoltaics: Also referred to as PV, these are solar electric systems that convert sunlight directly into electricity by using semiconductor materials. These materials do not create any pollution, noise, or other environmental impact [GAL200801].

Plenum: A receiving chamber for air used to direct airflow.

Point Of Presence (POP): A location (carrier-class building, e.g., a central office) where advanced communication services are aggregated using a variety of switching, routing, or gatewaying equipment.

Pole: A row of power receptacles with power supplied from a PDU.

Post-Occupancy Evaluation (POE): A process for evaluating the performance of a building once the building has been completed. This evaluation particularly focuses on energy and water use, occupant comfort, indoor air quality, and the proper operation of all building systems. The results of this evaluation can often lead to operational improvements [GAL200801].

Power Distribution Unit (PDU): The junction point between the UPS and the cabinets containing equipment.

Power Usage Effectiveness (PUE): Power usage effectiveness, a measure of data center energy efficiency calculated by dividing the total data center energy consumption by the energy consumption of the IT computing equipment. This measure is the inverse of DCiE.

Power, Firm: (electric utility industry) Power or power-producing capacity intended to be available at all times during the period covered by a commitment, even under adverse conditions [MGE200901].

Power, Interruptible: (electric utility industry) Power made available under agreements that permit curtailment or cessation of delivery by the supplier [MGE200901].

Power, Nonfirm: (electric utility industry) Power or power-producing capacity supplied or available under an arrangement that does not have the guaranteed continuous availability feature of firm power. Power supplied based on the availability of a generating unit is one type of such power [MGE200901].

Pressure Differential: The difference in pressure between two locations in the data center used to analyze air flow behaviors.

Primary Discount: (electric utility industry) A discount provision that is available to customers who can take delivery of electrical energy at primary distribution voltage levels. The transformer equipment discount is also available to customers taking primary voltage service who own their own transformers and transformer equipment [MGE200901].

Primary Distribution, Primary Distribution Feeder: A primary voltage distribution circuit, usually considered to be between a substation or point of supply and the distribution transformers, which supplies lower voltage distribution circuits or consumer service circuits [MGE200901].

Primary Voltage: The voltage of the circuit supplying power to a transformer is called the primary voltage, as opposed to the output voltage or load-supply voltage, which is called secondary voltage. In power supply practice, the primary is almost always the high-voltage side and the secondary is the low-voltage side of a transformer, except at generating stations [MGE200901].

Process: (gas utility industry) Gas used in appliances that were designed to burn a gaseous fuel so as to utilize those combustion characteristics of gaseous fuels, such as complete combustion, safe combustion products, flame geometry, ease of temperature control to precise levels, and optimum safety of heat application. Specifically excluded are steam and hot water boilers, gas turbines, space-heating equipment, and indirect air heaters, where, for all such equipment, alternate fuel-burning equipment is available. A process gas load by this definition is a load for which there is no usable alternate to a gaseous fuel [MGE200902].

Production: (electric utility industry) The act or process of generating electric energy [MGE200901].

Productivity: Worker efficiency gains are a major business benefit of green buildings. Numerous studies have identified a link between specific green features and higher employee productivity, which offsets the cost of installing or implementing green features [GAL200801].

Psychrometric Chart: A graph of the properties of air (temperature, relative humidity, etc.) used to determine how these properties vary as the amount of moisture in the air changes.

PU: Packaged unit, an air handler designed for outdoor use.

Public Interest Facility: (gas utility industry) Hospitals, schools, correctional facilities, nursing homes, hotels and motels, and any other facilities whose occupants are dependent upon operators of the facility for protection from the elements and whose continuing operations are necessary to ensure public safety [MGE200902].

Pump: Machine for imparting energy to a fluid, causing it to do work.

Purchased Gas Adjustment (PGA): (gas utility industry) The utility's monthly adjustment to adjust the average cost of gas sold through the utility's gas sales service rate schedules as set forth on Sheet G40 of this tariff [MGE200902].

Rack: A metal shelving structure for housing electronic equipment. These systems are generally specified in EIA (Electronic Industry Alliance) units such as 1U, 2U, 3U, etc., where 1U = 1.75 inches (44 millimeters). In the computing industry, a rack is an enclosed cabinet housing computer equipment. The front and back panels may be solid, perforated, or open, depending on the cooling requirements of the equipment within. In the telecom industry, a rack is a framework consisting of two vertical posts mounted to the floor and a series of open shelves upon which electronic equipment is placed. Typically, there are no enclosed panels on any side of the rack [GRE201001].

Rack-Mounted Equipment: Equipment that is to be mounted in an EIA (Electronic Industry Alliance) or similar cabinet.

Radial: An electrical circuit arranged like rays, radiating from or converging to a common center. An electric circuit that is not looped [MGE200901].

Raised Floor: Also known as access floor. A platform with removable panels (tiles) where equipment is installed, with the intervening space between it and the main building floor used to house the interconnecting cables. Often (but not always) the space under the floor is used as a means for supplying conditioned air to the information technology equipment and the room [GRE201001]. The floors utilize pedestals to support the floor panels (tiles). The cavity between the building floor slab and the finished floor can be used as an air distribution plenum to provide conditioned air throughout the raised floor area.

Raised Floor: Metal flooring on stanchions that creates a plenum for airflow and cabling, synonymous with RMF.

Raised Metal Floor (RMF): An alternate term for the more commonly used term "raised floor."

Rapidly renewable materials: Materials that can be grown and harvested for production in a short period of time. These materials are considered more sustainable because they reduce concerns of resource depletion. The U.S. Green Building Council uses a 10-year milestone as an evaluation point for deeming a material to be rapidly renewable [GAL200801].

Rate Case: The process in which a utility appears before its regulatory authority to determine the rates that can be charged to customers [MGE200901].

Rate Class: A group of customers identified as a class subject to a rate different from the rates of other groups [MGE200901].

Rate Level: The electric price a utility is authorized to collect [MGE200901].

Rate Structure: The design and organization of billing charges to customers.

Rates, Block: (electric utility industry) A certain specified price per unit is charged for all or any part of a block of such units, and reduced/increased prices per unit are charged for all or any part of succeeding blocks of such units, each such reduced/increased price per unit applying only to a particular block or portion thereof [MGE200901].

Rates, Demand: (electric utility industry) Any method of charge for electric service that is based upon, or is a function of, the rate of use, or size, of the customer's installation or maximum demand (expressed in kilowatts) during a given period of time like a billing period [MGE200901].

Rates, Flat: (electric utility industry) The price charged per unit is constant; it does not vary due to an increase or decrease in the number of units [MGE200901].

Rates, Seasonal: (electric utility industry) Rates vary depending upon the time of year. Charges are generally higher during the summer months when greater demand levels push up costs for generating electricity. Typically there are summer and winter seasonal rates. Summer rates are effective from 1 June through 30 September; during all other times of the year, winter rates are effective [MGE200901].

Rates, Step: (electric utility industry) A certain specified price per unit is charged for the entire consumption, the rate or price depending on the particular step within which the total consumption falls [MGE200901].

Rates, Time-of-Use: (electric utility industry) Prices for electricity that vary depending upon what time of day or night a customer uses it. Time-of-use rates are designed to reflect the different costs an electric company incurs in providing electricity during peak periods when electricity demand is high and off-peak periods when electricity demand is low. A power company may have two time periods defined for its time-of-use services: on-peak and off-peak. On-peak periods are defined as 10 a.m. through 9 p.m., Monday through Friday, excluding holidays. All other periods are off-peak. Whether customers benefit from time-of-use rates depends on the percentage of total consumption used during on-peak periods. Generally, customers who use less than 30% to 36% of their total consumption during on-peak periods may benefit from these rates. However, individual analysis of electricity usage habits is required to see if a time-of-use service would be of potential value [MGE200901].

Rates, Unbundled: (electric utility industry) The process of itemizing the rates for specific services that used to be covered under one rate [MGE200901].

Recirculation: Chilled airflow returning to cooling units without passing through IT equipment; also referred to as short cycling.

Recirculation Air Handler (RAH): A device that circulates air but does not cool the air.

Recycled-content materials: Products manufactured using post-consumer materials such as plastic, fiber, wood, and glass. Deconstruction of various structures can also produce a variety of "raw" materials to create new products from—everything from tiles to carpeting to composite flooring materials and beyond. Recycled-content materials help reduce the need for new raw materials and the accumulation and manufacturing processes involved [GAL200801].

"Reduce, Reuse, Recycle" EPA program: Any change in the design, manufacture, purchase, or use of materials or products to reduce the amount or toxicity before they become municipal solid waste.

Refrigerant: Any substance that transfers heat from one place to another, creating a cooling effect. The medium of heat transfer that picks up heat by evaporating at a low temperature and pressure and gives up heat when condensing at a higher temperature and pressure. Water is a refrigerant in absorption machines [TRA199901].

Regulated Utility: (electric utility industry) Utilities are distinguished as being a class of business "affected with a deep public interest" and therefore subject to regulation. Public utilities are further distinguished in that in most jurisdictions it is considered desirable for them to operate as controlled monopolies. As such, they are obligated to charge fair, nondiscriminatory rates and to render safe, reliable service to the public on demand. In return, they are generally free from substantial direct competition and are permitted, although not assured of or guaranteed to get, a fair return on investment [MGE200901].

Relative Humidity (RH): (1) Ratio of the partial pressure or density of water vapor to the saturation pressure or density, respectively, at the same dry-bulb temperature and barometric pressure of the ambient air; (2) ratio of the mole fraction of water vapor to the mole fraction of water vapor saturated at the same temperature and barometric pressure. At 100% relative humidity, the dry-bulb, wet-bulb, and dew-point temperatures are equal [GRE201001]. RH is dimensionless, and is usually expressed as a percentage.

Reliability: (electric utility industry) The guarantee of system performance at all times and under all reasonable conditions to assure constancy, quality, adequacy, and economy of electricity. It is also the assurance of the continuous supply of electricity for customers at the proper voltage and frequency [MGE200901]. More generally, a percentage value representing the frequency distribution (probability) that a piece of equipment or system will be operable throughout its mission duration.

Relief Valve: A valve that opens before a dangerously high pressure is reached.

Renewable Energy: Energy generated from natural resources that are inexhaustible. Renewable energy technologies include solar power, wind power, hydroelectricity and micro hydro, biomass, and biofuels [GAL200801].

Requirements Service: (electric utility industry) Service that the supplier plans to provide on an ongoing basis [MGE200901].

Reserve Margin: (electric utility industry) The difference between net system capability and system maximum load requirements (peak load or peak demand) [MGE200901].

Retail: (electric utility industry) Sales of electric energy to ultimate customers.

Retail Wheeling: (electric utility industry) An arrangement in which retail customers can purchase electricity from any supplier, as opposed to their local utility. The local utility would be required to allow the outside generating company to wheel the power over the local lines to the customer. Wisconsin does not currently allow retail wheeling [MGE200901].

Return Air: The heated air returning to air-conditioning equipment.

Return on Investment: A measure of the financial viability of a profit or loss on an investment, often expressed as a percentage.

Right-Sizing Systems: The practice of using smaller, more efficient building systems that are not "too big" for normal building operation. By using integrated design, a building design team can minimize the "extreme" requirements on a building system, allowing for the installation of smaller-capacity systems [GAL200801].

Rooftop Unit (RTU): An air handler designed for outdoor use mounted on a rooftop. A typical application of a PU (packaged unit).

Room Load Capacity: The point at which the equipment heat load in the room no longer allows the equipment to run within the specified temperature requirements of the equipment. The load capacity is influenced by many factors, the primary one being the room's theoretical capacity. Other factors, such as the layout of the room and load distribution, also influence the room load capacity [GRE201001].

Scheduling: (gas utility industry) A process by which nominations are first consolidated by receipt point, by contract, and verified with upstream/downstream parties. If the verified capacity is greater than or equal to the total nominated quantities, all nominated quantities are scheduled. If verified capacity is less than nominated quantities, nominated quantities will be allocated according to scheduling priorities [MGE200902].

Seasonal Energy Efficiency Ratio (SEER): A measure of equipment energy efficiency over the cooling season. It represents the total cooling of a central air conditioner or heat pump (in Btu) during the normal cooling season as compared to the total electric energy input (in watt-hours) consumed during the same period [ENE200901]. The values for SEER are determined through averaging readings of different air conditions, to represent air conditioner efficiency throughout the season. The units of SEER are Btu/hr/W.

Sensible Cooling Capacity: Cooling capacity related to dry-bulb temperature and objects that do not produce condensation.

Sensitivity: An equipment setting that bounds the set point range and triggers a change in device function when exceeded. Most commonly referring to CRAC/CRAH temperature and humidity setpoints [42U201001].

Service Area: (electric utility industry) Territory in which a utility system is required or has the right to supply electric service to ultimate consumers.

Service Drop: (electric utility industry) The overhead conductors between the electric supply, such as the last pole, and the building or structure being served [MGE200901].

Service Entrance: (electric utility industry) The equipment installed between the utility's service drop, or lateral, and the customer's conductors. Typically consists of the meter used for billing, switches and/or circuit breakers and/or fuses, and a metal housing [MGE200901].

Service Lateral: (electric utility industry) The underground service conductors between the street main and the first point of connection to the service entrance conductors [MGE200901].

Service Orientation: A way of thinking about business processes as linked, loosely coupled tasks supported by services. A new service can be created by composing a primitive group of services. This recursive definition is important because it enables the construction of more complex services above a set of existent ones [SOU200601].

Service-Oriented Architecture (SOA): SOA describes an IT architecture based around the concept of delivering reusable business services that are underpinned by IT components in such a way that the providers and the consumers of the business services are loosely coupled, with no knowledge of the technology, platform, location, or environment choices of each other [STA200501]. A business-driven IT architectural approach that supports integrating business as linked, repeatable business tasks, or services. SOA helps businesses innovate by ensuring that IT systems can adapt quickly, easily, and economically to support rapidly changing business needs [IBM200701]. In an SOA, resources are made available to other participants in the network as independent services that are accessed in a standardized way.

Service, Customer's: (electric utility industry) That portion of conductors usually between the last pole or manhole and the premises of the customer served.

Service-Oriented Computing (SOC): The computing paradigm that utilizes services as fundamental elements for developing applications [PAP200301].

Service-Oriented Network (SON): A service-oriented architecture for the development, deployment, and management of network services [SOU200601]. The application paradigm that utilizes services distributed across a network as fundamental functional elements.

Service-Oriented Network Architecture (SONA): Cisco Systems' architectural framework that aims at delivering business solutions to unify network-based services such as security, mobility, and location with the virtualization of IT resources.

Short Cycling: Chilled airflow returning to cooling units without passing through IT equipment; also referred to as recirculation.

Simple Object Access Protocol (SOAP): A standard of the W3C that provides a framework for exchanging XML-based information. A lightweight protocol for the exchange of XML-based information in a distributed environment.

Single-Phase Service: (electric utility industry) Service where the facility (e.g., house, office, warehouse, barn) has two energized wires coming into it. Typically serves smaller needs of 120 V/240 V. Requires less and simpler equipment and infrastructure to support and tends to be less expensive to install and to maintain [MGE200901].

SOA Infrastructure: A simplified, virtualized, and distributed application framework that supports SOA.

Solenoid Valve: A control device that is opened and closed by an electrically energized coil.

Solution Additive: Octyl alcohol added to lithium bromide to enhance operation.

Solution Pump: Pump that recirculates lithium bromide solution in the absorption cycle.

SPEC (Standard Performance Evaluation Corporation): An organization that develops standardized benchmarks and publishing reviewed results.

Specific Enthalpy: The sum of the internal (heat) energy of the moist air in question, including the heat of the air and water vapor within. Enthalpy is given in (SI) Joules per kilogram (J/kg) of air or Btu per pound (Btu/lb) of dry air. Specific enthalpy is also called heat content per unit mass.

Specific Heat: The amount of heat necessary to change the temperature of one pound of a substance 1 degree Fahrenheit.

Specific Volume: The volume per unit mass of the air sample. The SI units are cubic meters per kilogram (m^3/kg) of dry air; other units are cubic feet per pound (ft^3/lb) of dry air. Specific volume is also called inverse density.

SPECrate: A throughput metric based on the SPEC CPU benchmarks (such as SPEC CPU95). This metric measures a system's capacity for processing jobs of a specified type in a given amount of time.

SPECweb2005: SPECweb2005 is a standardized performance test for WWW servers, the successor to SPECweb99 and SPECweb99_SSL. The benchmark consists of different workloads (both SSL and non-SSL), such as banking and e-commerce, and writes dynamic content in scripting languages to model real-world deployments.

Spill Point: When the evaporator pan overflows into the absorber.

Spread: Numerical difference in the percentage concentration of the concentrated and dilute solutions [TRA199901].

Standard Cubic Feet of Air per Minute (SCFM): A measure free airflow into the intake filter on an air compressor. It is the volumetric flow rate of a gas representing a precise mass flow rate corrected to "standardized" conditions of temperature, pressure, and relative humidity. SCFM is used to designate flow in terms of some base or reference pressure, temperature, and relative humidity. Many standards are used, the most common being the Compressed Air & Gas Institute (CAGI) and the American Society of Mechanical Engineers (ASME) standards, which are 14.7 PSIA, 68°F and 36% relative humidity (RH). This converts to a density of 0.075 lb/ft^3 for air. If actual site conditions are different from the standard or reference conditions, corrections must be made to reflect the actual conditions of pressure, temperature, and relative humidity (i.e., convert to ACFM). Blower performance calculations, including head (used for centrifugal compressors) and horsepower, are based on actual (not standard) conditions existing at the inlet and outlet connections of the blower [REP201001].

Step-Down: (electric utility industry) To change electricity from a higher to a lower voltage.

Step-Up: (electric utility industry) To change electricity from a lower to a higher voltage.

Storage: (HVAC) Amount of refrigerant in the unit not in solution with the lithium bromide.

Storage: (IT) Infrastructure (typically in the form of appliances) used for the permanent or semi-permanent on-line retention of structured (e.g., databases) and unstructured (e.g., business/e-mail files) corporate information. Typically includes (1) a controller that manages incoming and outgoing communications as well as the data steering onto the physical storage medium (e.g., RAIDs [redundant arrays of independent disks], semiconductor memory, etc.); and (2) the physical storage medium itself. The communications mechanism could be a network interface (such as Gigabit Ethernet), a channel interface (such as Small Computer System Interface {SCSI]), or a SAN Interface (i.e., FC).

Storage Appliance: A storage platform designed to perform a specific task, such as NAS, routers, virtualization, etc.

Storage Virtualization: Software (sub)systems (typically middleware) that abstract the physical and logical storage assets from the host systems.

Strainer: Filter that removes solid particles from the liquid passing through it.

Sub-Cooling: Cooling of a liquid, at a constant pressure, below the point at which it was condensed [TRA199901].

Sub-floor: The open area underneath a raised computer floor, also called a subfloor plenum.

Submetering: (electric utility industry) Remetering of purchased energy by a customer for distribution to his tenants through privately owned or rented meters.

Substation: (electric utility industry) An assemblage of equipment for the purposes of switching and/or changing or regulating the voltage of electricity. Service equipment, line transformer installations, or minor distribution and transmission equipment are not classified as substations [MGE200901].

Summer Peak: (electric utility industry) The greatest load on an electric system during any prescribed demand interval in the summer (or cooling) season [MGE200901].

Superheat: The heat added to vapor after all liquid has been vaporized.

Supplier: (gas utility industry) A party that sells the commodity of natural gas [MGE200902].

Supply Air: The cooled airflow emitted from air-conditioning equipment.

Sustainability, in relation to the ICT industry: The ability to meet current needs without hindering the ability to meet the needs of future generations in terms of economic, environmental, and social challenges [ATI200902].

Sustainable: Refers not only to green physical attributes, as in a building, but also business processes, ethics, values, and social justice [GAL200801]. The United Nations World Commission on Environment and Development defines sustainability as "meeting the needs of the present without compromising the ability of future generations to meet their own needs."

Take or Pay (TOP): (gas utility industry) A surcharge collected by interstate pipeline customers permitted by FERC orders, designed to recover the interstate pipeline's cost of settling its historic take-or-pay liabilities [MGE200902].

Tariff: A schedule of prices or fees.

Tariff Schedule: (electric utility industry) A document filed with the regulatory authority(ies) specifying lawful rates, charges, rules, and conditions under which the utility provides service to the public [MGE200901].

Temperature: A measurement of heat intensity.

Temperature, Dew-Point (DPT): The temperature at which water vapor has reached the saturation point (100% relative humidity).

Temperature, Dry-Bulb (DBT): The temperature of air indicated by a thermometer.

Temperature, Wet-Bulb (WBT): The temperature indicated by a psychrometer when the bulb of one thermometer is covered with a water-saturated wick over which air is caused to flow at approximately 4.5 m/s (900 ft/min) to reach an equilibrium temperature of water evaporating into air, where the heat of vaporization is supplied by the sensible heat of the air [GRE201001].

Therm: The quantity of heat energy that is equivalent to one hundred thousand (100,000) Btu.

Thermal Effectiveness: Measure of the amount of mixing between hot and cold air streams before the supply air can enter the equipment and before the equipment discharge air can return to the air-handling unit.

Third-Party Natural Gas Supplies: Natural gas supplies that are purchased from an entity other than the utility.

Third-Party Natural Gas Supply Nomination Form: The utility-designated nomination form that is required to be completed and submitted to the utility in accordance with terms and provisions of the Company's Daily Balancing Service (Rate Schedule DBS-1).

Third-Party Pool: Customers who are aggregated, by a party other than the basic provider, for balancing purposes under Daily Balancing Service.

Three-Phase Service: (electric utility industry) Service where the facility (e.g., manufacturing plant, office building, warehouse, barn) has three energized wires coming into it. Typically serves larger power needs of greater than 120 V/240 V. Usually required for motors exceeding 10 horsepower or other inductive loads. Requires more sophisticated equipment and infrastructure to support and tends to be more expensive to install and maintain [MGE200901].

Tiered Storage: A process for the assignment of different categories of data to different types of storage media. The purpose is to reduce total storage cost and optimize accessibility. Organizations are reportedly finding cost savings and improved data management with a tiered storage approach. In practice, the assignment of data to particular media tends to be an evolutionary and complex activity. Storage categories may be based on a variety of design/architectural factors, including levels of protection required for the application or organization, performance requirements, and frequency of use. Software exists for automatically managing the process based on a company-defined policy. Tiered storage generally introduces more vendors into the environment and interoperability is important.

An example of tiered storage is as follows: Tier-1 data (e.g., mission-critical files) could be effectively stored on high-quality Directly Attached Storage (DAS) (but relatively expensive) media such as double-parity RAIDs (Redundant Arrays of Independent Disks). Tier-2 data (e.g., quarterly financial records) could be stored on media affiliated with a Storage Area Network (SAN); this media tends to be less expensive than DAS drives, but there may be network latencies associated with the access. Tier-3 data (e.g., e-mail backup files) could be stored on recordable compact discs (CD-Rs) or tapes. (Clearly, there could be more than three tiers, but the management of the multiple tiers then becomes fairly complex.)

Another example (in the medical field) is as follows: Real-time medical imaging information may be temporarily stored on DAS disks as a Tier-1, say, for a couple of weeks. Recent medical images and patient data may be kept on FC drives (Tier-2) for about a year. After that, less-frequently accessed images and patient records are stored on AT Attachment (ATA) drives (Tier-3) for 18 months or more. Tier-4 consists of a tape library for archiving.

Ton: (refrigeration) The amount of heat absorbed by melting 1 ton of ice in 24 hours. Equal to 288,000 Btu per day, 12,000 Btu per hour, or 200 Btu per minute.

Transformer: (electric utility industry) An electromagnetic device for changing the voltage level of alternating-current electricity.

Transformer: A coil or set of coils that increases or decreases voltage and current by induction.

Transmission: (electric utility industry) The act or process of transporting electric energy in bulk from a source or sources of supply to other principal parts of the system or to other utility systems [MGE200901].

Transmission Access: (electric utility industry) The ability of third parties to use transmission facilities owned by others (wheeling utilities) to deliver power to another utility [MGE200902].

Transportation Aggregation Group: A group of customers who have joined together for capacity release and nomination purposes under the Comprehensive Balancing Service Rate Schedule [MGE200902].

Trimming the Machine: Involves automatic adjustment of the generator solution flow to deliver the correct solution concentration at any operating condition and provide the right evaporator pan water storage capacity at design conditions [TRA199901].

Triple Bottom Line: A calculation of financial, environmental, and social performance. Often referred to as "profits, planet, and people." This calculation method contrasts with the traditional business bottom line, which considers only profits [GAL200801].

U: Rack mount unit, the standardized height of one unit is 1.75 inches.

Ultimate Customers (Consumers): (electric utility industry) Those customers purchasing electricity for their own use and not for resale.

Unbundling: (electric utility industry) Itemizing some of the different services a customer actually receives and charging for these services separately [MGE200901].

Undernomination: (gas utility industry) A nomination that is less than the amount of gas used by the customer during the period of time covered by the nomination (e.g., a Gas Day or a Gas Month). For balancing purposes, the difference between an undernomination and the usage volumes will be divided by the nomination to determine the undernomination percentage. Under Daily Balancing Service, the absolute difference between the nomination and the usage volumes will be subject to the appropriate Balancing Service Charges or Penalties. Usage volumes will be adjusted as necessary to be on a comparable basis with nomination volumes before an overnomination or undernomination is calculated [MGE200902].

Uninterruptible Power Supply (UPS): Device used to supply short-term power to computing equipment for brief outages or until an alternate power source, such as a generator, can begin supplying power.

Universal Discovery, Description and Integration (UDDI): A standardized method for publishing and discovering information about Web services. UDDI is an industry initiative that attempts to create a platform-independent, open framework for describing services, discovering businesses, and integrating business services. UDDI deals with the process of discovery in the SOA (WSDL is often used for service description, and SOAP for service invocation). Being a Web Service itself, UDDI is invoked using SOAP. In addition, UDDI also defines how to operate servers and how to manage replication among several servers.

U.S. Green Building Council (USGBC): A nonprofit trade organization that promotes sustainability in the way buildings are designed, built, and operated. The USGBC is best known for the development of the LEED™ rating system [GAL200801].

Utility Computing: (aka Cloud Computing, Grid Computing, Network Computing) The ability to access machine cycles and information services in a distributed fashion similar to those provided by an electric utility company.

Vacuum: A pressure below atmospheric pressure. A perfect vacuum is 30 inches Hg.

Variable Air Volume (VAV) AC: ACs where both the supply airflow and return airflow rates vary according to the thermal demands in the space.

Variable Costs: (electric utility industry) Costs that change or vary with usage, output, or production. Example: fuel costs [MGE200901].

Variable Frequency Drive (VFD): An electronic controller that adjusts the speed of an electric motor by regulating the power being delivered. VFDs provide continuous control, matching motor speed to the specific demands of the work being performed. VFDs allow operators to fine-tune processes while reducing costs for energy and equipment maintenance. For example, by lowering fan or pump speed by 15% to 20%, shaft power can be reduced by as much as 30%. VFDs reduce the capacity of centrifugal chillers, thus saving energy. They include electronic control of the fan motor's speed and torque to continually match fan speed with changing building-load conditions. Electronic control of the fan speed and airflow can replace inefficient mechanical controls, such as inlet vanes or outlet dampers [VAR201001].

Variable-Capacity Cooling: The application of newer technologies, such as Digital Scroll compressors and variable frequency drives in CRACs, which allow high efficiencies to be maintained at partial loads. Digital Scroll compressors allow the capacity of room air conditioners to be matched exactly to room conditions without turning compressors on and off. Typically, CRAC fans run at a constant speed and deliver a constant volume of air flow; converting these fans to variable frequency drive fans allows fan speed and power draw to be reduced as load decreases [EME201001].

Virtual Infrastructure: A virtual infrastructure is a dynamic mapping of physical resources to business needs [VMW200701].

Virtualization: An approach that allows several operating systems to run simultaneously on one (large) computer (e.g., IBM's z/VM operating system lets multiple instances of Linux coexist on the same mainframe computer).

More generally, it is the practice of making resources from diverse devices accessible to a user as if they were a single, larger, homogenous, appear-to-be-locally-available resource.

Dynamically shifting resources across platforms to match computing demands with available resources: the computing environment can become dynamic, enabling autonomic shifting applications between servers to match demand.

The abstraction of server, storage, and network resources to make them available dynamically for sharing by IT services, both internal to and external to an organization. In combination with other server, storage, and networking capabilities, virtualization offers customers the opportunity to build more efficient IT infrastructures. Virtualization is seen by some as a step on the road to utility computing.

Virtualization is a proven software technology that is rapidly transforming the IT landscape and fundamentally changing the way that people compute [VMW200701].

VOCs (Volatile Organic Compounds): Volatile organic compounds are emitted as gases from certain solids or liquids. They include a variety of chemicals, some of which may have short- and long-term adverse health effects. Concentrations of many VOCs are consistently higher indoors than outdoors [GAL200801].

Waterside Economizer: An economizer that redirects water flow to an external heat exchanger when the exterior ambient air temperature is at or below a temperature required to chill water to a given setpoint, simultaneously shutting down the mechanical chiller equipment [42U201001].

Watts Per Square Foot (WPSF): A measure used to normalize power and heat density on a per-cabinet basis.

Web Services (WSs): A software system designed to support interoperable machine-to-machine interaction over a network. It has an interface described in a machine-processable format (specifically, WSDL). Other systems interact with the Web service in a manner prescribed by its description using SOAP messages, typically conveyed using HTTP with an XML serialization in conjunction with other Web-related standards [IBM200701].

Web Services provide standard infrastructure for data exchange between two different distributed applications (grids provide an infrastructure for aggregation of high-end resources for solving large-scale problems). Web Services are expected to play a key constituent role in the standardized definition of Grid Computing, because Web Services have emerged as a standards-based approach for accessing network applications.

Web Services Description Language (WSDL): An XML-based language used to describe Web Services and how to locate them; it provides information on what the service is about, where it resides, and how it can be invoked.

Web Services Networking: Assembly of a more complex service from service modules that reside on different nodes connected to a network.

Weighted Average Cost of Gas (WACOG): Average cost of gas purchased during a given time period, usually a month. WACOG includes gas injected or withdrawn from storage [MGE200902].

Wet-Bulb Temperature (WBT): The temperature of an air sample after it has passed through a constant-pressure, ideal, adiabatic saturation process. Effectively, this is after the air has passed over a large surface of liquid water in an insulated channel. The temperature of the air measured using a wet-bulb thermometer, typically taken in conjunction with a dry-bulb reading to determine relative humidity [42U201001]. In the field, this is the reading of a thermometer whose sensing bulb is covered with a wet sock evaporating into a rapid stream of the sample air. Note: WBT = DBT when the air sample is saturated with water.

Wheeling Service: (electric utility industry) The use of the transmission facilities of one system to transmit power and energy by agreement of, and for, another system with a corresponding wheeling charge [MGE200901].

Wind Energy/Wind Generation: (electric utility industry) Electricity generated through wind-powered turbines [MGE200901].

Work Cell: The area of a rack and the related area immediately in front of and behind an equipment rack. Standard racks are 2 ft wide and 4 feet deep. Standard aisles are 4 ft wide, so half of that space is workspace for a given rack. This results in a standard work cell of 16 ft². Actual work cell size varies with data center design [42U201001].

References

[42U201001]   42U, Data Center Efficiency Glossary, Denver, CO. 42U provides data center efficiency solutions for data center and facilities managers.

[ATI200902]   ATIS Report On Environmental Sustainability, March 2009, A report by the ATIS Exploratory Group on Green, ATIS, 1200 G Street, NW, Suite 500, Washington, DC 20005.

[EME201001]   Emerson Network Power, Variable-Capacity Cooling, Whitepaper, Columbus, OH. Emerson Network Power is a business of Emerson.

[ENE200901]   ENERGY STAR® Program Requirements for Light Commercial HVAC Partner Commitments, Version 2.0 – DRAFT 1, 5 January 2009.

[GAL200801]   Galley Eco Capital LLC, Glossary, 2008, San Francisco, CA.

[GRE201001]   The Green Grid, Glossary of Terms, 2010.

[IBM200701]   IBM, Service Oriented Architecture — SOA, *Service Oriented Architecture Glossary,* IBM Corporation, 1 New Orchard Road, Armonk, NY 10504-1722.

[MGE200901]   Madison Gas and Electric, Electric Glossary, Madison, WI, 2009.

[MGE200902]   Madison Gas and Electric, Natural Gas Glossary, Madison, WI, 2009.

[MIN200501]   D. Minoli, *A Networking Approach to Grid Computing* (Wiley, 2005).

[MIT200601]   T. Mitra, Business-driven development, IBM DeveloptWorks, Online Magazine, January 13, 2006. http://www-128.ibm/developerworks/webservices/library/ws-bdd/

[OMG200701]   OMG, Object Management Group, 140 Kendrick Street, Building A Suite 300, Needham, MA 02494.

[PAP200301]   M.P. Papazoglou and D. Georgakopoulos, Service-oriented computing: Introduction, *Communications of the ACM,* 46(10): 24–28, October 2003.

[REP201001]   REP Inc., SCFM (STANDARD CFM) VS. ACFM (ACTUAL CFM), Whitepaper, 2405 Murphy Blvd., Gainesville, GA 30504.

[SOU200601]   V.A.S.M. de Souza and E. Cardozo, SOANet — A service oriented architecture for building compositional network services, *Journal of Software,* 1(2): 1–11, August 2006.

[STA200501]   Staff, Enterprise-Class SOA Management with webMethods Servicenet, Leveraging a Web Services Infrastructure Platform to deliver on the promise of SOA, January 2006, Whitepaper, December 2005, webMethods, Inc., 3877 Fairfax Ridge Road, South Tower, Fairfax, VA 22030.

[SYS200501]   Systems and Software Consortium, Glossary. http://www.software.org/pub/architecture/pzdefinitions.asp

[THO200801]   T. von Eicken, The three levels of cloud computing, *Cloud Computing Journal,* 31 July 2008.

[TOG200501]   TOGAF, Preliminary Phase: Framework and Principles, http://www.opengroup.org/architecture/togaf8-doc/arch/p2/p2_prelim.htm.

[TRA199901]   Trane, Glossary, ABS-M-11-M-11A Operations/Maintenance Manual. Ingersoll-Rand Co., Piscataway, NJ, April 27, 1999.

[TRI200201]   T. Tritt, Overview of various strategies and promising new bulk materials for potential thermoelectric applications, *Materials Research Society Symposium Proceedings,* Vol. 691, 2002, Materials Research Society.

[VAR201001]   Variable Frequency Drive Co., Variable Frequency Drive Basics, Fremont, CA, 94538.

[VMW200701]   VMWare Promotional Materials, VMware, Inc. World Headquarters, 3401 Hillview Ave, Palo Alto, CA 94304, http://www.vmware.com.

Index

Printed and bound by CPI Group (UK) Ltd, Croydon, CR0 4YY

21/10/2024

01777107-0017